T0353358

Continuous Symmetries,
Lie Algebras,
Differential Equations and
Computer Algebra

Continuous Symmetries, Lie Algebras, Differential Equations and Computer Algebra

Willi-Hans Steeb

Rand Afrikaans University
Johannesburg
South Africa

World Scientific
Singapore • New Jersey • London • Hong Kong

Published by

World Scientific Publishing Co. Pte. Ltd.

P O Box 128, Farrer Road, Singapore 912805

USA office: Suite 1B, 1060 Main Street, River Edge, NJ 07661

UK office: 57 Shelton Street, Covent Garden, London WC2H 9HE

Library of Congress Cataloging-in-Publication Data
Steeb, W.-H.
    Continuous symmetries, Lie algebras, differential equations and computer algebra
    /Willi-Hans Steeb.
        p.   cm.
    Includes bibliographical references and index.
    ISBN 9810228910
    1. Differential equations.   2. Differential equations, Partial.
3. Lie algebras.   4. Continuous groups.   5. Mathematical physics.
I. Title.
QC20.7.D5S74   1996
515'35--dc20   96-38283
                                                          CIP

British Library Cataloguing-in-Publication Data
A catalogue record for this book is available from the British Library.

This book is printed on acid-free paper.

Printed in Singapore by Uto-Print

Preface

The purpose of this book is to provide a comprehensive introduction to the application of continuous symmetries and their Lie algebras to ordinary and partial differential equations. The study of symmetries of differential equations provides important information about the behaviour of differential equations. The symmetries can be used to find exact solutions. They can be applied to verify and develop numerical schemes. One can also obtain conservation laws of a given differential equation with the help of the continuous symmetries. Gauge theory is also based on the continuous symmetries of certain relativistic field equations.

Apart from the standard techniques in the study of continuous symmetries, the book includes: the Painlevé test and symmetries, invertible point transformation and symmetries, Lie algebra valued differential forms, gauge theory, Yang-Mills theory and chaos, self-dual Yang-Mills equation and soliton equations, Bäcklund transformation, Lax representation, Bose operators and symmetries, discrete systems and invariants.

Each chapter includes computer algebra applications. Examples are the finding of the determining equation for the Lie symmetries, finding the curvature for a given metric tensor field and calculating the Killing vector fields for a metric tensor field.

The book is suitable for use by students and research workers whose main interest lies in finding solutions of differential equations. It therefore caters for readers primarily interested in applied mathematics and physics rather than pure mathematics. The book provides an application orientated text that is reasonably self-contained. A large number of worked examples have been included in the text to help the readers working independently of a teacher. The advance of algebraic computation has made it possible to write programs for the tedious calculations in this research field. Thus the last chapter gives a survey on computer algebra packages.

End of proofs are indicated by ♠. End of examples are indicated by ♣.

I wish to express my gratitude to Catharine Thompson for a critical reading of the manuscript.

Any useful suggestions and comments are welcome.

email address of the author: WHS@RAU3.RAU.AC.ZA

Home page of the author: http://zeus.rau.ac.za/steeb/steeb.html

Contents

Notation

$\emptyset$	empty set
$\mathbf{N}$	natural numbers
$\mathbf{Z}$	integers
$\mathbf{Q}$	rational numbers
$\mathbf{R}$	real numbers
$\mathbf{R}^+$	nonnegative real numbers
$\mathbf{C}$	complex numbers
$\mathbf{R}^n$	n-dimensional Euclidian space
$\mathbf{C}^n$	n-dimensional complex linear space
i	$:= \sqrt{-1}$
$\Re z$	real part of the complex number z
$\Im z$	imaginary part of the complex number z
$\mathbf{x} \in \mathbf{R}^n$	element $\mathbf{x}$ of $\mathbf{R}^n$
$A \subset B$	subset A of set B
$A \cap B$	the intersection of the sets A and B
$A \cup B$	the union of the sets A and B
$f \circ g$	composition of two mappings $(f \circ g)(x) = f(g(x))$
u	dependent variable
t	independent variable (time variable)
x	independent variable (space variable)
$\mathbf{x}^T = (x_1, x_2, \ldots, x_m)$	vector of independent variables, ^{T} means transpose
$\mathbf{u}^T = (u_1, u_2, \ldots, u_n)$	vector of dependent variables, ^{T} means transpose
$\|\cdot\|$	norm
$\mathbf{x} \cdot \mathbf{y}$	scalar product (inner product)
$\mathbf{x} \times \mathbf{y}$	vector product
$\otimes$	Kronecker product, tensor product
det	determinant of a square matrix
tr	trace of a square matrix
I	unit matrix
$[,]$	commutator
δ_{jk}	Kronecker delta with $\delta_{jk} = 1$ for $j = k$ and $\delta_{jk} = 0$ for $j \neq k$
d	exterior derivative
λ	eigenvalue
ϵ	real parameter
$\wedge$	Grassmann product (exterior product, wedge product)
H	Hamilton function
L	Lagrange function

Chapter 1

Introduction

Sophus Lie (1842–1899) and Felix Klein (1849–1925) studied mathematical systems from the perspective of those transformation groups which left the systems invariant. Klein, in his famous "Erlanger" program, pursued the role of finite groups in the studies of regular bodies and the theory of algebraic equations, while Lie developed his notion of continuous transformation groups and their role in the theory of differential equations. Today the theory of continuous groups is a fundamental tool in such diverse areas as analysis, differential geometry, number theory, atomic structure and high-energy physics. In this book we deal with Lie's theorems and extensions thereof, namely its applications to the theory of differential equations.

It is well known that many, if not all, of the fundamental equations of physics are nonlinear and that linearity is achieved as an approximation. One of the important developments in applied mathematics and theoretical physics over the recent years is that many nonlinear equations, and hence many nonlinear phenomena, can be treated as they are, without approximations, and be solved by essentially linear techniques.

One of the standard techniques for solving linear partial differential equations is the Fourier transform. During the past 25 years it was shown that a class of physically interesting nonlinear partial differential equations can be solved by a nonlinear extension of the Fourier technique, namely the inverse scattering transform. This reduces the solution of the Cauchy problem to a series of linear steps. This method, originally applied to the Korteweg-de Vries equation, is now known to be applicable to a large class of nonlinear evolution equations in one space and one time variable, to quite a few equations in 2 + 1 dimensions and also to some equations in higher dimensions.

Continuous group theory, Lie algebras and differential geometry play an important role in the understanding of the structure of nonlinear partial differential equations, in particular for generating integrable equations, finding Lax pairs, recursion operators, Bäcklund transformations and finding exact analytic solutions.

1

Most nonlinear equations are not integrable and cannot be treated via the inverse scattering transform, nor its generalizations. They can of course be treated by numerical methods, which are the most common procedures. Interesting qualitative and quantitative features are however often missed in this manner and it is of great value to be able to obtain, at least, particular exact analytic solutions of nonintegrable equations. Here group theory and Lie algebras play an important role. Indeed, Lie group theory was originally created as a tool for solving ordinary and partial differential equations, be they linear or nonlinear.

New developments have also occurred in this area. Some of them have their origins in computer science. The advent of algebraic computing and the use of such computer languages for symbolic computations such as REDUCE, MACSYMA, AXIOM, MAPLE, MATHEMATICA, SYMBOLICC++ etc., have made it possible (in principle) to write computer programs that construct the Lie algebra of the symmetry group of a differential equation. Other important advances concern the theory of infinite dimensional Lie algebras, such as loop algebras, Kac-Moody and Virasoro algebras which frequently occur as Lie algebras of the symmetry groups of integrable equations in $2 + 1$ dimensions such as the Kadomtsev-Petviashvili equation. Furthermore, practical and computerizable algorithms have been proposed for finding all subgroups of a given Lie group and for recognizing Lie algebras given their structure constants.

In chapter 2 we give an introduction into group theory. Both finite and infinite groups are discussed.

Lie and Lie transformation groups are introduced in chapter 3. In particular, the classical Lie groups are studied in detail.

Chapter 4 is devoted to the infinitesimal transformations (vector fields) of Lie transformation groups. In particular, the three theorems of Lie are discussed.

Chapter 5 gives a comprehensive introduction into Lie algebras. We also discuss representations of Lie algebras in details. Many examples are provided to clarify the definitions and theorems.

The form-invariance of partial differential equations under Lie transformation groups is illustrated by way of examples in chapter 6. This should be seen as an introduction to the development of the theory of invariance of differential equations by the jet bundle formalism. The Gauge transformation for the Schrödinger equation is also discussed. We also show how the electromagnetic field A_μ is coupled to the wave function ψ.

Chapter 7 deals with differential geometry. Theorems and definitions (with examples) are provided that are of importance in the application of Lie algebras to differential equations. A comprehensive introduction into differential forms and tensor fields is given.

The Lie derivative is of central importance for continuous symmetries. In chapter 8 we study invariance and conformal invariance of geometrical objects, i.e. functions, vector fields, differential forms, tensor fields, etc..

In chapter 9 the jet bundle formalism in connection with the prolongation of vector fields and (partial) differential equations is studied. The application of the Lie derivative in the jet bundle formalism is analysed to obtain the invariant Lie algebra. Explicit analytic solutions are then constructed by applying the invariant Lie algebra. These are the so-called similarity solutions which are of great theoretical and practical importance. The direct method is also introduced.

In chapter 10 the generalisation of the Lie point symmetry vector fields is considered. These generalised vector fields are known as the Lie-Bäcklund symmetry vector fields. Similarity solutions are constructed from the Lie-Bäcklund vector fields. The connection with gauge transformations is also discussed.

In chapter 11 the inverse problem is considered. This means that a partial differential equation is constructed from a given Lie algebra which is spanned by Lie point or Lie-Bäcklund symmetry vector fields.

A list of Lie symmetry vector fields of some important partial differential equations in physics is included in chapter 12. In particular the Lie symmetry vector fields for the Maxwell-Dirac equation have been calculated.

In chapter 13, the Gateaux derivative is defined. A Lie algebra is introduced using the Gateaux derivative. Furthermore, recursion operators are defined and applied. Then we can find hierarchies of integrable equations.

In chapter 14 we introduce Bäcklund transformations for partial and ordinary differential equations.

For soliton equations the Lax representations are the starting point for the inverse scattering method. In chapter 15 we discuss the Lax representation. Many illustrative examples are given.

The important concept of conservation laws is discussed in chapter 16. The connection between conservation laws and Lie symmetry vector fields is of particular interest. Extensive use is made of the definitions and theorems of exterior differential forms. The Cartan fundamental form plays an important role regarding the Lagrange density and Hamilton density.

In chapter 17 the Painlevé test is studied with regard to the symmetries of ordinary and partial differential equations. The Painlevé test provides an approach to study the integrability of ordinary and partial differential equations. This approach is studied and several examples are given. In particular a connection between the singularity manifold and similarity variables is presented.

In chapter 18 the extension of differential forms, discussed in chapter 7, to Lie algebra valued differential forms is studied. The covariant exterior derivative is defined. Then the Yang-Mills equations and self-dual Yang-Mills equations are introduced. It is conjectured that the self-dual Yang-Mills equations are the master equations of all integrable equations such as the Korteweg-de Vries equation.

The connection between nonlinear autonomous systems of ordinary differential equations, first integrals, Bose operators and Lie algebras is studied in chapter 19. It is shown that ordinary differential equations can be expressed with Bose operators. Then the time-evolution can be calculated using the Heisenberg picture. An extension to nonlinear partial differential equations is given where Bose field operators are considered.

Chapter 20 gives a survey of computer algebra packages. Of particular interest are the computer programs available for the calculation of symmetry vector fields.

The emphasis throughout this book is on differential equations that are of importance in physics and engineering. The examples and applications consist mainly of the following equations: the Korteweg-de Vries equation, the sine-Gordon equation, Burgers' equation, linear and nonlinear diffusion equations, the Schrödinger equation, the nonlinear Klein-Gordon equation, nonlinear Dirac equations, Yang-Mills equations, the Lorenz model, the Lotka-Volterra model and damped anharmonic oscillators.

Each chapter includes a section on computer algebra applications.

Chapter 2

Groups

2.1 Definitions and Examples

In this section we introduce some elementary definitions and fundamental concepts in general group theory. We present examples to illustrate these concepts and show how different structures form a group.

Let us define a group as an abstract mathematical entity Miller [84], Baumslag and Chandler [6].

Definition 2.1 *A group G is a set $e, g_1, g_2, \cdots \in G$ not necessarily countable, together with an operator, called group composition $(\cdot)$, such that*

1. *Closure: $g_i \in G, g_j \in G \implies g_i \cdot g_j \in G$.*

2. *Assosiativity: $g_i \cdot (g_j \cdot g_k) = (g_i \cdot g_j) \cdot g_k$.*

3. *Existence of identity $e \in G$: $e \cdot g_i = g_i = g_i \cdot e$ for all $g_i, e \in G$.*

4. *Existence of inverse $g_i^{-1} \in G$: $g_i \cdot g_i^{-1} = g_i^{-1} \cdot g_i = e$ for all $g_i \in G$.*

5. *A group that obeys a fifth postulate $g_i \cdot g_j = g_j \cdot g_i$ for all $g_i, g_j \in G$, in addition to the four listed above is called an* **abelian group** *or* **commutative group**.

The group composition in an abelian group is often written in the form $g_i + g_j$. The element $g_i + g_j$ is called the sum of g_i and g_j and G is called an **additive group**.

Definition 2.2 *If a group G consists of a finite number of elements, then G is called a* **finite group***; otherwise, G is called an* **infinite group**.

Example: The set of integers $\mathbf{Z}$ with addition as group composition is an infinite additive group with $e = 0$. ♣

Example: The set $\{1, -1\}$ with multiplication as group composition is a finite abelian group with $e = 1$. ♣

Definition 2.3 *Let G be a finite group. The number of elements of G is called the dimension or* **order** *of G.*

Definition 2.4 *A nonempty subset H of G is called a* **subgroup** *of G if H is a group with respect to the composition of the group G. We write $H < G$.*

Hence a nonempty subset H is a subgroup of G if and only if $h_i^{-1} \cdot h_j \in H$ for any $h_i, h_j \in H$. For a family $\{H_\lambda\}$ of subgroups of G, the intersection $\bigcap_\lambda H_\lambda$ is also a subgroup.

Theorem 2.1 *The identity element e is unique.*

Proof: Suppose $e' \in G$ such that $e' \cdot g_i = g_i \cdot e' = e$ for all $g_i \in G$. Setting $g_i = e$, we find $e \cdot e' = e' \cdot e = e$. But $e' \cdot e = e'$ since e is an identity element. Therefore, $e' = e$. ♠

Theorem 2.2 *The inverse element g_i^{-1} of g_i is unique.*

Proof: Suppose $g_i' \in G$ such that $g_i \cdot g_i' = e$. Multiplying on the left by g_i^{-1} and using the assosiative law, we get $g_i^{-1} = g_i^{-1} \cdot e = g_i^{-1} \cdot (g_i \cdot g_i') = (g_i^{-1} \cdot g_i) \cdot g_i' = e \cdot g_i' = g_i'$. ♠

Theorem 2.3 *The order of a subgroup of a finite group divides the order of the group.*

This theorem is called **Lagrange's theorem**. For the proof we refer to the literature (Miller [84]).

Definition 2.5 *Let H be a subgroup of G and $g \in G$. The set*

$$Hg := \{\ hg\ :\ h \in H\ \}$$

is called a right coset of H. The set

$$gH := \{\ gh\ :\ h \in H\ \}$$

is called a left coset of H.

Definition 2.6 *A subgroup N of G is called* **normal (invariant, self-conjugate)** *if $gNg^{-1} = N$ for all $g \in G$.*

If N is a normal subgroup we can construct a group from the cosets of N, called the **factor group** G/N. The elements of G/N are the cosets gN, $g \in G$. Of course, two cosets gN, $g'N$ containing the same elements of G define the same element $G/N : gN = g'N$. Since N is normal it follows that

$$(g_1 N)(g_2 N) = (g_1 N)(g_2 N) = g_1 N g_2 = g_1 g_2 N$$

as sets. Note that $NN = N$ as sets.

Consider an element g_i of a finite group G. If g_i is of order d, where the **order** of a group element is the smallest positive integer d with $g_i^d = g_1$ (identity), then the different powers of g_i are $g_i^0 (= g_1), g_i, g_i^2, \ldots g_i^{d-1}$. All the powers of g_i form a group $< g_i >$ which is a subgroup of G and is called a **cyclic group**. This is an abelian group where the order of the subgroup $< g_i >$ is the same as the order of the element g_i.

A way to partition G is by means of **conjugacy classes**.

Definition 2.7 *A group element h is said to be conjugate to the group element k, $h \sim k$, if there exists a $g \in G$ such that*

$$k = ghg^{-1}.$$

It is easy to show that conjugacy is an equivalence relation, i.e., (1) $h \sim h$ (reflexive), (2) $h \sim k$ implies $k \sim h$ (symmetric), and (3) $h \sim k, k \sim j$ implies $h \sim j$ (transitive). Thus, the elements of G can be divided into conjugacy classes of mutually conjugate elements. The class containing e consists of just one element since

$$geg^{-1} = e$$

for all $g \in G$. Different conjugacy classes do not necessarily contain the same number of elements.

Let G be an abelian group. Then each conjugacy class consists of one group element each, since

$$ghg^{-1} = h, \quad \text{for all} \quad g \in G.$$

Let us now give a number of examples to illustrate the definitions given above.

Example: A **field** is an (infinite) abelian group with respect to addition. The set of nonzero elements of a field forms a group with respect to multiplication, which is called a multiplicative group of the field. ♣

Example: A **linear vector space** over a field K (such as the real numbers **R**) is an abelian group with respect to the usual addition of vectors. The group composition of two elements (vectors) **a** and **b** is their vector sum **a** + **b**. The identity is the zero vector and the inverse of an element is its negative. ♣

Example: Let N be an integer with $N \geq 1$. The set

$$\{ e^{2\pi i n/N} \quad : \ n = 0, 1, \ldots, N-1 \}$$

is an abelian (finite) group under multiplication since

$$e^{2\pi i n/N} e^{2\pi i m/N} = e^{2\pi i(n+m)/N}$$

where $n, m = 0, 1, \ldots, N - 1$. Note that $e^{2\pi i n} = 1$ for $n \in \mathbf{N}$. We consider some special cases of N: For $N = 2$ we find the set $\{1, -1\}$ and for $N = 4$ we find $\{1, i, -1, -i\}$. These are elements on the unit circle in the complex plane. For $N \to \infty$ the number of points on the unit circle increases. As $N \to \infty$ we find the unitary group

$$U(1) := \left\{ \, e^{i\alpha} \; : \; \alpha \in \mathbf{R} \, \right\}. \qquad\qquad \clubsuit$$

Example: The two matrices

$$\left\{ \begin{pmatrix} 1 & 0 \\ 0 & 1 \end{pmatrix}, \begin{pmatrix} 0 & 1 \\ 1 & 0 \end{pmatrix} \right\}$$

form a finite abelian group of order two with matrix multiplication as group composition. The closure can easily be verified

$$\begin{pmatrix} 1 & 0 \\ 0 & 1 \end{pmatrix} \begin{pmatrix} 1 & 0 \\ 0 & 1 \end{pmatrix} = \begin{pmatrix} 1 & 0 \\ 0 & 1 \end{pmatrix}, \qquad \begin{pmatrix} 1 & 0 \\ 0 & 1 \end{pmatrix} \begin{pmatrix} 0 & 1 \\ 1 & 0 \end{pmatrix} = \begin{pmatrix} 0 & 1 \\ 1 & 0 \end{pmatrix}$$

$$\begin{pmatrix} 0 & 1 \\ 1 & 0 \end{pmatrix} \begin{pmatrix} 0 & 1 \\ 1 & 0 \end{pmatrix} = \begin{pmatrix} 1 & 0 \\ 0 & 1 \end{pmatrix}.$$

The identity element is the 2×2 unit matrix. $\qquad\qquad \clubsuit$

Example: Let $M = \{1, 2, \ldots, n\}$. The set $Bi(M, M)$ of bijective mappings $\sigma : M \to M$ so that

$$\sigma : \{1, 2, \ldots, n\} \to \{p_1, p_2, \ldots, p_n\}$$

forms a group S_n under the composition of functions. Let S_n be the set of all the permutations

$$\sigma = \begin{pmatrix} 1 & 2 & \cdots & n \\ p_1 & p_2 & \cdots & p_n \end{pmatrix}.$$

We say 1 is mapped into p_1, 2 into p_2, $\ldots$, n into p_n. The numbers $p_1, p_2, \ldots, p_n$ are a reordering of $1, 2, \ldots, n$ and no two of the p_j's $j = 1, 2 \ldots, n$ are the same. The inverse permutation is given by

$$\sigma^{-1} = \begin{pmatrix} p_1 & p_2 & \cdots & p_n \\ 1 & 2 & \cdots & n \end{pmatrix}.$$

The product of two permutations σ and τ, with

$$\tau = \begin{pmatrix} q_1 & q_2 & \cdots & q_n \\ 1 & 2 & \cdots & n \end{pmatrix}$$

is given by the permutation

$$\sigma \circ \tau = \begin{pmatrix} q_1 & q_2 & \cdots & q_n \\ p_1 & p_2 & \cdots & p_n \end{pmatrix}.$$

That is, the integer q_i is mapped to i by τ and i is mapped to p_i by σ, so q_i is mapped to p_i by $\sigma \circ \tau$. The identity permutation is

$$e = \begin{pmatrix} 1 & 2 & \cdots & n \\ 1 & 2 & \cdots & n \end{pmatrix}.$$

S_n has order $n!$. The group of all permutations on M is called the **symmetric group** on M which is non-abelian, if $n > 2$. ♣

Example: Let N be a positive integer. The set of all matrices

$$Z_{2\pi k/N} = \begin{pmatrix} \cos \frac{2k\pi}{N} & -\sin \frac{2k\pi}{N} \\ \sin \frac{2k\pi}{N} & \cos \frac{2k\pi}{N} \end{pmatrix}$$

where $k = 0, 1, 2, \ldots, N-1$, forms an abelian group under matrix multiplication. The elements of the group can be generated from the transformation

$$Z_{2k\pi/N} = \left(Z_{2\pi/N} \right)^k, \qquad k = 0, 1, 2, \ldots, N-1.$$

For example, if $N = 2$ the group consists of the elements $\{(Z_\pi)^0, (Z_\pi)^1\} \equiv \{-I, +I\}$ where I is the 2×2 unit matrix. This is an example of a cyclic group. ♣

Example: The set of all invertible $n \times n$ matrices form a group with respect to the usual multiplication of matrices. The group is called the **general linear group** over the real numbers $GL(n, \mathbf{R})$, or over the complex numbers $GL(n, \mathbf{C})$. This group together with its subgroups are the so-called **classical groups** which are Lie groups (see chapter 3). ♣

Example: Let

$$A(\alpha) = \begin{pmatrix} \cos \alpha & \sin \alpha \\ -\sin \alpha & \cos \alpha \end{pmatrix}$$

where $\alpha \in \mathbf{R}$. We show that the matrices $A(\alpha)$ form an abelian group under matrix multiplication, the so-called $SO(2)$ group which is a subgroup of the group $GL(2, \mathbf{R})$. Since

$$\begin{pmatrix} \cos \alpha & \sin \alpha \\ -\sin \alpha & \cos \alpha \end{pmatrix} \begin{pmatrix} \cos \beta & \sin \beta \\ -\sin \beta & \cos \beta \end{pmatrix} = \begin{pmatrix} \cos(\alpha + \beta) & \sin(\alpha + \beta) \\ -\sin(\alpha + \beta) & \cos(\alpha + \beta) \end{pmatrix},$$

the set is closed under multiplication. Here we have used the identities

$$\cos \alpha \cos \beta - \sin \alpha \sin \beta \equiv \cos(\alpha + \beta)$$

$$\sin \alpha \cos \beta + \cos \alpha \sin \beta \equiv \sin(\alpha + \beta).$$

For $\alpha = 0$ we obtain the identity element of the group, i.e. the unit matrix. Since $\det A(\alpha) = 1$ the inverse exists and is given by

$$A^{-1}(\alpha) = A(-\alpha) = \begin{pmatrix} \cos \alpha & -\sin \alpha \\ \sin \alpha & \cos \alpha \end{pmatrix}.$$

For arbitrary $n \times n$ matrices A, B, C the associative law holds, i.e.,

$$A(BC) = (AB)C.$$

Consequently the matrices $A(\alpha)$ form a group.   ♣

Example: Let $\mathbf{C}$ be the complex plane. Let $z \in \mathbf{C}$. The set of Möbius transformations in $\mathbf{C}$ form a group called the **Möbius group** denoted by M where $m : \mathbf{C} \to \mathbf{C}$,

$$M := \{ \, m(a,b,c,d) \; : \; a,b,c,d \in \mathbf{C}, \;\; ad - bc \neq 0 \, \}$$

and

$$m : z \mapsto z' = \frac{az + b}{cz + d}.$$

The condition $ad - bc \neq 0$ must hold for the transformation to be invertible. Here, $z = x + iy$, where $x, y \in \mathbf{R}$. This forms a group under the composition of functions: Let

$$m(z) = \frac{az + b}{cz + d}$$

$$\widetilde{m}(z) = \frac{ez + f}{gz + h}$$

where $ad - bc \neq 0$ and $eh - fg \neq 0$ ($e, f, g, h \in \mathbf{C}$). Consider the composition

$$
\begin{aligned}
m\left(\widetilde{m}(z)\right) &= \frac{a(ez + f)/(gz + h) + b}{c(ez + f)/(gz + h) + d} \\
&= \frac{aez + af + bgz + hb}{cez + cf + dgz + hd} \\
&= \frac{(ae + bg)z + (af + hb)}{(ce + dg)z + (cf + hd)}.
\end{aligned}
$$

Thus $m(\widetilde{m}(z))$ has the form of a Möbius transformation, since

$$(ae + bg)(cf + hd) - (af + hb)(ce + dg)$$

$$= aecf + aehd + bgcf + bghd - afce - afdg - hbce - hbdg$$

$$= ad(eh - fg) + bc(gf - eh)$$

$$= (ad - bc)(eh - fg) \neq 0.$$

Thus we conclude that m is closed under composition. Associativity holds since we consider the multiplication of complex numbers. The identity element is given by

$$m(1, 0, 0, 1) = z.$$

To find the inverse of $m(z)$ we assume that

$$m\left(\widetilde{m}(z)\right) = \frac{(ae + bg)z + (af + hb)}{(ce + dg)z + (cf + hd)} = z.$$

so that

$$ae + bg = 1, \qquad af + hb = 0$$
$$ce + dg = 0, \qquad cf + hd = 1$$

and we find

$$e = \frac{d}{ad - bc}, \qquad f = -\frac{b}{ad - bc}$$

$$g = -\frac{c}{ad - bc}, \qquad h = \frac{a}{ad - bc}.$$

The inverse is thus given by

$$(z')^{-1} = \frac{dz - b}{-cz + a}.$$ ♣

Example: Let **Z** be the abelian group of integers. Let E be the set of even integers. Obviously, E is an abelian group under addition and is a subgroup of **Z**. Let C_2 be the cyclic group of order 2. Then

$$\mathbf{Z}/E \cong C_2$$ ♣

We denote the mapping between two groups by ρ and present the following definitions:

Definition 2.8 *A mapping of a group G into another group G' is called a* **homomorphism** *if it preserves all combinatorial operations associated with the group G so that*

$$\rho(a \cdot b) = \rho(a) * \rho(b)$$

$a, b \in G$ *and* $\rho(a)$, $\rho(b) \in G'$. *Here* $\cdot$ *and* $*$ *is the group composition in G and G' respectively.*

Example: There is a homomorphism ρ from $GL(2, \mathbf{C})$ into the Möbius group M given by

$$\rho : \begin{pmatrix} a & b \\ c & d \end{pmatrix} \rightarrow m(z) = \frac{az + b}{cz + d}.$$

We now check that ρ is indeed a homomorphism: Consider

$$A = \begin{pmatrix} a & b \\ c & d \end{pmatrix}$$

where $a, b, c, d \in \mathbf{C}$ and $ad - bc \neq 0$. The matrices A form a group with matrix multiplication as group composition. We find

$$AB = \begin{pmatrix} a & b \\ c & d \end{pmatrix} \begin{pmatrix} e & f \\ g & h \end{pmatrix} = \begin{pmatrix} ae + bg & af + bh \\ ce + dg & cf + dh \end{pmatrix}$$

where $e, f, g, h \in \mathbf{C}$. Consider the mapping

$$\rho(AB) = \frac{(ae + bg)z + (af + bh)}{(ce + dg)z + (cf + dh)}$$

and

$$\rho(A) = \frac{az + b}{cz + d}$$

$$\rho(B) = \frac{ez + f}{gz + h}$$

so that

$$\rho\left(\rho(A)\right) = \frac{(ae + bg)z + (af + bh)}{(ce + dg)z + (cf + dh)}.$$

We have shown that $\rho(A \cdot B) = \rho(A) * \rho(B)$ and thus that ρ is a homomorphism.   ♣

Remark: An extension of the Möbius group is as follows: Consider the transformation

$$\mathbf{v} = \frac{A\mathbf{w} + B}{C\mathbf{w} + D}$$

where $\mathbf{v} = (v_1, \ldots, v_n)^T$, $\mathbf{w} = (w_1, \ldots, w_n)^T$ (T transpose). A is an $n \times n$ matrix, B an $n \times 1$ matrix, C a $1 \times n$ matrix and D a 1×1 matrix. The $(n+1) \times (n+1)$ matrix

$$\begin{pmatrix} A & B \\ C & D \end{pmatrix}$$

is invertible.

In the following let G and G' be groups.

Definition 2.9 *A mapping of all elements in G* **onto** *elements of G' is called* **surjective**.

Definition 2.10 *A* **one-to-one** *(or* **faithful***) mapping of elements in G to elements in G' is called an* **injection**.

Definition 2.11 *A map that is both one-to-one and onto is called a* **bijection**.

Definition 2.12 *If we have a mapping from G to G' that is a* **surjective homomorphism** *we say that G' is* **homomorphic** *to G.*

Definition 2.13 *If we have a mapping from G to G' that is a* **bijective homomorphism (isomorphism)** *we say that G' is* **isomorphic** *to G; we write $G \cong G'$.*

Example: An $n \times n$ permutation matrix is a matrix that has in each row and each column precisly one 1. There are $n!$ permutation matrices. The $n \times n$ permutation matrices form a group under matrix multiplication. Consider the symmetric group S_n given above. It is easy to see that the two groups are isomorphic. **Cayley's theorem** tells us that every finite group is isomorphic to a subgroup (or the group itself) of these permutation matrices. The six 3×3 permutation matrices are given by

$$A = \begin{pmatrix} 1 & 0 & 0 \\ 0 & 1 & 0 \\ 0 & 0 & 1 \end{pmatrix}, \qquad B = \begin{pmatrix} 1 & 0 & 0 \\ 0 & 0 & 1 \\ 0 & 1 & 0 \end{pmatrix}, \qquad C = \begin{pmatrix} 0 & 1 & 0 \\ 1 & 0 & 0 \\ 0 & 0 & 1 \end{pmatrix}$$

$$D = \begin{pmatrix} 0 & 1 & 0 \\ 0 & 0 & 1 \\ 1 & 0 & 0 \end{pmatrix}, \qquad E = \begin{pmatrix} 0 & 0 & 1 \\ 1 & 0 & 0 \\ 0 & 1 & 0 \end{pmatrix}, \qquad F = \begin{pmatrix} 0 & 0 & 1 \\ 0 & 1 & 0 \\ 1 & 0 & 0 \end{pmatrix}.$$

We have

$$AA = A \quad AB = B \quad AC = C \quad AD = D \quad AE = E \quad AF = F$$

$$BA = B \quad BB = A \quad BC = D \quad BD = C \quad BE = F \quad BF = E$$

$$CA = C \quad CB = E \quad CC = A \quad CD = F \quad CE = B \quad CF = D$$

$$DA = D \quad DB = F \quad DC = B \quad DD = E \quad DE = A \quad DF = C$$

$$EA = E \quad EB = C \quad EC = F \quad ED = A \quad EE = D \quad EF = B$$

$$FA = F \quad FB = D \quad FC = E \quad FD = B \quad FE = C \quad FF = A$$

For the inverse we find

$$A^{-1} = A, \quad B^{-1} = B, \quad C^{-1} = C, \quad D^{-1} = E, \quad E^{-1} = D, \quad F^{-1} = F.$$

The order of a finite group is the number of elements of the group. Thus our group has order 6. Lagrange's theorem tells us that the order of a subgroup of a finite group divides the order of the group. Thus the subgroups must have order 3, 2, 1. From the group table we find the subgroups

$$\{A, \ D, \ E\}$$

$$\{A, \ B\}, \qquad \{A, \ C\}, \qquad \{A, \ F\}$$

$$\{A\}$$

Cayley's theorem tells us that every finite group is isomorphic to a subgroup (or the group itself) of these permutation matrices. The **order of an element** $g \in G$ is the order of the cyclic subgroup generated by $\{g\}$, i.e. the smallest positive integer m such that

$$g^m = e$$

where e is the identity element of the group. The integer m divides the order of G. Consider, for example, the element D of our group. Then

$$D^2 = E, \qquad D^3 = A, \qquad A \quad \text{identity element.}$$

Thus $m = 3$.   ♣

Definition 2.14 *A homomorphism of G to itself is called an* **endomorphism** *of G.*

Definition 2.15 *An isomorphism of G to itself is called an* **automorphism** *of G.*

Definition 2.16 *The* **kernel** K *of a mapping ρ between two groups is the set*

$$K = \{\ g_i \in G : \rho(g_i) = e'\ \}$$

where e' is the identity element in G'.

Note that the set K is a subgroup of G.

Let us give examples of groups which are homomorphic :

Example: The group $\mathbf{R}$ and S^1 (unit circle) are homomorphic by the homomorphism $\rho : \mathbf{R} \to S^1$ defined by

$$\rho(x) = e^{2\pi i x}.$$

The transformation is locally one-to-one but globally it is infinite-to-one. For all points $x + n$ (with n an integer) the map ρ maps onto the same point $\exp(2\pi i x)$ in S^1 since

$$e^{2n\pi} = 1.$$

The kernel of ρ is $\mathbf{Z}$, the discrete group of integers.   ♣

Example: The group of positive real numbers $\mathbf{R}^+$ with ordinary multiplication being the group operation, is isomorphic to the additive group of the real numbers $\mathbf{R}$. The exponential function,

$$\rho(t) = e^t$$

with $t \in \mathbf{R}$, provides the isomorphism, $\rho : \mathbf{R} \to \mathbf{R}^+$.   ♣

2.2 Computer Algebra Applications

In the program we consider the permutation group S_3. We give the composition of the group elements. Then we evaluate the inverse of each group element. Finally we determine the conjugacy classes. The group consists of six elements which we denote by $a(0)$, $a(1)$, ..., $a(5)$. The neutral (identity) element is denoted by $a(0)$. Thus we have

$$a(0) * a(j) = a(j) * a(0) = a(j)$$

for $j = 0, 1, \ldots, 5$. The group is nonabelian.

```
%gr1.red;

operator a, g, res, test, cl1, cl2;
noncom a, g, res, test, cl1, cl2;

% a(0) is the neutral element;
for all j let a(j)*a(0) = a(j);
for all j let a(0)*a(j) = a(j);

let a(1)*a(1) = a(0);
let a(1)*a(2) = a(3); let a(2)*a(1) = a(4);
let a(1)*a(3) = a(2); let a(3)*a(1) = a(5);
let a(1)*a(4) = a(5); let a(4)*a(1) = a(2);
let a(1)*a(5) = a(4); let a(5)*a(1) = a(3);
let a(2)*a(2) = a(0);
let a(2)*a(3) = a(5); let a(3)*a(2) = a(1);
let a(2)*a(4) = a(1); let a(4)*a(2) = a(5);
let a(2)*a(5) = a(3); let a(5)*a(2) = a(4);
let a(3)*a(3) = a(4);
let a(3)*a(4) = a(0); let a(4)*a(3) = a(0);
let a(3)*a(5) = a(2); let a(5)*a(3) = a(1);
let a(4)*a(4) = a(3);
let a(4)*a(5) = a(1); let a(5)*a(4) = a(2);
let a(5)*a(5) = a(0);
res := a(0)*a(1)*a(2)*a(3)*a(4)*a(5);

% find the inverse;
g(0) := a(0);
for j:=0:5 do
begin
for k:=0:5 do
begin
test := a(j)*a(k);
if test = a(0) then g(j) := a(k);
```

```
end;
end;

for j:=0:5 do write g(j);

% conjugacy class of the group element a(1);
for j:= 0:5 do
cl1(j) := a(j)*a(1)*g(j);

for j:=0:5 do write cl1(j);

% conjugacy class of the group element a(3);
for j:=0:5 do
cl2(j) := a(j)*a(3)*g(j);

for j:=0:5 do write cl2(j);
```

The output is

```
res := a(2)
```

```
for j:=0:5 do write g(j);
a(0)
a(1)
a(2)
a(4)
a(3)
a(5)

for j:=0:5 do write cl1(j);
a(1)
a(1)
a(5)
a(2)
a(5)
a(2)

for j:=0:5 do write cl2(j);
a(3)
a(4)
a(4)
a(3)
a(3)
a(4)
```

Chapter 3

Lie Groups and Lie Transformation Groups

3.1 Lie Groups

First we introduce the definition of a Lie group, and then we give some examples (Gilmore [52], Von Westenholz [141], Sattinger and Weaver [103], Choquet-Bruhat *et al* [17]).

Definition 3.1 *An r-parameter Lie group G, which also carries the structure of an r-dimensional differentiable manifold in such a way that both the group composition*

$$c : G \times G \to G \qquad c(g_1, g_2) = g_1 \cdot g_2$$

with $g_1, g_2 \in G$, and the inversion

$$i : G \to G$$

are smooth maps between manifolds.

From the definition it follows that a Lie group carries the structure of a differentiable manifold so that group elements can be continuously varied.

Example: The additive group $\mathbf{R}$ is a Lie group since it is a differentiable manifold (see appendix A). ♣

Example: The general linear group $GL(m, \mathbf{R})$ is a Lie group. The manifold structure can be identified with the open subset

$$GL(m, \mathbf{R}) := \{ \; A \; : \; \det A \neq 0 \; \}$$

of the linear space of all $m \times m$ nonsingular matrices. This space is isomorphic to $\mathbf{R}^{m^2}$ with matrix entries A_{ij} of A. Thus $GL(m, \mathbf{R})$ is also an m^2-dimensional manifold. ♣

Note that in both cases mentioned above the group composition is analytic.

Definition 3.2 *The Lie group G is* **compact** *if its manifold is a closed and bounded submanifold of $\mathbf{R}^m$, for some m.*

A group is thus compact if all its elements are bounded. In particular, for a matrix group it must be true for all the matrices of the group that each entry of the matrices is bounded.

Example: Consider the unitary group $U(m)$ which is discussed in detail in the next section. The group is defined by

$$U(m) := \{A \in GL(m, \mathbf{C}) \;\; : \;\; A^\dagger A = I_m\}$$

where I_m is the $m \times m$ unit matrix and † denotes the transpose and complex conjugate. For $A \in U(m)$ it follows that

$$\sum_{k=1}^m A_{ki}^* A_{kj} = \delta_{ij}.$$

For $i = j$ this becomes

$$|A_{1j}|^2 + |A_{2j}|^2 + \cdots + |A_{mj}|^2 - 1,$$

so that $|A_{lj}| \leq 1$ for $1 \leq l, j \leq m$. Thus the entries of all the matrices in $U(m)$ are bounded by 1. This is also true for the identity element I_m (unit matrix) and so $U(m)$ is compact. ♣

In physics groups emerge as transformations. For applications of Lie groups to differential equations we consider the transformation of coordinates and fields. Such transformations form a Lie transformation group. Lie transformation groups such as the standard matrix groups, also known as classical groups, will be discussed in the next section. $GL(n, \mathbf{R})$, discussed above, is also a standard matrix group.

A Lie group is a special case of a topological group. A set G together with a group operation and a topology is said to be a **topological group** if the mappings which define its group structure are continuous, i.e. if the mappings

$$G \times G \to G \quad \text{by} \quad (g_1, g_2) \mapsto g_1 \cdot g_2 = g_1 g_2$$

$$G \mapsto G \quad \text{by} \quad g \mapsto g^{-1}$$

are continuous.

3.2 Lie Transformation Groups

Let M be a differential manifold and G be a Lie group.

Definition 3.3 *The group G is called a* **Lie transformation group** *of M if there is a differential map*

$$\varphi : G \times M \to M, \qquad \phi(g, \mathbf{x}) = g\mathbf{x}$$

such that

1. *$(g_1 \cdot g_2)\mathbf{x} = g_1 \cdot (g_2\mathbf{x})$ for $\mathbf{x} \in M$ and $g_1, g_2 \in G$,*

2. *$e\mathbf{x} = \mathbf{x}$ for the identity element e of G and $\mathbf{x} \in M$,*

are satisfied. Note that $\mathbf{x}$ is now transformed to $g\mathbf{x}$ by the transformation φ. This is also known as the **group action** *on $\mathbf{x}$ and will be discussed in more detail in section (3.2.3).*

The classical groups as well as the affine groups are examples of Lie transformation groups which will now be discussed in some detail.

3.2.1 Classical Groups

We consider the canonical m-dimensional vector space V^m of $m \times 1$ matrices with the standard basis

$$\mathbf{e}_1 = \begin{pmatrix} 1 \\ 0 \\ 0 \\ \vdots \\ 0 \end{pmatrix}, \quad \mathbf{e}_2 = \begin{pmatrix} 0 \\ 1 \\ 0 \\ \vdots \\ 0 \end{pmatrix}, \ldots, \quad \mathbf{e}_m = \begin{pmatrix} 0 \\ 0 \\ \vdots \\ 0 \\ 1 \end{pmatrix}.$$

The vector space V^m over the fields $\mathbf{R}$ and $\mathbf{C}$ is denoted by $\mathbf{R}^m$ and $\mathbf{C}^m$, respectively. Note that the choice of a set of basis vectors $\{\mathbf{e}_1, \mathbf{e}_2, \cdots, \mathbf{e}_m\}$ in a vector space V^m is not unique so that every set of basis vectors in V^m can be related to every other coordinate system by an $m \times m$ nonsingular matrix. Let $\mathbf{x}, \mathbf{y} \in V^m$ such that

$$\mathbf{x} = \begin{pmatrix} x_1 \\ x_2 \\ \vdots \\ x_m \end{pmatrix}, \quad \mathbf{y} = \begin{pmatrix} y_1 \\ y_2 \\ \vdots \\ y_m \end{pmatrix}.$$

A linear transformation of V^m can be represented by an $m \times m$ matrix A over $\mathbf{C}$ or $\mathbf{R}$ such that

$$A : \mathbf{x} \mapsto \mathbf{x}' = A\mathbf{x}$$

where

$$A(a\mathbf{x} + b\mathbf{y}) = a(A\mathbf{x}) + b(A\mathbf{y})$$

with $a, b \in \mathbf{C}$ (or $\mathbf{R}$) are constants.

The set of all $m \times m$ nonsingular matrices over $\mathbf{C}$ or $\mathbf{R}$ forms the **general linear group** $GL(m, \mathbf{C})$ and $GL(m, \mathbf{R})$, respectively, which is a linear invertible transformation of V^m. The group composition is given by matrix multiplication.

Remark: If we do not specify the field the results obtained are valid for real as well as complex numbers.

Let $A, B \in GL(m)$, so that $\det A \neq 0$ and $\det B \neq 0$. Since

$$\det(AB) \equiv (\det A)(\det B) \neq 0$$

we find that the set is closed by matrix multiplication and the inverse matrices exist. The unit matrix I is the identity element. Matrix multiplication is associative. $GL(m, \mathbf{C})$ has m^2 complex parameters and therefore $2m^2$ real parameters, we write $\dim\{GL(m, \mathbf{C})\} = 2m^2$ and $\dim\{GL(m, \mathbf{R})\} = m^2$.

The **special linear group**, defined by

$$SL(m) := \{ \ A \in GL(m) \ : \ \det A = 1 \ \}$$

is a subgroup of $GL(m)$ with the restriction that the determinant of the matrix elements be equal to one. For $A, B \in GL(m)$ and $\det A = 1$, $\det B = 1$. It follows that

$$\det(AB) = 1$$

so that $AB \in SL(m)$. The subgroups are as follows:

$$SL(m, \mathbf{R}) < GL(m, \mathbf{R}) < GL(m, \mathbf{C}) > SL(m, \mathbf{C})$$

where

$$SL(m, \mathbf{R}) < SL(m, \mathbf{C}).$$

We find $\dim\{SL(m, \mathbf{C})\} = 2(m^2 - 1)$ and $\dim\{SL(m, \mathbf{R})\} = m^2 - 1$.

The other classical groups leave certain **bilinear forms (metrics)** on the vector space V^m invariant.

Definition 3.4 *A bilinear metric function on a vector space V^m is a mapping of a pair of vectors into a number in the field F associated with the vector space:*

$$(\mathbf{v}_1, \mathbf{v}_2) = f, \qquad \mathbf{v}_1, \mathbf{v}_2 \in V^m, \qquad f \in F.$$

The mapping obeys

$$(\mathbf{v}_1, a\mathbf{v}_2 + b\mathbf{v}_3) = a(\mathbf{v}_1, \mathbf{v}_2) + b(\mathbf{v}_1, \mathbf{v}_3)$$

and

$$(a\mathbf{v}_1 + b\mathbf{v}_2, \mathbf{v}_3) = a(\mathbf{v}_1, \mathbf{v}_3) + b(\mathbf{v}_2, \mathbf{v}_3).$$

Let us now consider the metric preserving groups: Suppose V^m has a metric $(\mathbf{x}, \mathbf{y})$. Let $H \subset G$ be defined by

$$H := \{A \in GL(m, V) \; : \; (A\mathbf{x}, A\mathbf{y}) = (\mathbf{x}, \mathbf{y}) \text{ for all } \mathbf{x}, \mathbf{y} \in V^m\}$$

i.e., H preserves (leaves invariant) the metric. Then $H < GL(m, V)$ for any metric on V^m.

Orthogonal Groups

We consider the **Euclidean metric** or symmetric bilinear form in $\mathbf{R}^m$: Let $\mathbf{x}, \mathbf{y} \in \mathbf{R}^m$. Then

$$(\mathbf{x}, \mathbf{y}) = \mathbf{x}^T I_m \mathbf{y} = (x_1, x_2, \ldots, x_m) \begin{pmatrix} 1 & 0 & 0 & \ldots & 0 \\ 0 & 1 & 0 & \ldots & 0 \\ \vdots & & & & \vdots \\ 0 & 0 & \ldots & 0 & 1 \end{pmatrix} \begin{pmatrix} y_1 \\ y_2 \\ \vdots \\ y_m \end{pmatrix} = \sum_{i=1}^m x_i y_i$$

where $\mathbf{x}^T$ denotes the transpose of $\mathbf{x}$ and I_m is the $m \times m$ unit matrix. Thus $(\mathbf{x}, \mathbf{y})$ is the inner or scalar product in $\mathbf{R}^m$. We can now define the **orthogonal group**

$$O(m) := \{A \in GL(m, \mathbf{R}) \; : \; A \text{ preserves the Euclidean metric}\}.$$

Let $\mathbf{x}, \mathbf{y} \in \mathbf{R}^m$ so that

$$(A\mathbf{x}, A\mathbf{y}) = (A\mathbf{x})^T A\mathbf{y} = \mathbf{x}^T A^T A \mathbf{y} = \mathbf{x}^T I_m \mathbf{y} = \mathbf{x}^T \mathbf{y}$$

where A is such that $A^T A = I_m$. With the condition $\det A = 1$ we define

$$SO(m) := \{\, A \in O(m) \; : \; \det A = 1 \,\} < O(m)$$

which is called the **special orthogonal group**. The dimension of $O(m)$ and $SO(m)$ is $m(m-1)/2$. The group $O(m)$ is compact since every element is closed and bounded: From $A^T A = I_m$ it follows that

$$\sum_{k=1}^m (A^T)_{ik} A_{kj} = \delta_{ij}.$$

Therefore

$$\sum_{k=1}^m A_{ki} A_{kj} = \delta_{ij}$$

and consequently $|A_{ij}| \leq 1$. Obviously $SO(m)$ is also compact.

Unitary Groups

We consider the **Hermitian metric** or Hermitian symmetric form in $\mathbf{C}^m$: Let $\mathbf{x}, \mathbf{y} \in \mathbf{C}^m$, then

$$(\mathbf{x}, \mathbf{y}) = \mathbf{x}^\dagger I_m \mathbf{y} = (x_1^*, x_2^*, \ldots, x_m^*) \begin{pmatrix} 1 & 0 & 0 & \ldots & 0 \\ 0 & 1 & 0 & \ldots & 0 \\ \vdots & & & & \vdots \\ 0 & 0 & \ldots & 0 & 1 \end{pmatrix} \begin{pmatrix} y_1 \\ y_2 \\ \vdots \\ y_m \end{pmatrix} = \sum_{i=1}^m x_i^* y_i$$

where the dagger † denotes the transpose and complex conjugate. For example, $m = 2$ is the hermitian inner product defined on the two-component spinor. We can now define the unitary group

$$U(m) := \{\, A \in GL(m, \mathbf{C}) \;:\; A \text{ preserves the hermitian metric} \,\}.$$

Let $\mathbf{x}, \mathbf{y} \in \mathbf{C}^m$. Then

$$(A\mathbf{x}, A\mathbf{y}) = (A\mathbf{x})^\dagger I_m (A\mathbf{y}) = \mathbf{x}^\dagger A^\dagger A \mathbf{y} = \mathbf{x}^\dagger \mathbf{y}$$

for all $\mathbf{x}, \mathbf{y} \in \mathbf{C}^m$ where $A^\dagger A = I_m$. With the condition $\det A = 1$ we define

$$SU(m) := \{\, A \in U(m) \;:\; \det A = 1 \,\} < U(m),$$

which is called the **special unitary group**. $U(m)$ has the dimension m^2 and $SU(m)$ the dimension $m^2 - 1$. The group $U(m)$ is compact since every element is closed and bounded: From $A^T A = I_m$ it follows that

$$\sum_{k=1}^m (A^\dagger)_{ik} A_{kj} = \delta_{ij}.$$

Therefore

$$\sum_{k=1}^m A_{ki}^* A_{kj} = \delta_{ij}$$

and consequently $|A_{ij}| \leq 1$. Obviously $SU(m)$ is also compact.

Pseudo-Orthogonal Groups

We consider the **Lorentzian metric** in $\mathbf{R}^m$: Let $\mathbf{x}, \mathbf{y} \in \mathbf{R}^m$. Then

$$(\mathbf{x}, \mathbf{y}) = \mathbf{x}^T L \mathbf{y} = \mathbf{x}^T \begin{pmatrix} -I_P & 0 \\ 0 & I_{m-P} \end{pmatrix} \mathbf{y} = -\sum_{i=1}^P x_i y_i + \sum_{i=P+1}^m x_i y_i$$

where

$$L := \begin{pmatrix} -I_P & 0 \\ 0 & I_{m-P} \end{pmatrix}$$

and I_P is the $P \times P$ unit matrix. For example, the metric with $P = 1$ and $m = 4$ is important in the theory of special relativity. We can now define the **pseudo-orthogonal group**

$$O(P, m - P) := \{ A \in GL(m, \mathbf{C}) \ : \ A \ \text{preserves the Lorentzian metric} \}.$$

Let $\mathbf{x}, \mathbf{y} \in \mathbf{R}^m$. Then

$$(A\mathbf{x}, A\mathbf{y}) = (A\mathbf{x})^T L(A\mathbf{y}) = \mathbf{x}^T A^T L A\mathbf{y} = \mathbf{x}^T L\mathbf{y}$$

for all $\mathbf{x}, \mathbf{y} \in \mathbf{R}^m$, where $A^T L A = L$. With the condition $\det A = 1$ we write

$$SO(P, m - P) = \{A \in O(P, m - P) \ : \ \det A = 1\} < O(P, m - P),$$

which is called the **pseudo-special orthogonal group**. $O(P, m - P)$ has the dimension $\frac{1}{2}m(m - 1)$ which is the same as the dimension of $SO(P, m - P)$. $O(P, m - P)$ (as well as $SO(P, m - P)$) is not compact since $|A_{ij}|$ is not bounded by $A^T L A = L$.

Pseudo-Unitary group

Consider the **Hermitian-Lorentzian metric** in $\mathbf{C}^m$: Let $\mathbf{x}, \mathbf{y} \in \mathbf{C}^m$. Then

$$(\mathbf{x}, \mathbf{y}) = \mathbf{x}^\dagger L\mathbf{y} = \mathbf{x}^\dagger \begin{pmatrix} -I_P & 0 \\ 0 & I_{m-P} \end{pmatrix} \mathbf{y} = -\sum_{i=1}^{P} x_i^* y_i + \sum_{i=P+1}^{m} x_i^* y_i.$$

We can now define the **pseudo-unitary group**

$$U(P, m - P) := \{ A \in GL(m, \mathbf{C}) \ : \ A \ \text{preserves the hermitian-Lorentzian metric} \}.$$

Let $\mathbf{x}, \mathbf{y} \in \mathbf{C}^m$. Then

$$(A\mathbf{x}, A\mathbf{y}) = (A\mathbf{x})^\dagger L(A\mathbf{y}) = \mathbf{x}^\dagger A^\dagger L A\mathbf{y} = \mathbf{x}^\dagger L\mathbf{y} \quad \text{for all} \quad \mathbf{x}, \mathbf{y} \in \mathbf{C}^m$$

where $A^\dagger L A = L$ and L is defined as

$$L := \begin{pmatrix} -I_P & 0 \\ 0 & I_{m-P} \end{pmatrix}.$$

With the condition $\det A = 1$ we define

$$SU(P, m - P) := \{A \in U(P, m - P) \ : \ \det A = 1\} < U(P, m - P),$$

which is called the **pseudo-special unitary group**. $U(P, m - P)$ has the dimension m^2 and $SU(P, m - P)$ has the dimension $m^2 - 1$. $U(P, m - P)$ (as well as $SU(P, m - P)$) is not compact since $|A_{ij}|$ is not bounded by $A^\dagger L A = L$.

Symplectic Group

Consider the **symplectic metric** or skew-symmetric bilinear form: The elements are in a necessarily even dimensional vector space $\mathbf{C}^{2m}$ or $\mathbf{R}^{2m}$, so that

$$(\mathbf{x}, \mathbf{y}) = \mathbf{x}^\dagger J \mathbf{y} = \mathbf{x}^\dagger \begin{pmatrix} 0 & I_m \\ -I_m & 0 \end{pmatrix} \mathbf{y} = \sum_{i=1}^{m} x_i^* y_{m+i} - \sum_{i=1}^{m} x_{m+i}^* y_i$$

where we define

$$J := \begin{pmatrix} 0 & I_m \\ -I_m & 0 \end{pmatrix}.$$

An example is the group of canonical transformations in linear Hamilton systems. We can now define the **symplectic group**

$$Sp(2m) := \{ A \in GL(m) \ : \ A \ \text{preserves the symplectic metric} \}.$$

Let $\mathbf{x}, \mathbf{y} \in \mathbf{C}^{2m}$. Then

$$(A\mathbf{x}, A\mathbf{y}) = (A\mathbf{x})^\dagger J (A\mathbf{y}) = \mathbf{x}^\dagger A^\dagger J A \mathbf{y} = \mathbf{x}^\dagger J \mathbf{y}$$

for all $\mathbf{x}, \mathbf{y} \in \mathbf{C}^{2m}$ where $A^\dagger J A = J$. With the condition $\det A = 1$ we define

$$SSp(2m) := \{A \in Sp(2m) \ : \ \det A = 1\} < Sp(2m),$$

which is called the **special symplectic group**. $Sp(2m, \mathbf{R})$ has dimension $m(2m + 1)$ and $Sp(2m, \mathbf{C})$ has dimension $2m(m^2 + 1)$. The groups $SSp(2m, \mathbf{R})$ and $SSp(2m, \mathbf{C})$ have dimension $m(2m+1)$ and $2m(2m+1)-1$, respectively. $Sp(2m)$, as well as $SSp(2m)$, is not compact since $|A_{ij}|$ is not bounded by $A^\dagger J A = J$.

To summarize, we represent the classical groups and their properties in tabular form:

group	invariant metric	matrix condition	dimension	compact
$GL(m, \mathbf{C}/\mathbf{R})$	—	$\det A \neq 0$	$2m^2 \,/\, m^2$	no
$SL(m, \mathbf{C}/\mathbf{R})$	—	$\det A = 1$	$2(m^2 - 1) \,/\, m^2 - 1$	no
$O(m)$	$\mathbf{x}^T \mathbf{y}$	$A^T A = I$	$\frac{1}{2}m(m - 1)$	yes
$SO(m)$	$\mathbf{x}^T \mathbf{y}$	$A^T A = I$ and $\det A = 1$	$\frac{1}{2}m(m - 1)$	yes
$U(m)$	$\mathbf{x}^\dagger \mathbf{y}$	$A^\dagger A = I$	m^2	yes
$SU(m)$	$\mathbf{x}^\dagger \mathbf{y}$	$A^\dagger A = I$ and $\det A = 1$	$m^2 - 1$	yes
$O(P, m - P)$	$\mathbf{x}^T L \mathbf{y}$	$A^T L A = L$	$\frac{1}{2}m(m - 1)$	no
$SO(P, m - P)$	$\mathbf{x}^T L \mathbf{y}$	$A^T L A = L$ and $\det A = 1$	$\frac{1}{2}m(m - 1)$	no
$U(P, m - P)$	$\mathbf{x}^\dagger L \mathbf{y}$	$A^\dagger L A = L$	m^2	no
$SU(P, m - P)$	$\mathbf{x}^\dagger L \mathbf{y}$	$A^\dagger L A = L$ and $\det A = 1$	$m^2 - 1$	no
$Sp(2m, \mathbf{C}/\mathbf{R})$	$\mathbf{x}^\dagger J \mathbf{y}$	$A^\dagger J A = J$	$2m(2m + 1) \,/\, m(2m + 1)$	no
$SSp(2m, \mathbf{C}/\mathbf{R})$	$\mathbf{x}^\dagger J \mathbf{y}$	$A^\dagger J A = J$ and $\det A = 1$	$2m(2m + 1) - 1 \,/\, m(2m + 1)$	no

Parametrization of Classical Groups

Let us give examples of some well known classical group parametrizations.

Example: Recall that for the Pseudo-Orthogonal group $O(P, m - P)$, the condition on the matrix A is

$$A^T L A = L$$

with

$$L := \begin{pmatrix} -I_P & 0 \\ 0 & I_{m-P} \end{pmatrix}.$$

We now consider $P = 1$ and $m = 2$ which is the group $O(1,1)$. It follows that

$$\sum_{\alpha=1}^{2} \sum_{\beta=1}^{2} A_{\alpha\rho} L_{\alpha\beta} A_{\beta\sigma} = L_{\rho\sigma}$$

with ρ, $\sigma = 1, 2$ and

$$L = \begin{pmatrix} -1 & 0 \\ 0 & 1 \end{pmatrix}.$$

The conditions on A are then

$$-A_{11}^2 + A_{21}^2 = -1, \qquad -A_{12}^2 + A_{22}^2 = 1, \qquad -A_{11}A_{12} + A_{21}A_{22} = 0.$$

Note that $\dim O(1,1) = 1$ so that a solution is

$$A_{11} = \cosh\phi, \quad A_{21} = \sinh\phi, \quad A_{12} = \sinh\phi, \quad A_{22} = \cosh\phi$$

with one independent parameter ϕ. The group

$$\mathbf{x}' = A(\phi)\mathbf{x} = \begin{pmatrix} \cosh\phi & \sinh\phi \\ \sinh\phi & \cosh\phi \end{pmatrix} \mathbf{x}$$

is commonly known as the **Lorentz boost**. Here

$$\mathbf{x}' = \begin{pmatrix} x_1' \\ x_2' \end{pmatrix}, \qquad \mathbf{x} = \begin{pmatrix} x_1 \\ x_2 \end{pmatrix}$$

so that

$$x_1' = x_1 \cosh\phi + x_2 \sinh\phi, \qquad x_2' = x_1 \sinh\phi + x_2 \cosh\phi.$$

It follows that

$$x_1'^2 - x_2'^2 = x_1^2 - x_2^2.$$

In special relativity the parameter ϕ is considered as

$$\tanh\phi = \frac{v}{c}$$

where v is the relative velocity between the moving frames, c is the speed of light and $x_1 \equiv x$, $x_2 \equiv ct$ with $v < c$. ♣

Examples: We give the following classical group parametrizations:

i) The one-dimensional (one parameter) unitary group,

$$U(1) := \{\, e^{i\theta} \ : \ \theta \in \mathbf{R}\}.$$

ii) The one parameter special orthogonal group,

$$SO(2) := \left\{ \begin{pmatrix} \cos\theta & \sin\theta \\ -\sin\theta & \cos\theta \end{pmatrix} \ : \ \theta \in \mathbf{R} \right\}.$$

iii) The four parameter unitary group

$$
\begin{aligned}
U(2) \ &:= \ \left\{ \begin{pmatrix} e^{i\alpha}\cos\nu & e^{i\gamma}\sin\nu \\ -e^{i(\beta-\gamma)}\sin\nu & e^{i(\beta-\alpha)}\cos\nu \end{pmatrix} \ : \ \alpha,\beta,\gamma,\nu \in \mathbf{R} \right\} \\
&= \ \left\{ \begin{pmatrix} e^{i(\psi+\phi)/2}\cos\theta/2 & ie^{i(\psi-\phi)/2}\sin\theta/2 \\ ie^{i(\beta+\phi-\psi)/2}\sin\theta/2 & e^{i(\beta-\phi-\psi)/2}\cos\theta/2 \end{pmatrix} \ : \ \beta,\theta,\phi,\psi \in \mathbf{R} \right\}.
\end{aligned}
$$

This parametrization is called the **Euler parametrization**.

iv) For the three parameter special unitary group $SU(2)$ we can set $\beta = 0$ in the above parametrization for $U(2)$.

v) The three parameter special orthogonal group,

$$
\begin{aligned}
SO(3) = & \\
\left\{ \begin{pmatrix} \cos\psi\cos\phi - \cos\theta\sin\phi\sin\psi & -\cos\psi\sin\phi - \cos\theta\cos\phi\sin\psi & \sin\psi\sin\theta \\ \sin\psi\cos\phi + \cos\theta\sin\phi\cos\psi & -\sin\psi\sin\phi + \cos\theta\cos\phi\cos\psi & -\cos\phi\sin\theta \\ \sin\theta\sin\phi & \sin\theta\cos\phi & \cos\theta \end{pmatrix} \right. & \\
: \ \phi,\psi,\theta \in \mathbf{R} \ \} &
\end{aligned}
$$

where ϕ, ψ, θ are known as the **Euler angles**.   ♣

Homomorphisms between some classical groups

Here we give a list of homomorphisms (which include some isomorphisms) between some classical groups.

Dimension	Homomorphism
1	$U(1) \sim SO(2)$
3	$SU(2) \sim SO(3)$
	$SO(1,2) \sim SU(1,1) \sim Sp(2,\mathbf{R})$
6	$SO(1,3) \sim SL(2,\mathbf{C})$
10	$SO(2,3) \sim Sp(4,\mathbf{R})$
15	$SO(6) \sim SU(4)$
	$SO(2,4) \sim SU(2,2)$
	$SO(3,3) \sim SL(4,\mathbf{R})$

Classical Groups in Physics

Let us give a list of some important classical groups in physics:

$SO(2)$	—	Axial rotation.
$SO(3)$	—	Full rotation.
$SO(4)$	—	Kepler problem; Quantum gravity.
$SO(10)$	—	Great unified theories (GUT's).
$SO(32)$	—	Heterotic superstring theory.
$U(1)$	—	Gauge group in electromagnetism.
$SU(2)$	—	Isospin; gauge group in Yang-Mills theory.
$SU(3)$	—	Gauge group in quantum electrodynamics.
$U(3)$	—	Symmetry group for 3 dimensional harmonic oscillator.
$O(1,3)$	—	Lorentz group in special relativity.
$Sp(2n)$	—	Hamiltonian canonical transformations.

3.2.2   Affine Groups

The affine groups are also examples of Lie transformation groups. Before we can define affine groups we must consider the translation groups.

Translation Group

We denote by $T(m)$ the translation group on the vector space $\mathbf{R}^m$:

$$\tau(\mathbf{a}) : \mathbf{x} \mapsto \mathbf{x}' = \mathbf{x} + \mathbf{a}.$$

Thus the dimension of $T(m)$ is m and

$$T(m) = \{\tau(\mathbf{a}) \ : \ \mathbf{a} \in \mathbf{R}^m\}.$$

$T(m)$ is an abelian group since

$$\tau(\mathbf{a}') \cdot \tau(\mathbf{a}) = (\mathbf{x} + \mathbf{a}) + \mathbf{a}' = \mathbf{x} + (\mathbf{a} + \mathbf{a}') = \tau(\mathbf{a} + \mathbf{a}') = \tau(\mathbf{a}' + \mathbf{a}) = \tau(\mathbf{a}) \cdot \tau(\mathbf{a}').$$

The inverse is given by

$$\tau(\mathbf{a})^{-1} = \tau(-\mathbf{a})$$

and the identity is $e = \tau(\mathbf{0})$.

Affine Groups

Definition 3.5 *An* **affine group** *on* $\mathbf{R}^m$ *is a combination of (pseudo-) orthogonal and translation transformations in the vector space* $\mathbf{R}^m$ *with a Lorentzian or Euclidean metric.*

We denote the general transformation by

$$g(A, \mathbf{a}) : \mathbf{x} \mapsto \mathbf{x}' = A\mathbf{x} + \mathbf{a}$$

where $A \in O(P, m - P)$ or $A \in O(m)$. We can show that the transformation is closed:

$$g(A', \mathbf{a}') \cdot g(A, \mathbf{a}) = A'(A\mathbf{x} + \mathbf{a}) + \mathbf{a}' = (A'A)\mathbf{x} + A'\mathbf{a} + \mathbf{a}' = g(A'A, A'\mathbf{a} + \mathbf{a}').$$

We thus find the composition law for the general transformation to be

$$g(A', \mathbf{a}') \cdot g(A, \mathbf{a}) = g(A'A, A'\mathbf{a} + \mathbf{a}').$$

This composition law is characteristic of the **semidirect product** $\bigwedge$, where

$$\text{Affine group} = O(P, m - P) \bigwedge T(m).$$

Furthermore the identity e is given by $e = g(I_m, 0)$ and we can deduce the inverse $g(A, \mathbf{a})^{-1}$ from the composition law: Let $g(A, \mathbf{a})^{-1} = g(A', \mathbf{a}')$ so that

$$g(A'A, A'\mathbf{a} + \mathbf{a}') = g(I, 0).$$

Therefore, $A'A = I$ and $A'\mathbf{a} + \mathbf{a}' = 0$, and we find $A' = A^{-1}$ and $\mathbf{a}' = -A^{-1}\mathbf{a}$, i.e.,

$$g(A, \mathbf{a})^{-1} = g(A^{-1}, -A^{-1}\mathbf{a}).$$

The dimension of an affine group is as follows

$$
\begin{aligned}
\text{Dimension of an affine group} \quad &= \quad \dim O(P, m - P) \, + \, \dim T(m) \\
&= \quad \frac{1}{2}m(m - 1) + m \\
&= \quad \frac{1}{2}m(m + 1).
\end{aligned}
$$

Euclidean Group $E(m)$

The m-dimensional Euclidean space E^m is the space $\mathbf{R}^m$ with the Euclidean metric. The affine group $E(m)$ known as the Euclidean group is defined by

$$E(m) := \{\ g(A, \mathbf{a}) \quad : \quad A \in O(m), \mathbf{a} \in \mathbf{R}^m\ \}$$

so that

$$E(m) = O(m) \bigwedge T(m).$$

The transformation is

$$g(A, \mathbf{a}) : \mathbf{x} \mapsto \mathbf{x}' = A\mathbf{x} + \mathbf{a}.$$

The dimension is as follows: $\dim E(m) = m(m + 1)/2$. We say that $E(m)$ is the symmetry group of E^m, i.e., E^m is isotropic about $\mathbf{0}$, meaning all directions are equivalent $(O(m) < E(m))$, and E^m is homogeneous, meaning all points are equivalent $(T(m) < E(m))$.

Example: We consider the affine group on E^2, denoted by $E(2)$. Its dimension is three so that there are three independent parameters. We have

$$
\begin{aligned}
x_1' &= x_1 \cos\theta - x_2 \sin\theta \\
x_2' &= x_1 \sin\theta + x_2 \cos\theta
\end{aligned}
$$

with

$$x_1'^2 + x_2'^2 = x_1^2 + x_2^2.$$

Adding the translations we find that the transformation group $E(2)$ is given by

$$\begin{pmatrix} x_1' \\ x_2' \end{pmatrix} = \begin{pmatrix} \cos\theta & -\sin\theta \\ \sin\theta & \cos\theta \end{pmatrix} \begin{pmatrix} x_1 \\ x_2 \end{pmatrix} + \begin{pmatrix} a_1 \\ a_2 \end{pmatrix}.$$

The three independent parameters of the group are θ, a_1 and a_2. ♣

Poincaré Group $P(m)$

The affine group $P(m)$ is the symmetry group of the space M^m which is the space $\mathbf{R}^m$ with the Minkowski metric. $P(m)$ is defined by

$$P(m) := \{\ g(A, \mathbf{a})\ :\ A \in O(P, m - P), \mathbf{a} \in \mathbf{R}^m\ \}$$

so that

$$P(m) := O(P, m - P) \bigwedge T(m).$$

The dimension is as follows: dim $P(m) = m(m + 1)/2$. $P(m)$ is also known as the **inhomogeneous Lorentz group**. The transformation is

$$g(A, \mathbf{a}) :\ \mathbf{x} \mapsto \mathbf{x}' = A\mathbf{x} + \mathbf{a}.$$

Example: We consider the group $P(2) = O(1, 1) \bigwedge T(2)$ on the space $M^2 = \{x_1, x_2\}$. Note that dim $P(2) = 3$. The transformation is

$$P(2) :\ x_\alpha \to x'_\alpha = \sum_{\beta=1}^{2} A_{\alpha\beta} x_\beta + a_\alpha$$

where $\alpha = 1, 2$ and $A_{\alpha\beta} \in O(1, 1)$. Recall that for the one dimensional pseudo-orthogonal group $O(1, 1)$ we found the Lorentz boost

$$\mathbf{x}' = A(\phi)\mathbf{x} = \begin{pmatrix} \cosh\phi & \sinh\phi \\ \sinh\phi & \cosh\phi \end{pmatrix} \mathbf{x}.$$

Adding the translations we find the transformation group

$$\begin{pmatrix} x'_1 \\ x'_2 \end{pmatrix} = \begin{pmatrix} x_1 \cosh\phi + x_2 \sinh\phi \\ x_1 \sinh\phi + x_2 \cosh\phi \end{pmatrix} + \begin{pmatrix} a_1 \\ a_2 \end{pmatrix}$$

with the three independent parameters ϕ, a_1 and a_2.   ♣

Galilean Group - A Non-Affine Group

The group of Galilean transformations expresses the invariance of Newtonian non-gravitational dynamics according to the Galilean principle of relativity. The Lie transformation group is defined by

$$G(m) := \{\ g(A, \mathbf{a}, \tau, \mathbf{u})\ :\ A \in O(m),\ \mathbf{u}, \mathbf{a} \in \mathbf{R}^m,\ \tau \in \mathbf{R}\ \}.$$

Here $\mathbf{u}$ is the velocity between frames. The transformation is

$$g(A, \mathbf{a}, \tau, \mathbf{u}) := \begin{cases} t & \mapsto & t' = t + \tau \\ \mathbf{x} & \mapsto & \mathbf{x}' = A\mathbf{x} + \mathbf{u}t + \mathbf{a}. \end{cases}$$

The group is also known as the maximal symmetry of Newton's second law. Note that $G(m)$ is not affine since we cannot write $\mathbf{x}' = A\mathbf{x} + \mathbf{a}$ with $A \in O(m)$ or $A \in O(P, m - P)$. The composition law for the Galilean group is

$$g(A', \mathbf{a}', \tau', \mathbf{u}') \cdot g(A, \mathbf{a}, \tau, \mathbf{u}) = g(A'A, A'\mathbf{a} + \mathbf{u}'\tau + \mathbf{a}', \tau + \tau', A'\mathbf{u} + \mathbf{u}')$$

and the inverse

$$g(A, \mathbf{a}, \tau, \mathbf{u})^{-1} = g(A^{-1}, -A^{-1}\mathbf{a} + A^{-1}\mathbf{u}\tau, -\tau, -A^{-1}\mathbf{u}).$$

3.2.3 Group Actions and Group Representations

We now consider the general group action in more detail.

Definition 3.6 *The* **action** *or* **realization** *of a group G on a set M is defined as a map φ_g*

$$\varphi_g : \mathbf{x} \mapsto \mathbf{x}' = \varphi_g(\mathbf{x})$$

where $\mathbf{x}, \mathbf{x}' \in M$ and $g \in G$.

With each g a function is identified and

$$\varphi_e(\mathbf{x}) = \mathbf{x}$$
$$\varphi_{g_2} \circ \varphi_{g_1} = \varphi_{g_2 g_1},$$

for all $\mathbf{x} \in M$, where g_1, g_2, e (identity element) $\in G$.

This action, as defined, is also known as the **left action**, whereas, for the **right action** we have $\varphi_{g_2} \circ \varphi_{g_1} = \varphi_{g_1 g_2}$. We conclude that

$$GM := \{\varphi_g \ : \ g \in G\}$$

forms a group with identity φ_e and inverse $\varphi_{g^{-1}} = (\varphi_g)^{-1}$. Thus the mapping φ_g is 1-1. In fact, G is the transformation group of G on M.

Definition 3.7 *The action is* **effective** *if the only element of G leaving all M fixed is the group identity element e, i.e., $\varphi_g(\mathbf{x}) = \mathbf{x}$, for all $\mathbf{x} \in M$ and so $g = e$.*

Note that for an effective action, GM is isomorphic to G; $GM \cong G$.

Example: Consider the group $SO(3)$ on $\mathbf{R}^3$ where

$$\varphi_A(\mathbf{x}) \equiv A\mathbf{x}$$

with $A \in SO(3)$. For an effective action we have

$$A\mathbf{x} = \mathbf{x} \quad \text{for all} \quad \mathbf{x} \in \mathbf{R}^3$$

and so $A = I$. Hence $SO(3)\mathbf{R}^3 \cong SO(3)$. ♣

Definition 3.8 *The action is* **free** *if*

$$\varphi_g(\mathbf{x}) = \mathbf{x} \quad \text{for any} \quad \mathbf{x} \in M.$$

Example: The left translation

$$\varphi_g(\mathbf{x}) = g\mathbf{x} = L_g\mathbf{x},$$

or self action, is free. ♣

Definition 3.9 *An element of the manifold M is a* **fixed point** *if*

$$\varphi_g(\mathbf{x}) = \mathbf{x} \quad for\ all \quad g \in G.$$

If the action is free, then there are no fixed points.

Example: For $SO(3)$ on $\mathbf{R}^3$ the only fixed point is $\mathbf{0}$ and for $SO(3)$ on $M = S^2$ there are no fixed points. ♣

Definition 3.10 *An action is* **transitive** *if any* $\mathbf{y} \in M$ *can be reached from any* $\mathbf{x} \in M$.

Definition 3.11 *An action is* **simply transitive** *if* $g \in G$ *is unique, otherwise the action is* **multiply transitive**.

Example: $E(2)$ on $\mathbf{R}^2$ is transitive, for we have

$$\mathbf{x}' = \varphi_{\mathbf{a}}(\mathbf{x}) = A\mathbf{x} + \mathbf{a}$$

where $A \in O(2)$ and $\mathbf{a} \in \mathbf{R}^2$. $SO(3)$ on $\mathbf{R}^3$ is not transitive while $SO(3)$ on S^2 is multiply transitive. ♣

Definition 3.12 *The* **orbit** *of the point* $\mathbf{x} \in M$ *under the group* G *is the set* $G(\mathbf{x})$ *of all points* $y \in M$ *that can be reached by applying some group operation* $g \in G$ *to the point* $\mathbf{x}$.

If an action is transitive we find that

$$G(\mathbf{x}) = M \quad for\ any \quad \mathbf{x} \in M.$$

Example: Consider $SO(2)$ on $\mathbf{R}^2$. With the transformation

$$\begin{aligned}
x_1'(x_1, x_2) &= x_1 \cos\theta - x_2 \sin\theta \\
x_2'(x_1, x_2) &= x_1 \sin\theta + x_2 \cos\theta
\end{aligned}$$

we find that

$$x_1'^2 + x_2'^2 = x_1^2 + x_2^2.$$

Thus $x_1^2 + x_2^2$ is invariant. ♣

Example: $SO(1,1)$ on $\mathbf{R}^2$. With the transformation

$$\begin{aligned}
x_1'(x_1, x_2) &= x_1 \cosh\theta + x_2 \sinh\theta \\
x_2'(x_1, x_2) &= x_1 \sinh\theta + x_2 \cosh\theta
\end{aligned}$$

we find that

$$x_1'^2 - x_2'^2 = x_1^2 - x_2^2.$$

Thus $x_1^2 - x_2^2$ is invariant. ♣

Definition 3.13 *A group* **representation** *of G is a realization of G as a group of linear transformations on a vector space, i.e., a representation of G is a linear (left) action of G on a linear space.*

Consider the homomorphism

$$A : G \to GL(V)$$

where $A : g \to A(g)$ and $A(g)\mathbf{x} = \mathbf{y}$ is linear with

$$
\begin{aligned}
A(e) &= I \\
A(g_2)A(g_1) &= A(g_2 g_1).
\end{aligned}
$$

In view of our notation for realizations we now set

$$
\begin{aligned}
M &\equiv V \\
\varphi_g &\equiv A(g),
\end{aligned}
$$

where

$$
\begin{aligned}
V : \quad & \text{representation space which may be infinite,} \\
A(g) : \quad & \text{(matrix) representation of } g.
\end{aligned}
$$

By

$$D = \{A(g) \quad \text{for all } g \in G\}$$

we denote the representation of G, where $D < GL(n)$.

Definition 3.14 *The representation is known to be* **faithful** *if $D \cong G$, i.e., the realization is faithful or the action is effective.*

Definition 3.15 *The representation is* **unitary** *if $A(g)$ is unitary for all $g \in G$.*

Definition 3.16 *The* **dimension (degree)** *of the representation is the dimension of the vector space V.*

Example: The classical groups as matrix groups on $\mathbf{R}^m$ or $\mathbf{C}^m$ are all faithful representations of dimension m. ♣

Example: $SO(3)$ on $\mathbf{R}^3$ is a three dimensional faithful representation of $SO(3)$, but $SO(3)$ on S^2 is only a realization since S^2 is not a vector space. ♣

Example: A unitary representation of $SO(2)$ is

$$A : R \to A(R) = \exp(i\alpha) \quad \text{on } \mathbf{C}^1.$$

This representation is faithful. ♣

Definition 3.17 *Two representations D and D' of G are* **equivalent** *if there exists a fixed linear transformation $P : D \to D'$, such that*

$$A'(g) = P^{-1}A(g)P \quad \text{for all } g \in G.$$

For a given linear vector space V a basis transformation

$$\mathbf{e}' = P\mathbf{e}$$

induces a transformation

$$
\begin{aligned}
P \;:\; & V \to V' \\
\;:\; & A(g) \to A'(g) = P^{-1}A(g)P.
\end{aligned}
$$

Thus $\{A'(g)\}$ is equivalent to $\{A(g)\}$.

Let V_1 be a linear subspace of V.

Definition 3.18 *If V_1 is invariant under G, i.e.,*

$$A(g)\mathbf{x} \in V_1 \quad \text{for all} \quad \mathbf{x} \in V_1 \quad \text{and all} \quad g \in G$$

then V_1 is called **invariant**.

Definition 3.19 *A representation of G is* **irreducible** *if there are no non-trivial subspaces.*

Trivial invariant subspaces of V are $\{\mathbf{0}\}$ and V, so that an irreducible representation means that the size of the representation space cannot be decreased.

Suppose V_1 is an m-dimensional invariant subspace in an n-dimensional V. Since $A(g)\mathbf{x} \in V_1$ for all $\mathbf{x} \in V_1$ and all $g \in G$, it follows that $A(g)$ is equivalent to

$$A(g) = \begin{pmatrix} A_1(g) & Q \\ 0 & S \end{pmatrix}$$

where $A_1(g)$ is an $m \times m$ matrix and $A(g)$ an $n \times n$ matrix. By choosing a basis

$$\{\, \mathbf{e}_1, \ldots, \mathbf{e}_m, \mathbf{e}_{m+1}, \ldots, \mathbf{e}_n \,\}$$

of V such that $\{\, \mathbf{e}_1, \ldots, \mathbf{e}_m \,\}$ is a basis of V_1:

$$
\begin{pmatrix}
A_1(g) & \vdots & Q \\
\cdots & \vdots & \cdots \quad \cdots \\
 & \vdots & \\
0 & \vdots & S
\end{pmatrix}
\begin{pmatrix}
x_1 \\ \vdots \\ x_m \\ 0 \\ \vdots \\ 0
\end{pmatrix}
=
\begin{pmatrix}
y_1 \\ \vdots \\ y_m \\ 0 \\ \vdots \\ 0
\end{pmatrix}.
$$

Thus $\{A_1(g)\}$ is an induced lower dimensional representation of G on V_1 called D_1. D_1 may be reducible or not.

Definition 3.20 *A representation of G on V is* **completely reducible** *if*

$$V = V_1 \oplus V_2 \oplus \cdots \oplus V_m$$

where each vector subspace V_i has an irreducible induced representation D_i, (i.e., V_i contains no invariant subspaces), so that

$$D = D_1 + D_2 + \cdots + D_m.$$

A completely irreducible representation can be given by

$$A(g) = \text{diag}\,(A_1(g),\ A_2(g),\ \ldots,\ A_m(g)) = A_1(g) \oplus A_2(g) \oplus \cdots \oplus A_m(g).$$

Example: $SO(3)$ on $\mathbf{R}^3$ is irreducible since the only invariant linear subspaces are $\{0\}$ and $\mathbf{R}^3$. ♣

Theorem 3.1 *Every unitary representation is completely reducible.*

The importance of complete reducibility in quantum mechanics is that the label i of each irreducible subspace V_i, and the induced representations D_i, is a quantum number.

3.3 General Remarks on Lie Groups

In this section we introduce cosets and the centre of a group.

Definition 3.21 *Any element $g \in G$ defines the* **left \ right translations** *by*

$$Lg : x \mapsto gx$$

$$Rg : x \mapsto xg$$

for all $x \in G$.

If the group is Abelian Rg and Lg are identical.

Definition 3.22 *The* **centre** *of a group G is*

$$
\begin{aligned}
Z(G) \ &:= \ \{\, z \in G \ : \ zg = gz \quad \text{for all } g \in G \,\} \\
&= \ \{\, z \in G \ : \ Lz = Rz \,\}.
\end{aligned}
$$

Clearly the identity element $e \in Z(G)$. For G Abelian the centre is $Z(G) = G$.

Definition 3.23 *For $H < G$ and $g \in G$, we define the* **left \ right coset** *of H with* **coset representative** *g:*

$$
\begin{aligned}
gH = Lg(H) = \{\, gh \ : \ h \in H \,\} \\
Hg = Rg(H) = \{\, hg \ : \ h \in H \,\}.
\end{aligned}
$$

Note that

(i) if $g \in H$ then $gH = Hg = H$

(ii) if $g \notin H$ then gH as well as Hg is not a subgroup of G, and in general $gH \neq Hg$.

The **coset partition** G is:

$$G = \bigcup_g gH$$

where any $x \in G$ lies in a unique coset.

Examples: Consider the matrix

$$S = \begin{pmatrix} 1 & 0 \\ 0 & -1 \end{pmatrix}$$

(parity reversal) where $S \in O(2)$ but $S \notin SO(2)$. Then every $A \in O(2)$, $A \notin SO(2)$ may be expressed as

$$A = \tilde{A} S$$

with $\tilde{A} \in SO(2)$. Thus

$$O(2) = I \cdot SO(2) \cup S \cdot SO(2),$$

($\cdot$ denotes matrix multiplication) i.e., cosets of $SO(2)$ with representations I, S, form a complete partition of $O(2)$.   ♣

Example: Consider the Lorentz group $O(1,3)$. We have the condition

$$A^T L A = L \quad \Longleftrightarrow \quad \sum_{\alpha=1}^{4} \sum_{\beta=1}^{4} A_{\alpha\rho} L_{\alpha\beta} A_{\beta\sigma} = L_{\rho\sigma}$$

with ρ, $\sigma = 1, 2, 3, 4$ and

$$L = \mathrm{diag}\,(-1, 1, 1, 1).$$

The condition on $A_{\alpha\beta}$ is then given by

$$-A_{11}^2 + A_{21}^2 + A_{31}^2 + A_{41}^2 = -1,$$

which means that $A_{11}^2 \geq 1$ or $\mathrm{sgn} A_{11} = \pm 1$. We can associate the following physical meaning with A:

$$
\begin{aligned}
\mathrm{sgn} A_{11} &= +1, \quad \text{preserves time direction} \\
\mathrm{sgn} A_{11} &= -1, \quad \text{time inversion,}
\end{aligned}
$$

also

$$A_P = \text{diag}\,(1, -1, -1, -1), \quad \text{spatial reversal or parity reversal}$$
$$\det A_P = -1, \quad \text{sgn}A_{11} = 1$$

$$A_T = \text{diag}\,(-1, 1, 1, 1), \quad \text{time inversion}$$
$$\det A_T = -1, \quad \text{sgn}A_{11} = -1$$

$$A_{PT} = \text{diag}\,(-1, -1, -1, -1), \quad \text{time and parity reversal}$$
$$\det A_{PT} = +1, \quad \text{sgn}A_{11} = -1.$$

Thus, the full Lorentz group $O(1,3)$ can be partitioned with respect to the subgroup

$$SO(1,3)^\uparrow := \{\, A \in SO(1,3) \ : \ \text{sgn}A_{11} = +1 \,\}$$

where $A_{PT} \in SO(1,3)$ but $A_{PT} \notin SO(1,3)^\uparrow$. The group $SO(1,3)^\uparrow$ is called the **proper Lorentz group**. The partition of $O(1,3)$ is

$$O(1,3) = I \cdot SO(1,3)^\uparrow \cup A_P \cdot SO(1,3)^\uparrow \cup A_T \cdot SO(1,3)^\uparrow \cup A_{PT} \cdot SO(1,3)^\uparrow.$$

The special orthogonal group $SO(1,3)$ can also be partitioned with the proper Lorentz group

$$SO(1,3) = I \cdot SO(1,3)^\uparrow \cup A_{PT} \cdot SO(1,3)^\uparrow. \qquad \clubsuit$$

Definition 3.24 *A* **conjugation map** *(also called* **inner automorphism***) can be defined by*

$$I_g : \quad x \mapsto gxg^{-1} \quad \text{for all } x \in G$$

where $g \in G$.

Every element of the form $I_g x$ is conjugate to x. The **conjugacy class** of x is

$$(x) := \{I_g x \ : \ g \in G\} = \text{all elements of the form } gxg^{-1}.$$

Note that (x) is not a subgroup of G unless $x = e$. The conjugacy classes partition the group G.

Example: Consider the group $GL(m)$. The conjugacy class is given by

$$(A) = \{\, I_B A \ : \ B \in GL(m)\} = \{BAB^{-1} \ : \ B \in GL(m) \,\}$$

for all $A \in GL(m)$. This means that every $C \in (A)$ has the same determinant, trace and eigenvalues as A. $\qquad \clubsuit$

Definition 3.25 *For* $H < G$ *and*

$$I_g H = H \quad \text{for all } g \in G$$

we call H *an* **invariant subgroup**.

Definition 3.26   *1. A group G is called* **simple** *if G has no invariant subgroups.*

2. G is called **semisimple** *if G has no Abelian invariant subgroups.*

3. G is called non-simple if G has an invariant subgroup.

4. G is called non-semisimple if G has an Abelian invariant subgroup.

Example: Consider $SL(m)$, which is a subgroup of $GL(m)$. Let $A \in GL(m)$ and $S \in SL(m)$. We show that $I_A SL(m) = SL(m)$ for all $A \in GL(m)$: Since $\det S = 1$ for all $S \in SL(m)$ and

$$\det(ASA^{-1}) = \det(A)\det(A^{-1})\det(S) = \det(AA^{-1})\det(S) = 1$$

which is true for all $A \in GL(m)$ and all $S \in SL(m)$, we have

$$I_A SL(m) = SL(m) \quad \text{for all } A \in GL(m).$$

Thus $SL(m)$ is an invariant subgroup of $GL(m)$ and $GL(m)$ is non-simple.   ♣

Example: Consider the group $T(3)$ which is a subgroup of $E(3)$ where

$$
\begin{aligned}
(A',\mathbf{a}') \cdot (A,\mathbf{a}) &= (A'A, A'\mathbf{a} + \mathbf{a}') \\
(A,\mathbf{a})^{-1} &= (A^{-1}, -A^{-1}\mathbf{a})
\end{aligned}
$$

and $A,\ A' \in O(3)$ and $\mathbf{a},\ \mathbf{a}' \in T(3)$. We consider a translation $(I,\mathbf{a}) \in T(3)$ and calculate

$$(A,\mathbf{b}) \cdot (I,\mathbf{a}) \cdot (A,\mathbf{b})^{-1}$$

for $(A,\mathbf{b}) \in E(3)$:

$$
\begin{aligned}
(A,\mathbf{b})(I,\mathbf{a})(A,\mathbf{b})^{-1} &= (AI, B + A\mathbf{a})(A^{-1}, -A^{-1}\mathbf{b}) \\
&= (AA^{-1}, \mathbf{b} + A\mathbf{a} + A(-A^{-1}\mathbf{b})) \\
&= (I, A\mathbf{a})
\end{aligned}
$$

Since $(I, A\mathbf{a}) \in T(3)$ for all $(A,\mathbf{b}) \in E(3)$ and all $(I,\mathbf{a}) \in T(3)$ it follows that $T(3)$ is an invariant subgroup of $E(3)$. $T(3)$ is Abelian so that $E(3)$ is a non-semisimple group. ♣

Let N be an invariant subgroup of G and by G/N we denote the set of cosets of N in G (there is no need to specify whether the cosets are left or right - for since N is an invariant subgroup they coincide). Since

$$(gN)(g'N) = gg'N,$$

the set G/N is closed under elementwise multiplication of its cosets. In fact G/N forms a group with respect to coset multiplication. The identity element is the coset N and the coset gH is the coset $g^{-1}N$.

Definition 3.27 *The group G/N is called the* **quotient** *or* **factor group** *of the group G by the invariant subgroup N.*

Example: $SO(3)$ is an invariant subgroup of $O(3)$. The factor group is then given by

$$O(3)/SO(3) = \{ SO(3), \ -I \cdot SO(3) \} \equiv \{I, \ -I\}. \qquad \clubsuit$$

Example: Consider $SO(1,3)^\dagger$ is an invariant subgroup of $O(1,3)$. The factor group is

$$
\begin{aligned}
O(1,3)/SO(1,3)^\dagger &= \{ SO(1,3)^\dagger, \ A_P \cdot SO(1,3)^\dagger, \ A_T \cdot SO(1,3)^\dagger, A_{PT} \cdot SO(1,3)^\dagger \} \\
&\equiv \{ I, \ A_P, \ A_T, \ A_{PT} \}.
\end{aligned}
$$

$\clubsuit$

Definition 3.28 *A* **direct product** $\otimes$ *between two groups G_1 and G_2 is a group*

$$G_1 \otimes G_2 = \{ (g_1, g_2) \}$$

with composition law

$$(g_1', g_2') \cdot (g_1, g_2) := (g_1' g_1, g_2' g_2).$$

Example: An obvious example is

$$O(3) = SO(3) \otimes \{I, -I\},$$

where the composition law is

$$(A', \pm I)(A, \pm I) = (A'A, \pm I).$$

As we have already seen, for the Euclidean group the direct product cannot be used so that

$$E(3) \neq O(3) \otimes T(3)$$

since the subgroup $O(3)$ acts on the subgroup $T(3)$, where the composition law is given by

$$(A', a') \cdot (A, a) = (A'A, a' + A'a). \qquad \clubsuit$$

3.4   Computer Algebra Applications

In the program we implement the Lie groups $SO(2)$ and and $SO(1,1)$ and show that $x_1^2 + x_2^2$ and $x_1^2 - x_2^2$ are invariant, respectively. Furthermore we show that the the determinant of the matrix of the four parameter representation of the unitary group $U(2)$ is given by $\exp(i\beta)$, where $\beta \in \mathbf{R}$.

```
% transf.red

% group SO(1,1)
matrix A(2,2);
A(1,1) := cosh(x); A(1,2) := sinh(x);
A(2,1) := sinh(x); A(2,2) := cosh(x);

for all q let cosh(q)**2 = 1 + sinh(q)**2;

matrix v(2,1);
v(1,1) := x1; v(2,1) := x2;

matrix w(2,1);
w := A*v;
w(1,1)**2 - w(2,1)**2;

% group SO(2)
matrix B(2,2);
B(1,1) := cos(x); B(1,2) := -sin(x);
B(2,1) := sin(x); B(2,2) := cos(x);

for all q let cos(q)**2 = 1 - sin(q)**2;

matrix r(2,1);
r(1,1) := x1; r(2,1) := x2;

matrix s(2,1);
s := B*r;
s(1,1)**2 + s(2,1)**2;

% group U(2), calculating the determinant
matrix U(2,2);
U(1,1) := exp(i*alpha)*cos(nu);
U(1,2) := exp(i*gamma)*sin(nu);
U(2,1) := - exp(i*(beta - gamma))*sin(nu);
U(2,2) := exp(i*(beta - alpha))*cos(nu);

det(U);
```

Chapter 4

Infinitesimal Transformations

4.1 Introduction

We consider a one-parameter Lie transformation group in the form (Bluman and Kumei [7], Gilmore [52], Olver [87])

$$\mathbf{x}' = \varphi(\mathbf{x}, \epsilon) \tag{1}$$

where $\mathbf{x}' = (x'_1, \ldots, x'_m)$ and $\mathbf{x} = (x_1, \ldots, x_m)$ lie in an open domain $D \subset \mathbf{R}^m$. The group composition is denoted by $\phi(\varepsilon, \eta)$, where ϵ and η are the group parameters in an open interval $S \subset \mathcal{R}$ including 0. If $\mathbf{x}' = \varphi(\mathbf{x}, \varepsilon)$ and $\mathbf{x}'' = \varphi(\mathbf{x}', \eta)$, then

$$\mathbf{x}'' = \varphi(\mathbf{x}, \phi(\varepsilon, \eta))$$

where $\mathbf{x}'$ and $\mathbf{x}''$ lie in D. φ is analytic in its domain of definition. In this section we show how any such group of transformations can be re-parametrized such that the group composition is given by

$$\phi(\varepsilon, \eta) = \varepsilon + \eta,$$

the identity of the transformation group by $\varepsilon = 0$ and the inverse by $-\varepsilon$. This will be known as the standard form of a one-parameter Lie transformation group.

Example: Consider the **scaling group** in the plane $\mathbf{R}^2$, given by

$$x'_1 = \alpha x_1, \qquad x'_2 = \alpha^2 x_2$$

where $0 < \alpha < \infty$. The group composition is $\phi(\alpha, \beta) = \alpha\beta$ and the identity element $\alpha = 1$. This group of transformations can be re-parametrized by $\varepsilon = \alpha - 1$ so that the identity element is $\varepsilon = 0$

$$x'_1 = (1 + \varepsilon)x_1, \qquad x'_2 = (1 + \varepsilon)^2 x_2$$

where $-1 < \varepsilon < \infty$. The group composition is found by considering

$$x_1'' = (1+\eta)x_1' = (1+\eta)(1+\varepsilon)x_1 = (1+\varepsilon+\eta+\varepsilon\eta)x_1$$

so that $\phi(\varepsilon,\eta) = \varepsilon + \eta + \varepsilon\eta$. The inverse is obtained by finding η such that

$$(1+\varepsilon+\eta+\varepsilon\eta)x_1 = x_1.$$

The inverse is thus given by

$$-\frac{\varepsilon}{1+\varepsilon}. \qquad\qquad \clubsuit.$$

In the example above we note that, although the identity is given by $\varepsilon = 0$, due to the re-parametrization the group composition and inverse are still not in the standard form as mentioned before. The first theorem of Lie provides us with an algorithmic method to re-parametrize a one-parameter transformation group such that it is of the standard form. Before we can state Lie's first theorem we have to define an infinitesimal transformation.

Expanding (1) about $\varepsilon = 0$ we obtain, in a neighbourhood of $\varepsilon = 0$,

$$\mathbf{x}' = \mathbf{x} + \varepsilon \left.\frac{\partial\varphi}{\partial\varepsilon}(\mathbf{x},\varepsilon)\right|_{\varepsilon=0} + \frac{\varepsilon^2}{2}\left.\frac{\partial^2\varphi}{\partial\varepsilon^2}(\mathbf{x},\varepsilon)\right|_{\varepsilon=0} + O(\varepsilon^3). \tag{2}$$

We define

$$\boldsymbol{\xi}(\mathbf{x}) := \left.\frac{\partial\varphi}{\partial\varepsilon}(\mathbf{x},\varepsilon)\right|_{\varepsilon=0}. \tag{3}$$

Definition 4.1 *The transformation*

$$\mathbf{x}' = \mathbf{x} + \varepsilon\boldsymbol{\xi}(\mathbf{x}) \tag{4}$$

is called the **infinitesimal transformation** *of the Lie transformation group (1) and the components of $\boldsymbol{\xi}$ are called the infinitesimals of (1).*

Theorem 4.1 *(Lie's first fundamental theorem.) There exists a parametrization $\tau(\varepsilon)$ such that the Lie group of transformation (1) is equivalent to the solution of the initial value problem for the autonomous system of first order ordinary differential equations*

$$\frac{d\mathbf{x}'}{d\tau} = \boldsymbol{\xi}(\mathbf{x}') \tag{5}$$

with $\mathbf{x}' = \mathbf{x}$ when $\tau = 0$. In particular

$$\tau(\varepsilon) = \int_0^\varepsilon \Gamma(\varepsilon')d\varepsilon' \tag{6}$$

where

$$\Gamma(\varepsilon) = \left.\frac{\partial\phi(a,b)}{\partial b}\right|_{(a,b)=(\varepsilon^{-1},\varepsilon)} \tag{7}$$

and $\Gamma(0) = 1$. Here ε^{-1} denotes the inverse parameter. In terms of ε the one-parameter group is given by the solution of the initial value problem

$$\frac{d\mathbf{x}'}{d\varepsilon} = \Gamma(\varepsilon)\boldsymbol{\xi}(\mathbf{x}'). \tag{8}$$

For the proof we refer to Bluman and Kumei [7].

Example: We consider the scaling group from the previous example

$$x_1' = (1+\varepsilon)x_1, \qquad x_2' = (1+\varepsilon)^2 x_2$$

where $-1 < \varepsilon < \infty$. The law of composition is

$$\phi(a,b) = a + b + ab$$

(where a and b denote the group parameters) and the inverse is given by

$$\frac{-\varepsilon}{1+\varepsilon}.$$

Thus

$$\Gamma(\varepsilon) = \left. \frac{\partial \phi(a,b)}{\partial b} \right|_{(a,b)=(\varepsilon^{-1},\varepsilon)} = (1+a)|_{(\varepsilon^{-1},\varepsilon)} = 1 + \left(\frac{-\varepsilon}{1+\varepsilon} \right) = \frac{1}{1+\varepsilon}$$

and

$$\boldsymbol{\xi}(\mathbf{x}) = \left. \frac{\partial \boldsymbol{\varphi}}{\partial \varepsilon}(\mathbf{x},\varepsilon) \right|_{\varepsilon=0} = (x_1, 2(1+\varepsilon)x_2)|_{\varepsilon=0} = (x_1, 2x_2)$$

where

$$\boldsymbol{\varphi}(\mathbf{x},\varepsilon) = ((1+\varepsilon)x_1, (1+\varepsilon)^2 x_2)$$

and $\boldsymbol{\xi} = (\xi_1, \xi_2)$, $\mathbf{x} = (x_1, x_2)$. From

$$\frac{d\mathbf{x}'}{d\varepsilon} = \Gamma(\varepsilon)\boldsymbol{\xi}(\mathbf{x}')$$

with $\mathbf{x}'(0) = \mathbf{x}$, it follows that

$$\frac{dx_1'}{d\varepsilon} = \left(\frac{1}{1+\varepsilon} \right) x_1', \qquad \frac{dx_2'}{d\varepsilon} = \left(\frac{1}{1+\varepsilon} \right) 2x_2'.$$

The solution of this initial value problem will result in the given scaling group. The parametrization is now given by

$$\tau(\varepsilon) = \int_0^\varepsilon \Gamma(\varepsilon')d\varepsilon' = \int_0^\varepsilon \left(\frac{1}{1+\varepsilon'} \right) d\varepsilon'$$

so that $\tau(\varepsilon) = \ln(1 + \varepsilon)$ or $\varepsilon = e^\tau - 1$. The parametrized group is then

$$x_1' = e^\tau x_1, \qquad x_2' = e^{2\tau} x_2$$

where $-\infty < \tau < \infty$. The law of composition for this parametrized group is

$$\phi(a, b) = a + b. \qquad\qquad \clubsuit$$

Example: We consider the group of translations in the plane

$$x_1' = x_1 + \varepsilon, \qquad x_2' = x_2$$

with law of composition $\phi(a, b) = a + b$ and inverse $\varepsilon^{-1} = -\varepsilon$. Thus

$$\Gamma(\varepsilon) = \left.\frac{\partial\phi(a, b)}{\partial b}\right|_{(a,b)=(\varepsilon^{-1},\varepsilon)} = 1$$

and the parametrization is

$$\tau(\varepsilon) = \int_0^\varepsilon \Gamma(\varepsilon')d\varepsilon' = \varepsilon.$$

We now find the infinitesimals

$$\boldsymbol{\xi}(\mathbf{x}') = \left.\frac{\partial\boldsymbol{\varphi}(\mathbf{x}, \varepsilon)}{\partial\varepsilon}\right|_{\varepsilon=0}$$

where $\boldsymbol{\xi} = (\xi_1, \xi_2)$ and $\mathbf{x} = (x_1, x_2)$. We have

$$\boldsymbol{\varphi}(\mathbf{x}, \varepsilon) = (x_1 + \varepsilon, x_2)$$

so that

$$\left.\frac{\partial\boldsymbol{\varphi}}{\partial\varepsilon}\right|_{\varepsilon=0} = (1, 0).$$

The system of differential equations

$$\frac{d\mathbf{x}'}{d\varepsilon} = \Gamma(\varepsilon)\boldsymbol{\xi}(\mathbf{x}')$$

is thus given by

$$\frac{dx_1'}{d\varepsilon} = 1, \qquad \frac{dx_2'}{d\varepsilon} = 0.$$

With the initial conditions $\mathbf{x}'(0) = \mathbf{x}$ the given translation group can be found by solving this autonomous system of differential equations. $\qquad \clubsuit$

4.2 Infinitesimal Generators

We now introduce a representation of a local one-parameter Lie group of transformations by way of a group generator. This will lead us to the discussion of Lie algebras.

Definition 4.2 *The* **infinitesimal generators** *also known as the* **Lie symmetry vector field** *of the one-parameter Lie group of transformations (1) is defined by the linear differential operator*

$$Z \equiv Z(\mathbf{x}) := \boldsymbol{\xi}(\mathbf{x}) \cdot \boldsymbol{\nabla} \equiv \sum_{j=1}^{m} \xi_j(\mathbf{x}) \frac{\partial}{\partial x_j} \tag{9}$$

where

$$\boldsymbol{\xi} := (\xi_1, \xi_2, \ldots, \xi_m)$$

and $\boldsymbol{\nabla}$ is the gradient operator

$$\boldsymbol{\nabla} := \left(\frac{\partial}{\partial x_1}, \frac{\partial}{\partial x_2}, \cdots, \frac{\partial}{\partial x_m} \right)^T$$

where T denotes transpose. For a differentiable function f we have

$$Zf(\mathbf{x}) = \boldsymbol{\xi}(\mathbf{x}) \cdot \boldsymbol{\nabla} f(\mathbf{x}) = \sum_{j=1}^{m} \xi_j(\mathbf{x}) \frac{\partial f(\mathbf{x})}{\partial x_j}.$$

In particular, for $f(\mathbf{x}) = x_j$ we find

$$Zx_j = \xi_j(\mathbf{x}).$$

We are now interested in the relation between the Lie symmetry vector and an one-parameter Lie group of transformations.

The following theorem shows that use of the Lie symmetry vector field leads to an algorithm to find the explicit solution of the initial value problem stated in Lie's first theorem.

Theorem 4.2 *The one-parameter Lie group of transformations (1) can be written as*

$$
\begin{aligned}
\mathbf{x}'(\mathbf{x}, \varepsilon) &= e^{\varepsilon Z} \mathbf{x} = \mathbf{x} + \varepsilon Z \mathbf{x} + \frac{\varepsilon^2}{2!} Z^2 \mathbf{x} + O(\varepsilon^3) \mathbf{x} \\
&= \left(1 + \varepsilon Z + \frac{\varepsilon^2}{2!} Z^2 + O(\varepsilon^3) \right) \mathbf{x} \\
&= \sum_{k=0}^{\infty} \frac{\varepsilon^k}{k!} Z^k \mathbf{x}
\end{aligned}
$$

where the linear operator Z is defined by (9) and

$$Z^k f(\mathbf{x}) = Z(Z^{k-1} f(\mathbf{x})), \qquad k = 1, 2, \ldots .$$

Proof: We consider

$$Z(\mathbf{x}) = \sum_{j=1}^{m} \xi_j(\mathbf{x}) \frac{\partial}{\partial x_j}$$

and

$$Z(\mathbf{x}') = \sum_{j=1}^{m} \xi_j(\mathbf{x}') \frac{\partial}{\partial x_j'}.$$

The function φ in the one-parameter Lie group of transformations

$$\mathbf{x}' = \varphi(\mathbf{x}, \varepsilon)$$

can be expanded about $\varepsilon = 0$ as

$$\mathbf{x}' = \sum_{k=0}^{\infty} \frac{\varepsilon^k}{k!} \left. \frac{\partial^k \varphi(\mathbf{x}, \varepsilon)}{\partial \varepsilon^k} \right|_{\varepsilon=0} = \sum_{k=0}^{\infty} \frac{\varepsilon^k}{k!} \left. \frac{d^k \mathbf{x}'}{d\varepsilon^k} \right|_{\varepsilon=0}.$$

Note that for any differentiable function $\mathbf{f}(\mathbf{x}')$

$$\frac{d\mathbf{f}(\mathbf{x}')}{d\varepsilon} = \sum_{j=1}^{m} \frac{\partial \mathbf{f}(\mathbf{x}')}{\partial x_j'} \frac{dx_j'}{d\varepsilon} = \sum_{j=1}^{m} \xi_j(\mathbf{x}') \frac{\partial \mathbf{f}(\mathbf{x}')}{\partial x_j'} = Z(\mathbf{x}')\mathbf{f}(\mathbf{x}').$$

We now consider the case $\mathbf{f}(\mathbf{x}') = \mathbf{x}'$, so that

$$\frac{d\mathbf{x}'}{d\varepsilon} = Z(\mathbf{x}')\mathbf{x}'$$

and

$$\frac{d^2\mathbf{x}'}{d\varepsilon^2} = \frac{d}{d\varepsilon} Z(\mathbf{x}')\mathbf{x}' = Z(\mathbf{x}')Z(\mathbf{x}')\mathbf{x}' = Z^2(\mathbf{x}')\mathbf{x}'.$$

In general we have

$$\frac{d^k\mathbf{x}'}{d\varepsilon^k} = Z(\mathbf{x}')Z^{k-1}(\mathbf{x}')\mathbf{x}' = Z^k(\mathbf{x}')\mathbf{x}'$$

where $k = 1, 2, \ldots$. Consequently

$$\left. \frac{d^k\mathbf{x}'}{d\varepsilon^k} \right|_{\varepsilon=0} = Z^k(\varphi(\mathbf{x}, \varepsilon))\varphi(\mathbf{x}, \varepsilon)|_{\varepsilon=0} = Z^k(\mathbf{x})\mathbf{x},$$

so that

$$\mathbf{x}' = \sum_{k=0}^{\infty} \frac{\varepsilon^k}{k!} Z^k(\mathbf{x})\mathbf{x} = e^{\varepsilon Z}\mathbf{x}. \tag{10}$$

The series (10) is known as a **Lie series**. ♠

From the proof of theorem 4.2 we have the following

Corollary 4.1 *Let f be an analytic function. For a Lie group of transformations (1) with infinitesimal generator (9) it follows that*

$$f(\mathbf{x}') = f(e^{\varepsilon Z}\mathbf{x}) = e^{\varepsilon Z}f(\mathbf{x}).$$

Proof: From the proof of Theorem 4.2 it follows that

$$\left.\frac{df(\mathbf{x}')}{d\varepsilon}\right|_{\varepsilon=0} = Zf(\mathbf{x}).$$

We now consider the Taylor expansion of the function

$$
\begin{aligned}
f(\mathbf{x}') = f(e^{\varepsilon Z}\mathbf{x}) &= f(\mathbf{x}) + \varepsilon\left.\frac{df(\mathbf{x}')}{d\varepsilon}\right|_{\varepsilon=0} + O(\varepsilon^2) \\
&= f(\mathbf{x}) + \varepsilon Zf(\mathbf{x}) + O(\varepsilon^2) \\
&= \sum_{k=0}^{\infty}\frac{\varepsilon^k}{k!}Z^k f(\mathbf{x}) \\
&= e^{\varepsilon Z}f(\mathbf{x}).
\end{aligned}
$$

♠

Summary: There are two ways to find explicitly a one-parameter Lie group of transformations with identity $\varepsilon = 0$, group composition $\phi(\varepsilon, \delta) = \varepsilon + \delta$, and inverse $-\varepsilon$ from its infinitesimal transformation

$$\mathbf{x}' = \mathbf{x} + \varepsilon\boldsymbol{\xi}(\mathbf{x})$$

namely:

i) Solve the initial value problem

$$\frac{d\mathbf{x}'}{d\varepsilon} = \boldsymbol{\xi}(\mathbf{x}')$$

with $\mathbf{x}' = \mathbf{x}$ at $\varepsilon = 0$.

ii) Express the group in terms of the Lie series

$$\mathbf{x}' = \sum_{k=0}^{\infty}\frac{\varepsilon^k}{k!}Z^k\mathbf{x}$$

which is found from the infinitesimal generator

$$Z = \sum_{j=1}^{m}\xi_j(\mathbf{x})\frac{\partial}{\partial x_j}$$

corresponding to the infinitesimal transformation.

It thus follows that the Lie series in (ii) provides the solution of the autonomous system of differential equations in (i).

Example: Consider the pseudo-orthogonal group $O(1,1)$, which is a one-parameter group given by

$$A(\varepsilon) = \begin{pmatrix} \cosh \varepsilon & \sinh \varepsilon \\ \sinh \varepsilon & \cosh \varepsilon \end{pmatrix}.$$

The transformation group is given by

$$\mathbf{x}' = A(\varepsilon)\mathbf{x}$$

so that

$$x_1' = x_1 \cosh \varepsilon + x_2 \sinh \varepsilon, \qquad x_2' = x_1 \sinh \varepsilon + x_2 \cosh \varepsilon.$$

We thus have

$$\boldsymbol{\varphi}(\mathbf{x}, \varepsilon) = (x_1 \cosh \varepsilon + x_2 \sinh \varepsilon, \, x_1 \sinh \varepsilon + x_2 \cosh \varepsilon)$$

so that the infinitesimals are

$$\boldsymbol{\xi}(\mathbf{x}) = \left.\frac{\partial \boldsymbol{\varphi}}{\partial \varepsilon}\right|_{\varepsilon=0} = (x_2, x_1).$$

The infinitesimal generator is

$$Z = x_2 \frac{\partial}{\partial x_1} + x_1 \frac{\partial}{\partial x_2}$$

with corresponding Lie series

$$(x_1', x_2') = (e^{\varepsilon Z} x_1, e^{\varepsilon Z} x_2).$$

For the first component we have

$$x_1' = e^{\varepsilon Z} x_1 = x_1 + \varepsilon Z x_1 + \frac{\varepsilon^2}{2!} Z^2 x_1 + O(\varepsilon^3) = x_1 + \varepsilon Z x_1 + \frac{\varepsilon^2}{2!} Z(Z x_1) + O(\varepsilon^3)$$

where $Z x_1 = x_2$, $Z(Z x_1) = x_1$, $Z(Z(Z x_1)) = Z x_1 = x_2$ etc. It follows that

$$x_1' = x_1 + \varepsilon x_2 + \frac{\varepsilon^2}{2!} x_1 + \frac{\varepsilon^3}{3!} x_2 + O(\varepsilon^4) = x_1 \cosh \varepsilon + x_2 \sinh \varepsilon.$$

Analogously

$$x_2' = e^{\varepsilon Z} x_2 = x_1 \sinh \varepsilon + x_2 \cosh \varepsilon. \qquad \clubsuit$$

Example: Consider the nonlinear differential equation

$$\frac{dx}{d\varepsilon} = x^2, \qquad x(\varepsilon = 0) = x_0 > 0.$$

We find the solution of the initial value problem using the Lie series

$$\exp(\varepsilon Z)x$$

where

$$Z := x^2 \frac{d}{dx}.$$

We find

$$\exp(\varepsilon Z)x = (1 + \varepsilon Z + \frac{\varepsilon^2 Z^2}{2!} + \frac{\varepsilon^3 Z^3}{3!} + \ldots)x.$$

Thus

$$\exp(\varepsilon Z)x = x + \varepsilon x^2 + \varepsilon^2 x^3 + \ldots = x(1 + \varepsilon x + \varepsilon^2 x^2 + \ldots).$$

We can write

$$\exp(\varepsilon Z)x = x \sum_{j=0}^{\infty} (\varepsilon x)^j.$$

Direct integration of the differential equation yields

$$x(\varepsilon) = \frac{x_0}{1 - \varepsilon x_0} = x_0 \left(\frac{1}{1 - \varepsilon x_0}\right)$$

For $\varepsilon x_0 < 1$ we find the expansion

$$x(\varepsilon) = x_0 \left(1 + \varepsilon x_0 + \varepsilon^2 x_0^2 + \ldots\right).$$

Thus the Lie series gives us the solution of the initial value problem

$$\exp(\varepsilon Z)x|_{x=x_0} = x_0 \sum_{j=0}^{\infty} (\varepsilon x_0)^j.$$

The solution has a singularity at

$$1 = \epsilon x_0.$$

♣

4.3   Multi-Parameter Lie Groups of Transformations

We now generalise the first fundamental theorem of Lie by considering an r-**parameter Lie group of transformations**, which we denote by

$$\mathbf{x}' = \boldsymbol{\varphi}(\mathbf{x}, \boldsymbol{\varepsilon}). \tag{11}$$

Here $\mathbf{x}' = (x'_1, \ldots, x'_m)$, $\mathbf{x} = (x_1, \ldots, x_m)$, $\boldsymbol{\varphi} = (\varphi_1, \ldots, \varphi_m)$, and the r-parameters are denoted by $\boldsymbol{\varepsilon} = (\varepsilon_1, \varepsilon_2, \ldots, \varepsilon_r)$. The law of composition of parameters is given by

$$\boldsymbol{\phi}(\boldsymbol{\varepsilon}, \boldsymbol{\delta}) = (\phi_1(\boldsymbol{\varepsilon}, \boldsymbol{\delta}), \phi_2(\boldsymbol{\varepsilon}, \boldsymbol{\delta}), \ldots, \phi_r(\boldsymbol{\varepsilon}, \boldsymbol{\delta}))$$

where $\boldsymbol{\delta} = (\delta_1, \ldots, \delta_r)$. $\boldsymbol{\phi}(\boldsymbol{\varepsilon}, \boldsymbol{\delta})$ satisfies the group axioms with $\boldsymbol{\varepsilon} = 0$ corresponding to the identity

$$\varepsilon_1 = \varepsilon_2 = \cdots = \varepsilon_r = 0.$$

$\boldsymbol{\phi}(\boldsymbol{\varepsilon}, \boldsymbol{\delta})$ is assumed to be analytic in its domain of definition.

The **infinitesimal matrix** $\Xi(\mathbf{x})$ is the $r \times m$ matrix with entries

$$\xi_{\alpha j}(\mathbf{x}) := \left. \frac{\partial \varphi_j(\mathbf{x}, \boldsymbol{\varepsilon})}{\partial \varepsilon_\alpha} \right|_{\boldsymbol{\varepsilon}=0} \tag{12}$$

where $\alpha = 1, \ldots, r$ and $j = 1, \ldots, m$.

Theorem 4.3 *(Lie's first Theorem for an r-parameter group.) Let $\Theta(\boldsymbol{\varepsilon})$ be the $r \times r$ matrix with entries*

$$\Theta_{\alpha\beta}(\boldsymbol{\varepsilon}) := \left. \frac{\partial \phi_\beta(\boldsymbol{\varepsilon}, \boldsymbol{\delta})}{\partial \delta_\alpha} \right|_{\boldsymbol{\delta}=0}, \tag{13}$$

and let

$$\Psi(\boldsymbol{\varepsilon}) = \Theta^{-1}(\boldsymbol{\varepsilon}), \tag{14}$$

be the inverse of the matrix $\Theta(\boldsymbol{\varepsilon})$. The Lie transformations group (11) is equivalent to the solution of the initial value problem for the system of $m \times r$ first order partial differential equations (in some neighbourhood of $\boldsymbol{\varepsilon} = 0$):

$$\begin{pmatrix} \dfrac{\partial x'_1}{\partial \varepsilon_1} & \dfrac{\partial x'_2}{\partial \varepsilon_1} & \cdots & \dfrac{\partial x'_m}{\partial \varepsilon_1} \\[2mm] \dfrac{\partial x'_1}{\partial \varepsilon_2} & \dfrac{\partial x'_2}{\partial \varepsilon_2} & \cdots & \dfrac{\partial x'_m}{\partial \varepsilon_2} \\[2mm] \vdots & \vdots & \cdots & \vdots \\[2mm] \dfrac{\partial x'_1}{\partial \varepsilon_r} & \dfrac{\partial x'_2}{\partial \varepsilon_r} & \cdots & \dfrac{\partial x'_m}{\partial \varepsilon_r} \end{pmatrix} = \Psi(\boldsymbol{\varepsilon})\Xi(\mathbf{x}'), \tag{15}$$

with

$$\mathbf{x}' = \mathbf{x} \quad at \quad \boldsymbol{\varepsilon} = 0.$$

Definition 4.3 *The infinitesimal generator Z_α, corresponding to the parameter ε_α of the r-parameter Lie group of transformations (11) is*

$$Z_\alpha := \sum_{j=1}^{m} \xi_{\alpha j}(\mathbf{x}) \frac{\partial}{\partial x_j}, \quad \alpha = 1, 2, \ldots, r \tag{16}$$

so that there are r infinitesimal generators denoted by $\mathbf{Z} = (Z_1, \ldots, Z_r)$.

The r-parameter Lie group of transformations can be written as

$$\mathbf{x}' = \prod_{j=1}^{r} \exp(\varepsilon_j Z_j) \mathbf{x}.$$

We now discuss the arguments of Theorem (4.3) in terms of a concrete example.

Example: Let us consider the Poincaré group $P(2)$. Recall that

$$P(2) = O(1,1) \bigwedge T(2)$$

where $\dim P(2) = 3$ so that we have a three-parameter group. The transformation group is given by

$$\begin{pmatrix} x_1' \\ x_2' \end{pmatrix} = \begin{pmatrix} \cosh \varepsilon_1 & \sinh \varepsilon_1 \\ \sinh \varepsilon_1 & \cosh \varepsilon_1 \end{pmatrix} \begin{pmatrix} x_1 \\ x_2 \end{pmatrix} + \begin{pmatrix} \varepsilon_2 \\ \varepsilon_3 \end{pmatrix}$$

and for $\boldsymbol{\varepsilon} = (\varepsilon_1, \varepsilon_2, \varepsilon_3) = (0,0,0)$ we find

$$x_1'(0) = x_1, \qquad x_2'(0) = x_2.$$

To find the law of composition $\phi(\boldsymbol{\varepsilon}, \boldsymbol{\delta})$ we consider

$$x_1'' = x_1' \cosh \delta_1 + x_2' \sinh \delta_1 + \delta_2, \qquad x_2'' = x_1' \sinh \delta_1 + x_2' \cosh \delta_1 + \delta_3.$$

It follows that

$$x_1'' = x_1 \cosh \phi_1(\boldsymbol{\varepsilon}, \boldsymbol{\delta}) + x_2 \sinh \phi_1(\boldsymbol{\varepsilon}, \boldsymbol{\delta}) + \phi_2(\boldsymbol{\varepsilon}, \boldsymbol{\delta})$$

so that

$$\begin{aligned} \phi_1(\boldsymbol{\varepsilon}, \boldsymbol{\delta}) &= \varepsilon_1 + \delta_1 \\ \phi_2(\boldsymbol{\varepsilon}, \boldsymbol{\delta}) &= \varepsilon_2 \cosh \delta_1 + \varepsilon_3 \sinh \delta_1 + \delta_2. \end{aligned}$$

Also,

$$x_2'' = x_1 \sinh \phi_1(\boldsymbol{\varepsilon}, \boldsymbol{\delta}) + x_2 \cosh \phi_1(\boldsymbol{\varepsilon}, \boldsymbol{\delta}) + \phi_3(\boldsymbol{\varepsilon}, \boldsymbol{\delta})$$

so that

$$\phi_3(\boldsymbol{\varepsilon}, \boldsymbol{\delta}) = \varepsilon_2 \sinh \delta_1 + \varepsilon_3 \cosh \delta_1 + \delta_3.$$

We now find the system of differential equations that represents the given transformation group. The entries of the infinitesimal matrix $\Xi(\mathbf{x})$ are given by

$$\xi_{\alpha j}(\mathbf{x}) = \frac{\partial x'_j}{\partial \varepsilon_\alpha}\bigg|_{\boldsymbol{\varepsilon}=0}$$

where $\alpha = 1, 2, 3$ and $j = 1, 2$. We find

$$\xi_{11}(\mathbf{x}) = \frac{\partial x'_1}{\partial \varepsilon_1}\bigg|_{\boldsymbol{\varepsilon}=0} = x_2, \qquad \xi_{12}(\mathbf{x}) = \frac{\partial x'_2}{\partial \varepsilon_1}\bigg|_{\boldsymbol{\varepsilon}=0} = x_1$$

$$\xi_{21}(\mathbf{x}) = \frac{\partial x'_1}{\partial \varepsilon_2}\bigg|_{\boldsymbol{\varepsilon}=0} = 1, \qquad \xi_{22}(\mathbf{x}) = \frac{\partial x'_2}{\partial \varepsilon_2}\bigg|_{\boldsymbol{\varepsilon}=0} = 0$$

$$\xi_{31}(\mathbf{x}) = \frac{\partial x'_1}{\partial \varepsilon_3}\bigg|_{\boldsymbol{\varepsilon}=0} = 0, \qquad \xi_{32}(\mathbf{x}) = \frac{\partial x'_2}{\partial \varepsilon_3}\bigg|_{\boldsymbol{\varepsilon}=0} = 1$$

so that

$$\Xi(\mathbf{x}) = \begin{pmatrix} x_2 & x_1 \\ 1 & 0 \\ 0 & 1 \end{pmatrix}.$$

To determine $\Psi(\boldsymbol{\varepsilon})$ we first calculate $\Theta(\boldsymbol{\varepsilon})$:

$$\Theta_{11} = \frac{\partial \phi_1}{\partial \delta_1}\bigg|_{\boldsymbol{\delta}=0} = 1, \qquad \Theta_{12} = \frac{\partial \phi_2}{\partial \delta_1}\bigg|_{\boldsymbol{\delta}=0} = -\varepsilon_3, \qquad \Theta_{13} = \frac{\partial \phi_3}{\partial \delta_1}\bigg|_{\boldsymbol{\delta}=0} = -\varepsilon_2,$$

$$\Theta_{21} = \frac{\partial \phi_1}{\partial \delta_2}\bigg|_{\boldsymbol{\delta}=0} = 0, \qquad \Theta_{22} = \frac{\partial \phi_2}{\partial \delta_2}\bigg|_{\boldsymbol{\delta}=0} = 1, \qquad \Theta_{23} = \frac{\partial \phi_3}{\partial \delta_2}\bigg|_{\boldsymbol{\delta}=0} = 0,$$

$$\Theta_{31} = \frac{\partial \phi_1}{\partial \delta_3}\bigg|_{\boldsymbol{\delta}=0} = 0, \qquad \Theta_{32} = \frac{\partial \phi_2}{\partial \delta_3}\bigg|_{\boldsymbol{\delta}=0} = 0, \qquad \Theta_{33} = \frac{\partial \phi_3}{\partial \delta_3}\bigg|_{\boldsymbol{\delta}=0} = 1$$

so that

$$\Theta(\boldsymbol{\varepsilon}) = \begin{pmatrix} 1 & -\varepsilon_3 & -\varepsilon_2 \\ 0 & 1 & 0 \\ 0 & 0 & 1 \end{pmatrix}$$

and

$$\Phi(\boldsymbol{\varepsilon}) = \Theta^{-1}(\boldsymbol{\varepsilon}) = \begin{pmatrix} 1 & \varepsilon_3 & \varepsilon_2 \\ 0 & 1 & 0 \\ 0 & 0 & 1 \end{pmatrix}.$$

Then we find that

$$\Psi(\boldsymbol{\varepsilon})\Xi(\mathbf{x}') = \begin{pmatrix} 1 & \varepsilon_3 & \varepsilon_2 \\ 0 & 1 & 0 \\ 0 & 0 & 1 \end{pmatrix} \begin{pmatrix} x'_2 & x'_1 \\ 1 & 0 \\ 0 & 1 \end{pmatrix}.$$

From Lie's Theorem it follows that

$$
\begin{pmatrix}
\dfrac{\partial x_1'}{\partial \varepsilon_1} & \dfrac{\partial x_2'}{\partial \varepsilon_1} \\[2mm]
\dfrac{\partial x_1'}{\partial \varepsilon_2} & \dfrac{\partial x_2'}{\partial \varepsilon_2} \\[2mm]
\dfrac{\partial x_1'}{\partial \varepsilon_3} & \dfrac{\partial x_2'}{\partial \varepsilon_3}
\end{pmatrix}
=
\begin{pmatrix}
x_2' + \varepsilon_3 & x_1' + \varepsilon_2 \\[2mm]
1 & 0 \\[2mm]
0 & 1
\end{pmatrix}
$$

so that the initial value problem is

$$
\frac{\partial x_1'}{\partial \varepsilon_1} = x_2' + \varepsilon_3, \quad
\frac{\partial x_1'}{\partial \varepsilon_2} = 1, \quad
\frac{\partial x_1'}{\partial \varepsilon_3} = 0
$$

$$
\frac{\partial x_2'}{\partial \varepsilon_1} = x_1' + \varepsilon_2, \quad
\frac{\partial x_2'}{\partial \varepsilon_2} = 0, \quad
\frac{\partial x_2'}{\partial \varepsilon_3} = 1
$$

with initial value $x_1'(0) = x_1$ and $x_2'(0) = x_2$. We now show that the solution of this initial value problem gives the transformation group $P(2)$. It follows from the second and third of each set of equations that

$$
x_1'(\varepsilon_1, \varepsilon_2) = \varepsilon_2 + f_1(\varepsilon_1), \qquad
x_2'(\varepsilon_1, \varepsilon_3) = \varepsilon_3 + f_2(\varepsilon_1)
$$

where $f_1(\varepsilon_1)$ and $f_2(\varepsilon_1)$ are arbitrary smooth functions. Inserting these results into the first of each set of equations we have

$$
\frac{df_1}{d\varepsilon_1} = f_2(\varepsilon_1), \qquad
\frac{df_2}{d\varepsilon_1} = f_1(\varepsilon_1)
$$

with solution

$$
f_2(\varepsilon_1) = a_1 \cosh \varepsilon_1 + a_2 \sinh \varepsilon_1, \qquad
f_1(\varepsilon_1) = a_1 \sinh \varepsilon_1 + a_2 \cosh \varepsilon_1.
$$

From the initial values we have that

$$
x_1'(0) = f_1(0) = x_1, \qquad
x_2'(0) = f_2(0) = x_2
$$

so that

$$
f_1(0) = a_2 = x_1, \qquad
f_2(0) = a_1 = x_2.
$$

It follows that

$$
f_1(\varepsilon_1) = x_2 \sinh \varepsilon_1 + x_1 \cosh \varepsilon_1, \qquad
f_2(\varepsilon_1) = x_2 \cosh \varepsilon_1 + x_1 \sinh \varepsilon_1
$$

and

$$
x_1' = x_2 \sinh \varepsilon_1 + x_1 \cosh \varepsilon_1 + \varepsilon_2, \qquad
x_2' = x_2 \cosh \varepsilon_1 + x_1 \sinh \varepsilon_1 + \varepsilon_3.
$$

Let us now find the infinitesimal generators $\boldsymbol{Z} = (Z_1, Z_2, Z_3)$ for $P(2)$ to show that the representation

$$
\mathbf{x}' = e^{\varepsilon_1 Z_1} e^{\varepsilon_2 Z_2} e^{\varepsilon_3 Z_3} \mathbf{x}
$$

also leads to the group $P(2)$. The infinitesimal generators are

$$Z_1 = \xi_{11}(\mathbf{x})\frac{\partial}{\partial x_1} + \xi_{12}(\mathbf{x})\frac{\partial}{\partial x_2} = x_2\frac{\partial}{\partial x_1} + x_1\frac{\partial}{\partial x_2}$$

$$Z_2 = \xi_{21}(\mathbf{x})\frac{\partial}{\partial x_1} + \xi_{22}(\mathbf{x})\frac{\partial}{\partial x_2} = \frac{\partial}{\partial x_1}$$

$$Z_3 = \xi_{31}(\mathbf{x})\frac{\partial}{\partial x_1} + \xi_{32}(\mathbf{x})\frac{\partial}{\partial x_2} = \frac{\partial}{\partial x_2}.$$

We find

$$
\begin{aligned}
e^{\varepsilon_1 Z_1}e^{\varepsilon_2 Z_2}e^{\varepsilon_3 Z_3}(x_1, x_2) &= e^{\varepsilon_1 Z_1}e^{\varepsilon_2 Z_2}\left(e^{\varepsilon_3 Z_3}x_1, e^{\varepsilon_3 Z_3}x_2\right) \\
&= e^{\varepsilon_1 Z_1}e^{\varepsilon_2 Z_2}(x_1, x_2 + \varepsilon_3) \\
&= e^{\varepsilon_1 Z_1}\left(e^{\varepsilon_2 Z_2}x_1, e^{\varepsilon_2 Z_2}(x_2 + \varepsilon_3)\right) \\
&= e^{\varepsilon_1 Z_1}(x_1 + \varepsilon_2, x_2 + \varepsilon_3) \\
&= \left(e^{\varepsilon_1 Z_1}(x_1 + \varepsilon_2), e^{\varepsilon_1 Z_1}(x_2 + \varepsilon_3)\right) \\
&= (x_1 \cosh\varepsilon_1 + x_2 \sinh\varepsilon_1 + \varepsilon_2, x_1 \sinh\varepsilon_1 + x_2 \cosh\varepsilon_1 + \varepsilon_3).
\end{aligned}
$$

♣

Each parameter of an r-parameter Lie group of transformations leads to an infinitesimal generator. These infinitesimal generators belong to an r-dimensional linear vector space on which there is an additional structure, called the commutator. This special vector space is called a **Lie algebra** and will be discussed in the next chapter. For our purposes the study of a local r-parameter Lie group of transformations is equivalent to the study of its infinitesimal generators and the structure of its corresponding Lie algebra.

4.4 Computer Algebra Applications

In Lie1.red we calculate the Lie series

$$\exp(\varepsilon Z)x, \qquad Z := (x - x^2)\frac{d}{dx}$$

up to order four, i.e.

$$\exp(\varepsilon Z) \approx 1 + \varepsilon Z + \frac{\varepsilon^2 Z^2}{2!} + \frac{\varepsilon^3 Z^3}{3!} + \frac{\varepsilon^4 Z^4}{4!}.$$

In Lie2.red we calculate the Lie series

$$\exp(\varepsilon Z)\begin{pmatrix} x_1 \\ x_2 \end{pmatrix}, \qquad Z := (x_1 - x_1 x_2)\frac{\partial}{\partial x_1} + (-x_2 + x_1 x_2)\frac{\partial}{\partial x_2}$$

```
%lie1.red

n  := 4;
operator Z;
depend Q, x;
depend V, x;

for all P let Z(P) = V*df(P,x);

operator SD;
SD(Q)  := ep*Z(Q);
operator RES;
RES(Q)  := Q + ep*Z(Q);

for j  := 1:n do
begin
SD(Q)  := ep*Z(SD(Q))/(j+1);
RES(Q)  := RES(Q) + SD(Q);
end;

RES(Q);

V  := x - x**2;

depend F, x;
F  := RES(Q);

Q  := x;

F;
```

```
%lie2.red

n := 4;
operator Z;
depend Q1, x1, x2;
depend Q2, x1, x2;
depend V1, x1, x2;
depend V2, x1, x2;

for all P let Z(P) = V1*df(P,x1) + V2*df(P,x2);

operator SD;
SD(Q1) := ep*Z(Q1);
SD(Q2) := ep*Z(Q2);
operator RES;
RES(Q1) := Q1 + SD(Q1);
RES(Q2) := Q2 + SD(Q2);

for j := 1:n do
begin
SD(Q1) := ep*Z(SD(Q1))/(j+1);
SD(Q2) := ep*Z(SD(Q2))/(j+1);
RES(Q1) := RES(Q1) + SD(Q1);
RES(Q2) := RES(Q2) + SD(Q2);
end;

RES(Q1);
V1 := x1 - x1*x2;
RES(Q2);
V2 := -x2 + x1*x2;

depend F1, x1, x2;
F1 := RES(Q1);
depend F2, x1, x2;
F2 := RES(Q2);

Q1 := x1;
F1;
Q2 := x2;
F2;
```

Chapter 5

Lie Algebras

5.1 Definition and Examples

In this section we give the definition of a Lie algebra (Jacobson [62], Miller [84], Gilmore [52], Olver [87], Sattinger and Weaver [103]) and consider a few examples.

Definition 5.1 *A* **Lie algebra** *L is a vector space over a field F on which a product $[\ ,\ ]$ called the Lie bracket or commutator, is defined with the properties*
A1. Closure: For $X,\ Y\ \in\ L$ it follows that $[X,Y]\ \in\ L$
A2. Bilinearity: $[X, \alpha Y + \beta Z] = \alpha[X,Y] + \beta[X,Z]$ for $\alpha, \beta \in F$ and $X,\ Y,\ Z \in L$
A3. Antisymmetry: $[X,Y] = -[Y,X]$
A4. Derivation property known as the Jacobi identity:

$$[X,[Y,Z]] + [Y,[Z,X]] + [Z,[X,Y]] = 0.$$

L is a **real Lie algebra** if F is the field of real numbers **R** and a **complex Lie algebra** if F is the field of complex numbers **C**.

We now give a few examples of Lie algebras.

Example: Let L be the real vector space $\mathbf{R}^3$. Consider the definition

$$[\mathbf{x}, \mathbf{y}] := \mathbf{x} \times \mathbf{y}$$

where $\mathbf{x} = (x_1, x_2, x_3)^T$, $\mathbf{y} = (y_1, y_2, y_3)^T \in L$ and $\times$ denotes the vector cross product. We now prove that L satisfies all the properties of a Lie algebra. Property A1 is satisfied since

$$\mathbf{x} \times \mathbf{y} = \begin{pmatrix} x_2 y_3 - x_3 y_2 \\ x_3 y_1 - x_1 y_3 \\ x_1 y_2 - x_2 y_1 \end{pmatrix} \in L.$$

For the bilinearity property A2 we consider $\mathbf{x}, \mathbf{x}', \mathbf{y}, \mathbf{y}' \in L$ with the constants $\alpha, \beta \in \mathbf{R}$. It follows that

$$
\begin{aligned}
[\alpha\mathbf{x} + \beta\mathbf{x}', \mathbf{y}] &= (\alpha\mathbf{x} + \beta\mathbf{x}') \times \mathbf{y} \\
&= \begin{pmatrix} (\alpha x_2 + \beta x_2')y_3 - (\alpha x_3 + \beta x_3')y_2 \\ (\alpha x_3 + \beta x_3')y_1 - (\alpha x_1 + \beta x_1')y_3 \\ (\alpha x_1 + \beta x_1')y_2 - (\alpha x_2 + \beta x_2')y_1 \end{pmatrix} \\
&= \begin{pmatrix} \alpha(x_2 y_3 - x_3 y_2) + \beta(x_2' y_3 - x_3' y_2) \\ \alpha(x_3 y_1 - x_1 y_3) + \beta(x_3' y_1 - x_1' y_3) \\ \alpha(x_1 y_2 - x_2 y_1) + \beta(x_1' y_2 - x_2' y_1) \end{pmatrix} \\
&= \alpha[\mathbf{x}, \mathbf{y}] + \beta[\mathbf{x}', \mathbf{y}].
\end{aligned}
$$

Analogously,

$$[\mathbf{x}, \alpha\mathbf{y} + \beta\mathbf{y}'] = \alpha[\mathbf{x}, \mathbf{y}] + \beta[\mathbf{x}, \mathbf{y}'].$$

The antisymmetry property A3 is obviously satisfied since

$$\mathbf{x} \times \mathbf{y} = -\mathbf{y} \times \mathbf{x}.$$

To show that the Jacobi identity is satisfied we consider $\mathbf{x}, \mathbf{y}, \mathbf{z} \in L$. We find

$$\mathbf{x} \times (\mathbf{y} \times \mathbf{z}) + \mathbf{y} \times (\mathbf{z} \times \mathbf{x}) + \mathbf{z} \times (\mathbf{x} \times \mathbf{y})$$

$$
\begin{aligned}
&= \begin{pmatrix} x_2(y_1 z_2 - y_2 z_1) - x_3(y_3 z_1 - y_1 z_3) \\ x_3(y_2 z_3 - y_3 z_2) - x_1(y_1 z_2 - y_2 z_1) \\ x_1(y_3 z_1 - y_1 z_3) - x_2(y_2 z_3 - y_3 z_2) \end{pmatrix} + \begin{pmatrix} y_2(z_1 x_2 - z_2 x_1) - y_3(z_3 x_1 - z_1 x_3) \\ y_3(z_2 x_3 - z_3 x_2) - y_1(z_1 x_2 - z_2 x_1) \\ y_1(z_3 x_1 - z_1 x_3) - y_2(z_2 x_3 - z_3 x_2) \end{pmatrix} \\
&\quad + \begin{pmatrix} z_2(x_1 y_2 - x_2 y_1) - z_3(x_3 y_1 - x_1 y_3) \\ z_3(x_2 y_3 - x_3 y_2) - z_1(x_1 y_2 - x_2 y_1) \\ z_1(x_3 y_1 - x_1 y_3) - z_2(x_2 y_3 - x_3 y_2) \end{pmatrix} \\
&= \begin{pmatrix} 0 \\ 0 \\ 0 \end{pmatrix}.
\end{aligned}
$$

It follows that L is a Lie algebra. For the elements of the standard basis $\{\, \mathbf{e}_1, \mathbf{e}_2, \mathbf{e}_3 \,\}$ we find the commutator

$$[\mathbf{e}_1, \mathbf{e}_2] = \mathbf{e}_3, \qquad [\mathbf{e}_3, \mathbf{e}_1] = \mathbf{e}_2, \qquad [\mathbf{e}_2, \mathbf{e}_3] = \mathbf{e}_1. \qquad \clubsuit$$

Example: The set of all $m \times m$ matrices form a Lie algebra with the commutator defined by

$$[A, B] := AB - BA. \qquad \clubsuit$$

Example: The set of all traceless $m \times m$ matrices form a Lie algebra under the commutator, since

$$\text{tr}([A, B]) = 0$$

for two arbitrary $m \times m$ matrices. ♣

Example: The set of all upper (lower) triangular matrices form a Lie algebra under the commutator.

Example: The set of diagonal matrices form a Lie algebra under the commutator. ♣

Example: Let L be a Lie algebra with $x, y, h \in L$. Let

$$[x, h] = 0, \qquad [y, h] = 0.$$

From the Jacobi identity it follows that

$$[[x, y], h] = 0.$$

If h describes a Hamilton operator we say that the operators x and y commute with h. Thus the operator $[x, y]$ also commutes with h. ♣

Example: Consider C^∞-functions which depend on $2n$ variables $(q_1, \ldots, q_n, p_1, \ldots, p_n)$. These variables are known as the phase space variables where q_i denotes the position and q_i the momentum. The commutator of two such functions f and g is defined by the **Poisson bracket**

$$[f, g] := \sum_{i=1}^{n} \left(\frac{\partial f}{\partial q_i} \frac{\partial g}{\partial p_i} - \frac{\partial f}{\partial p_i} \frac{\partial g}{\partial q_i} \right).$$

The vector space of C^∞-functions $f(q_1, \ldots, q_n, p_1, \ldots, p_n)$ form a Lie algebra under the Poisson bracket. Let

$$f(\mathbf{p}, \mathbf{q}) = \frac{p_1^2}{2} + \frac{p_2^2}{2} + \frac{1}{12}(q_1^4 + q_2^4) + \frac{1}{2}q_1^2 q_2^2$$

and

$$g(\mathbf{p}, \mathbf{q}) = 3p_1 p_2 + q_1 q_2(q_1^2 + q_2^2).$$

Then we have

$$[f, g] = 0.$$

The function f can be considered as a Hamilton function, where $p_1^2/2 + p_2^2/2$ is the kinetic term. Since $[f, g] = 0$ we say that g is a first integral of the Hamiltion system.

5.2   Lie Algebras and Bose and Fermi Operators

Consider a family of linear operators b_i, $b_j^\dagger$ $(i, j = 1, 2, \ldots, N)$ on an inner product space V, satisfying the **commutation relations**

$$[b_i, b_j] = 0, \qquad [b_i^\dagger, b_j^\dagger] = 0, \qquad [b_i, b_j^\dagger] = \delta_{ij} I$$

where I is the identity operator. $b_i^\dagger$ denotes the creation operator and b_i the annihilation operator. The operators

$$\{\ b_i^\dagger b_j\ : i, j = 1, 2, \ldots, N\ \}$$

form a Lie algebra. We have

$$[b_i^\dagger b_j, b_k^\dagger b_l] = -\delta_{li} b_k^\dagger b_j + \delta_{jk} b_i^\dagger b_l.$$

The vector space V must be infinite-dimensional. For, if A and B are finite-dimensional square matrices such that

$$[A, B] = \lambda I$$

then

$$\mathrm{tr}([A, B]) = 0$$

implies $\lambda = 0$.

Consider a family of linear operators c_i, $c_j^\dagger$ $(i, j = 1, 2, \ldots, N)$ on a finite-dimensional vector space V satisfying the **anticommutation relations**

$$[c_i, c_j^\dagger]_+ = \delta_{ij} I$$

$$[c_i, c_j]_+ = [c_i^\dagger, c_j^\dagger]_+ = 0.$$

The linear operators c_i, $c_j^\dagger$ are called annihilation and creation operators for fermions. The operators

$$\{\ c_i^\dagger c_j\ : i, j = 1, 2, \ldots, N\ \}$$

form a Lie algebra under the commutator. We have

$$[c_i^\dagger c_j, c_k^\dagger c_l] = -\delta_{li} c_k^\dagger c_j + \delta_{kj} c_i^\dagger c_l.$$

Exercise: Do the operators

$$\{\ c_i^\dagger c_j^\dagger c_k c_l\ :\ i, j, k, l = 1, 2, \ldots, N\ \}$$

form a Lie algebra under the commutator ?

5.3 Lie Algebras and Lie Groups

We recall that a Lie group is a differential manifold (see Appendix A) so that it makes sense to consider the tangent space at a point in the manifold. In particular we are interested in the tangent space at the identity of the Lie group.

Let G be an r-dimensional matrix Lie group where $A(\varepsilon)$ is a smooth curve through the identity of G. In the neighbourhood of the identity $I = A(0)$ we consider the expansion

$$A(\varepsilon) = I + \varepsilon \frac{dA}{d\varepsilon}\Big|_{\varepsilon=0} + O(\varepsilon^2)$$

where

$$X := \frac{dA}{d\varepsilon}\Big|_{\varepsilon=0}$$

is a tangent vector to $A(\varepsilon)$ at $\varepsilon = 0$. X is called the infinitesimal generator for a matrix Lie group. Note that this is in correspondence with the previous definition of the infinitesimal $\boldsymbol{\xi}$ for the Lie transformation group which will be discussed later in this chapter. The set of all X, i.e., the set of all tangents to all curves through I in G, is a linear vector space of dimension r, where

$$L := \left\{ X = \frac{dA}{d\varepsilon}\Big|_{\varepsilon=0} \ : \ A(\varepsilon) \text{ curve through } I \right\}.$$

L is an r-dimensional Lie algebra which is a local representation of the corresponding Lie group. We can recover all the properties of the local Lie group from the Lie algebra in the neighbourhood of the identity element via the exponentiation of the Lie algebra

$$A = \exp(\varepsilon X).$$

Example: Consider the group $SO(3)$ which is the counter clockwise rotation of $\mathbf{R}^3$ about the coordinate axis:

$$A_1(\varepsilon_1) = \begin{pmatrix} 1 & 0 & 0 \\ 0 & \cos\varepsilon_1 & -\sin\varepsilon_1 \\ 0 & \sin\varepsilon_1 & \cos\varepsilon_1 \end{pmatrix}$$

$$A_2(\varepsilon_2) = \begin{pmatrix} \cos\varepsilon_2 & 0 & \sin\varepsilon_2 \\ 0 & 1 & 0 \\ -\sin\varepsilon_2 & 0 & \cos\varepsilon_2 \end{pmatrix}$$

$$A_3(\varepsilon_3) = \begin{pmatrix} \cos\varepsilon_3 & -\sin\varepsilon_3 & 0 \\ \sin\varepsilon_3 & \cos\varepsilon_3 & 0 \\ 0 & 0 & 1 \end{pmatrix}.$$

These are all curves through the identity in $SO(3)$. We have

$$A_1(0) = A_2(0) = A_3(0) = I.$$

The derivatives at the identity

$$X_j := \left. \frac{dA_j(\varepsilon_j)}{d\varepsilon_j} \right|_{\varepsilon_j=0}$$

are given by

$$X_1 = \begin{pmatrix} 0 & 0 & 0 \\ 0 & 0 & -1 \\ 0 & 1 & 0 \end{pmatrix}, \qquad X_2 = \begin{pmatrix} 0 & 0 & 1 \\ 0 & 0 & 0 \\ -1 & 0 & 0 \end{pmatrix}, \qquad X_3 = \begin{pmatrix} 0 & -1 & 0 \\ 1 & 0 & 0 \\ 0 & 0 & 0 \end{pmatrix}.$$

We find

$$[X_1, X_2] = X_3, \qquad [X_2, X_3] = X_1, \qquad [X_3, X_1] = X_2.$$

It follows that the matrices X_j form the basis for the Lie algebra $so(3)$. By exponentiating this Lie algebra we find the group $SO(3)$:

$$A_j = e^{\varepsilon_j X_j}. \qquad\qquad \clubsuit$$

Let us now consider the r-parameter Lie transformation group as given in the previous chapter

$$\mathbf{x}' = \boldsymbol{\varphi}(\mathbf{x}, \boldsymbol{\varepsilon}).$$

Here $\mathbf{x}' = (x'_1, \ldots, x'_m)$, $\mathbf{x} = (x_1, \ldots, x_m)$, $\boldsymbol{\varepsilon} = (\varepsilon_1, \ldots, \varepsilon_r)$ and $\boldsymbol{\varphi} = (\varphi_1, \ldots, \varphi_m)$. The infinitesimal generators are given by

$$Z_\alpha = \sum_{i=1}^m \xi_{\alpha i}(\mathbf{x}) \frac{\partial}{\partial x_i}$$

where

$$\xi_{\alpha j}(\mathbf{x}) = \left. \frac{\partial \varphi_j(\mathbf{x}, \boldsymbol{\varepsilon})}{\partial \varepsilon_\alpha} \right|_{\boldsymbol{\varepsilon}=0}$$

with $\alpha = 1, \ldots, r$ and $j = 1, \ldots, m$. We have

$$\mathbf{x}' = \prod_{\alpha=1}^r \exp(\varepsilon_\alpha Z_\alpha)\mathbf{x}.$$

Consider the autonomous system of ordinary differential equations

$$\frac{d\mathbf{x}}{d\varepsilon_\alpha} = \boldsymbol{\xi}_\alpha(\mathbf{x})$$

($\boldsymbol{\xi}_\alpha = (\xi_{\alpha 1}, \ldots, \xi_{\alpha m})$) with corresponding infinitesimal generators

$$Z_\alpha = \sum_{i=1}^m \xi_{\alpha i}(\mathbf{x}) \frac{\partial}{\partial x_i}$$

and the autonomous system of ordinary differential equations

$$\frac{d\mathbf{x}}{d\varepsilon_\beta} = \boldsymbol{\xi}_\beta(\mathbf{x})$$

$(\boldsymbol{\xi}_\beta = (\xi_{\beta 1}, \ldots, \xi_{\beta m}))$ with corresponding infinitesimal generators

$$Z_\beta = \sum_{j=1}^{m} \xi_{\beta j}(\mathbf{x}) \frac{\partial}{\partial x_j}.$$

We show that these infinitesimal generators are closed under the commutator:

$$[Z_\alpha, Z_\beta]f := Z_\alpha(Z_\beta f) - Z_\beta(Z_\alpha f)$$

where f is a C^∞-function. It follows that

$$
\begin{aligned}
[Z_\alpha, Z_\beta]f &= \sum_{i=1}^{m}\sum_{j=1}^{m} \left(\left(\xi_{\alpha i}(\mathbf{x})\frac{\partial}{\partial x_i}\right)\left(\xi_{\beta j}(\mathbf{x})\frac{\partial f(\mathbf{x})}{\partial x_j}\right) - \left(\xi_{\beta j}(\mathbf{x})\frac{\partial}{\partial x_j}\right)\left(\xi_{\alpha i}(\mathbf{x})\frac{\partial f(\mathbf{x})}{\partial x_i}\right) \right) \\
&= \sum_{i=1}^{m}\sum_{j=1}^{m} \left(\xi_{\alpha i}\frac{\partial \xi_{\beta j}}{\partial x_i}\frac{\partial f(\mathbf{x})}{\partial x_j} + \xi_{\alpha i}\xi_{\beta j}\frac{\partial^2 f(\mathbf{x})}{\partial x_i \partial x_j} - \xi_{\beta j}\frac{\partial \xi_{\alpha i}}{\partial x_j}\frac{\partial f(\mathbf{x})}{\partial x_i} - \xi_{\beta j}\xi_{\alpha i}\frac{\partial^2 f(\mathbf{x})}{\partial x_j \partial x_i} \right) \\
&= \sum_{i=1}^{m}\sum_{j=1}^{m} \left(\xi_{\alpha i}\frac{\partial \xi_{\beta j}}{\partial x_i}\frac{\partial f(\mathbf{x})}{\partial x_j} - \xi_{\beta j}\frac{\partial \xi_{\alpha i}}{\partial x_j}\frac{\partial f(\mathbf{x})}{\partial x_i} \right)
\end{aligned}
$$

where $\partial^2 f(\mathbf{x})/\partial x_i \partial x_j = \partial^2 f(\mathbf{x})/\partial x_j \partial x_i$ was used. It follows that

$$[Z_\alpha, Z_\beta] = \sum_{j=1}^{m} \eta_j(\mathbf{x})\frac{\partial}{\partial x_j}$$

where

$$\eta_j(\mathbf{x}) := \sum_{i=1}^{m} \left(\xi_{\alpha i}(\mathbf{x})\frac{\partial \xi_{\beta j}(\mathbf{x})}{\partial x_i} - \xi_{\beta i}(\mathbf{x})\frac{\partial \xi_{\alpha j}(\mathbf{x})}{\partial x_i} \right).$$

We conclude that the infinitesimal generators of a Lie group are closed under the commutator. The bilinearity, antisymmetry and Jacobi identity can easily be proven for the infinitesimal generators. It follows that the infinitesimal generators of a Lie group form a Lie algebra.

For any Lie algebra with basis elements $\{\, Z_\alpha \;:\; \alpha = 1, \ldots, r \,\}$ the commutator may be expanded in terms of the basis

$$[Z_\alpha, Z_\beta] = \sum_{\gamma=1}^{r} C_{\alpha\beta}^{\gamma} Z_\gamma \tag{1}$$

$(\, \alpha, \beta = 1, \ldots, r \,)$ where $C_{\alpha\beta}^{\gamma}$ are called the structure constants.

Definition 5.2 *Equation (1) is called the* **commutation relation** *of an r-parameter Lie algebra.*

Theorem 5.1 *(Second Fundamental Theorem of Lie). The structure constants $C_{\alpha\beta}^{\gamma}$ in (1) are constants.*

The proof is straightforward and left as an exercise.

Theorem 5.2 *(Third Fundamental Theorem of Lie). The structure constants, defined by the commutation relations (1) satisfy the relations*

$$C_{\alpha\beta}^{\gamma} = -C_{\beta\alpha}^{\gamma},$$

$$\sum_{\rho=1}^{r} \left(C_{\alpha\beta}^{\rho} C_{\rho\gamma}^{\delta} + C_{\beta\gamma}^{\rho} C_{\rho\alpha}^{\delta} + C_{\gamma\alpha}^{\rho} C_{\rho\beta}^{\delta} \right) = 0$$

where α, β, $\gamma = 1, \ldots, r$.

Proof: Since $\{ Z_\alpha \ : \ \alpha = 1, \ldots, m \}$ form a basis of a Lie algebra it follows that $[Z_\alpha, Z_\beta] = -[Z_\beta, Z_\alpha]$ where

$$[Z_\alpha, Z_\beta] = \sum_{\rho=1}^{r} C_{\alpha\beta}^{\rho} Z_\rho$$

and

$$[Z_\beta, Z_\alpha] = \sum_{\rho=1}^{r} C_{\beta\alpha}^{\rho} Z_\rho$$

so that $C_{\alpha\beta}^{\mu} = -C_{\alpha\beta}^{\ell}$. From the Jacobi identity we obtain

$$[Z_\gamma, [Z_\alpha, Z_\beta]] + [Z_\beta, [Z_\gamma, Z_\alpha]] + [Z_\alpha[Z_\beta, Z_\gamma]] = 0$$

$$\Rightarrow \quad [Z_\gamma, \sum_{\rho=1}^{r} C_{\alpha\beta}^{\rho} Z_\rho] + [Z_\beta, \sum_{\rho=1}^{r} C_{\gamma\alpha}^{\rho} Z_\rho] + [Z_\alpha, \sum_{\rho=1}^{r} C_{\beta\gamma}^{\rho} Z_\rho] = 0$$

$$\Rightarrow \quad \sum_{\rho=1}^{r} C_{\alpha\beta}^{\rho} [Z_\gamma, Z_\rho] + \sum_{\rho=1}^{r} C_{\gamma\alpha}^{\rho} [Z_\beta, Z_\rho] + \sum_{\rho=1}^{r} C_{\beta\gamma}^{\rho} [Z_\alpha, Z_\rho] = 0$$

$$\Rightarrow \quad \sum_{\rho=1}^{r} C_{\alpha\beta}^{\rho} \sum_{\delta=1}^{r} C_{\gamma\rho}^{\delta} Z_\delta + \sum_{\rho=1}^{r} C_{\gamma\alpha}^{\rho} \sum_{\delta=1}^{r} C_{\beta\rho}^{\delta} Z_\delta + \sum_{\rho=1}^{r} C_{\beta\gamma}^{\rho} \sum_{\delta=1}^{r} C_{\alpha\rho}^{\delta} Z_\delta = 0$$

$$\Rightarrow \quad \sum_{\delta=1}^{r} (\sum_{\rho=1}^{r} C_{\alpha\beta}^{\rho} C_{\gamma\rho}^{\delta} + \sum_{\rho=1}^{r} C_{\gamma\alpha}^{\rho} C_{\beta\rho}^{\delta} + \sum_{\rho=1}^{r} C_{\beta\gamma}^{\rho} C_{\alpha\rho}^{\delta}) Z_\delta = 0.$$

Since the Z_δ's are linearly independent the linear combination can only be zero if

$$\sum_{\rho=1}^{r} C_{\alpha\beta}^{\rho} C_{\gamma\rho}^{\delta} + \sum_{\rho=1}^{r} C_{\gamma\alpha}^{\rho} C_{\beta\rho}^{\delta} + \sum_{\rho=1}^{r} C_{\beta\gamma}^{\rho} C_{\alpha\rho}^{\delta} = 0. \qquad \spadesuit$$

Example: Consider the three parameter Euclidean group $E(2)$. Recall that this is an affine group of rigid motions in $\mathbf{R}^2$ given by

$$\begin{pmatrix} x_1' \\ x_2' \end{pmatrix} = \begin{pmatrix} \cos \varepsilon_1 & -\sin \varepsilon_1 \\ \sin \varepsilon_1 & \cos \varepsilon_1 \end{pmatrix} \begin{pmatrix} x_1 \\ x_2 \end{pmatrix} + \begin{pmatrix} \varepsilon_2 \\ \varepsilon_3 \end{pmatrix}.$$

The corresponding infinitesimal generators are

$$Z_1 = -x_2 \frac{\partial}{\partial x_1} + x_1 \frac{\partial}{\partial x_2}, \qquad Z_2 = \frac{\partial}{\partial x_1}, \qquad Z_3 = \frac{\partial}{\partial x_2}. \qquad \clubsuit$$

It is convenient to display the commutation relation of a Lie algebra through its commutator table whose (i, j)-th entry is $[Z_i, Z_j]$. From the definition of the commutator it follows that the table is antisymmetric with its diagonal elements all zero. The structure constants are easily obtained from the commutator table. For the given example the commutator table is as follows

	Z_1	Z_2	Z_3
Z_1	0	$-Z_3$	Z_2
Z_2	Z_3	0	0
Z_3	$-Z_2$	0	0.

We have thus found a three dimensional Lie algebra $e(2)$ for the three dimensional affine group $E(2)$.

Example: Consider

$$Z_1 = x_1 \frac{\partial}{\partial x_2} + x_2 \frac{\partial}{\partial x_1}, \qquad Z_2 = x_1 \frac{\partial}{\partial x_1} + x_2 \frac{\partial}{\partial x_2}.$$

Then we have

$$[Z_1, Z_2] = 0$$

and therefore

$$\exp(\varepsilon Z_1 + \varepsilon Z_2) = \exp(\varepsilon Z_1) \exp(\varepsilon Z_2).$$

Thus Z_1 and Z_2 form a basis of an abelian Lie algebra. $\clubsuit$

Remark: Let V, W be to infinitesimal generators. Then, in general,

$$\exp(V + W) \neq \exp(V) \exp(W).$$

5.4 Classical Lie Algebras

The classical Lie algebras are associated with the classical Lie groups which were discussed in chapter 3. Note that the dimensions of the associated Lie algebras are the same as for the classical Lie groups, as listed in the table in chapter 3.

The Lie algebras associated with the general linear groups $GL(m, \mathbf{C})$ and $GL(m, \mathbf{R})$ which consist of all $m \times m$ nonsingular matrices, are given by the set of all $m \times m$ complex and real matrices, respectively. The corresponding Lie algebras are denoted by $gl(m, \mathbf{C})$ and $gl(m, \mathbf{R})$, respectively. The Lie algebra $gl(m, \mathbf{C})$ thus consists of all linear mappings from $\mathbf{C}^m$ to $\mathbf{C}^m$, where this set is denoted by $L(\mathbf{C}^m, \mathbf{C}^m)$, and $gl(m, \mathbf{R})$ consists of all linear mappings from $\mathbf{R}^m$ to $\mathbf{R}^m$ denoted by $L(\mathbf{R}^m, \mathbf{R}^m)$.

The Lie algebra associated with the special linear group $SL(m, \mathbf{R})$ is denoted by $sl(m, \mathbf{R})$ and defined by

$$sl(m, \mathbf{R}) := \{\ X \in L(\mathbf{R}^m, \mathbf{R}^m)\ :\ \operatorname{tr} X = 0\ \}.$$

The condition on X is found as follows: For any arbitrary $m \times m$ matrix X we have the identity

$$\det(\exp(X)) \equiv \exp(\operatorname{tr} X).$$

Let $A \in SL(m)$ and $X \in gl(m)$. It follows that $\det A = \det(\exp(\varepsilon X)) = \exp(\varepsilon \operatorname{tr} X) = 1$ so that $\operatorname{tr} X = 0$.

Example: For $m = 2$ a basis for the three dimensional Lie algebra $sl(2, \mathbf{R})$ is given by

$$\left\{ \begin{pmatrix} 0 & 0 \\ 1 & 0 \end{pmatrix}, \begin{pmatrix} 0 & 1 \\ 0 & 0 \end{pmatrix}, \begin{pmatrix} 1 & 0 \\ 0 & -1 \end{pmatrix} \right\}. \qquad \clubsuit$$

Similarly, for the complex group $SL(m, \mathbf{C})$ the associated Lie algebra is defined by

$$sl(m, \mathbf{C}) := \{\ X \in L(\mathbf{C}^m, \mathbf{C}^m)\ :\ \operatorname{tr} X = 0\ \}.$$

Example: A basis for the six dimensional Lie algebra $sl(2, \mathbf{C})$ is given by

$$\{\ E_1, E_2, E_3, H_1, H_2, H_3\ \}$$

where

$$E_1 = \frac{1}{2} \begin{pmatrix} 0 & -i \\ -i & 0 \end{pmatrix}, \qquad E_2 = \frac{1}{2} \begin{pmatrix} 0 & -1 \\ 1 & 0 \end{pmatrix}, \qquad E_3 = \frac{1}{2} \begin{pmatrix} -i & 0 \\ 0 & i \end{pmatrix}$$

$$H_1 = \frac{1}{2} \begin{pmatrix} 0 & 1 \\ 1 & 0 \end{pmatrix}, \qquad H_2 = \frac{1}{2} \begin{pmatrix} 0 & -i \\ i & 0 \end{pmatrix}, \qquad H_3 = \frac{1}{2} \begin{pmatrix} 1 & 0 \\ 0 & -1 \end{pmatrix}.$$

The commutation relations are

$$[E_i, E_j] = \sum_{k=1}^{3} \epsilon_{ijk} E_k$$

$$[H_i, H_j] = -\sum_{k=1}^{3} \epsilon_{ijk} H_k$$

$$[E_i, H_j] = \sum_{k=1}^{3} \epsilon_{ijk} H_k$$

with $i, j = 1, 2, 3$. Here $\epsilon_{123} = +1$ is the totally antisymmetric tensor.   ♣

The Lie algebra $o(m)$ associated with the orthogonal group $O(m)$ is given by

$$o(m) := \left\{ X \in L(\mathbf{R}^m, \mathbf{R}^m) \ : \ X^T = -X \right\}.$$

Recall that the condition on $A \in O(m)$ is $A^T A = I$. To find the associated condition on the Lie algebra elements we consider

$$\frac{d}{d\varepsilon} \left(A^T A \right)\Big|_{\varepsilon=0} = \left(\frac{dA^T}{d\varepsilon} A + A^T \frac{dA}{d\varepsilon} \right)\Big|_{\varepsilon=0} = X^T A(0) + A^T(0)X = X^T + X = 0.$$

This condition is equivalent to the bilinear metric form

$$(X\mathbf{x}, \mathbf{y}) + (\mathbf{x}, X\mathbf{y}) = 0$$

where $\mathbf{x}, \mathbf{y} \in \mathbf{R}^m$. The special orthogonal group $SO(m)$ has the additional condition $\det A = 1$ so that the Lie algebra is defined by

$$so(m) := \{ X \in L(\mathbf{R}^m, \mathbf{R}^m) \ : \ X^T = -X, \ \mathrm{tr}X = 0 \}.$$

The Lie algebra consists of all skew-symmetric real matrices.

Example: For $m = 2$, a basis for the one-dimensional Lie algebra $so(m)$ is provided by

$$\begin{pmatrix} 0 & -1 \\ 1 & 0 \end{pmatrix}.$$ ♣

The Lie algebra $u(m)$ associated with the unitary group $U(m)$ is defined by

$$u(m) := \{ X \in L(\mathbf{C}^n, \mathbf{C}^n) \ : \ X^\dagger = -X \}.$$

It thus consists of all skew-hermitian matrices.

The special unitary group $SU(m)$ has the additional condition $\det A = 1$ $(A \in U(m))$ so that the associated Lie algebra is defined by

$$su(m) := \{ X \in L(\mathbf{C}^m, \mathbf{C}^m) \ : \ X^\dagger = -X, \ \mathrm{tr}X = 0 \}.$$

It thus consists of all traceless skew-hermitian matrices.

Example: A basis of the three dimensional Lie algebra $su(2)$ is given by $\{\sigma_1, \sigma_2, \sigma_3\}$, where

$$\sigma_1 := \begin{pmatrix} 0 & 1 \\ 1 & 0 \end{pmatrix}, \qquad \sigma_2 := \begin{pmatrix} 0 & -i \\ i & 0 \end{pmatrix}, \qquad \sigma_3 := \begin{pmatrix} 1 & 0 \\ 0 & -1 \end{pmatrix}.$$

σ_1, σ_2 and σ_3 are known as the **Pauli spin matrices**.   ♣

Example: Consider $su(2)$ with the basis $\{ J_j : j = 1, 2, 3 \}$ where

$$[J_j, J_k] = \sum_{l=1}^{3} i\epsilon_{jkl} J_l, \qquad j, k = 1, 2, 3.$$

If we define

$$J_\pm := J_1 \pm iJ_2, \qquad J_0 := J_3$$

we find the following commutation relations

$$
\begin{aligned}
[J_0, J_\pm] &= J_0 J_\pm - J_\pm J_0 \\
&= J_0(J_1 \pm iJ_2) - (J_1 \pm iJ_2)J_0 \\
&= J_0 J_1 \pm iJ_0 J_2 - (J_1 J_0 \pm iJ_2 J_0) \\
&= [J_0, J_1] \pm i[J_0, J_2] \\
&= [J_3, J_1] \pm i[J_3, J_2] \\
&= i\epsilon_{312} J_2 \pm i(\epsilon_{321} J_1) \\
&= iJ_2 \pm J_1 \\
&= \pm(J_1 \pm iJ_2) \\
&= \pm J_\pm
\end{aligned}
$$

$$
\begin{aligned}
[J_+, J_-] &= (J_1 + iJ_2)(J_1 - iJ_2) - (J_1 - iJ_2)(J_1 + iJ_2) \\
&= J_1 J_1 + i(J_2 J_1 - J_1 J_2) + J_2 J_2 - J_1 J_1 + i(J_2 J_1 - J_1 J_2) - J_2 J_2 \\
&= 2i[J_2, J_1] \\
&= 2i(i\epsilon_{213} J_3) \\
&= 2J_0.
\end{aligned}
$$

We define the following operators:

$$J_+ := b_1^\dagger b_2, \qquad J_- := b_2^\dagger b_1, \qquad J_0 := \frac{1}{2}(b_1^\dagger b_1 - b_2^\dagger b_2)$$

where b_1, b_2, $b_1^\dagger$, $b_2^\dagger$ are Bose annihilation and creation operators. Thus

$$
\begin{aligned}
[J_+, J_-] &= b_1^\dagger b_2 b_2^\dagger b_1 - b_2^\dagger b_1 b_1^\dagger b_2 \\
&= b_1^\dagger (I + b_1^\dagger b_2) b_1 - b_2^\dagger (I + b_1^\dagger b_1) b_2 \\
&= b_1^\dagger b_1 - b_2^\dagger b_2 \\
&= 2J_0
\end{aligned}
$$

$$
\begin{aligned}
[J_0, J_+] &= \frac{1}{2}(b_1^\dagger b_1 - b_2^\dagger b_2)b_1^\dagger b_2 - b_1^\dagger b_2 \frac{1}{2}(b_1^\dagger b_1 - b_2^\dagger b_2) \\
&= \frac{1}{2}b_1^\dagger b_1 b_1^\dagger b_2 - \frac{1}{2}(b_2 b_1^\dagger - I)b_1^\dagger b_2 - \frac{1}{2}(b_1^\dagger b_2 b_1 b_1^\dagger - b_1^\dagger b_2) + \frac{1}{2}b_1^\dagger b_2 b_2^\dagger b_2 \\
&= b_1^\dagger b_2 + \frac{1}{2}b_1^\dagger b_1 b_1^\dagger b_2 - \frac{1}{2}b_1^\dagger b_2 b_1 b_1^\dagger + \frac{1}{2}b_1^\dagger b_2 b_2^\dagger b_2 - \frac{1}{2}b_2 b_2^\dagger b_1^\dagger b_2 \\
&= b_1^\dagger b_2 \\
&= J_+ \\
[J_0, J_-] &= \frac{1}{2}(b_1^\dagger b_1 - b_2^\dagger b_2)b_2^\dagger b_1 - b_2^\dagger b_1 \frac{1}{2}(b_1^\dagger b_1 - b_2^\dagger b_2) \\
&= \frac{1}{2}(b_1 b_1^\dagger - I)b_2^\dagger b_1 - \frac{1}{2}b_2^\dagger b_2 b_2^\dagger b_1 - \frac{1}{2}b_2^\dagger b_1 b_1^\dagger b_1 + \frac{1}{2}b_2^\dagger b_1 (b_2 b_2^\dagger - I) \\
&= -b_2^\dagger b_1 + \frac{1}{2}b_1 b_1^\dagger b_2^\dagger b_1 - \frac{1}{2}b_2^\dagger b_2 b_2^\dagger b_1 - \frac{1}{2}b_2^\dagger b_1 b_1^\dagger b_1 + \frac{1}{2}b_2^\dagger b_1 b_2 b_2^\dagger \\
&= -b_2^\dagger b_1 \\
&= -J_-.
\end{aligned}
$$

This is known as a Bose realisation of the Lie algebra $su(2)$.   ♣

Let us finally give a list of the classical matrix Lie algebras with the conditions on the matrix X. The Lie algebras of the pseudo-classical groups discussed in chapter 3 are also given in the following list:

Lie Algebra	Conditions on X
$gl(m)$	-
$sl(m)$	$\operatorname{tr} X = 0$
$o(m)$	$X^T = -X$
$so(m)$	$X^T = -X,\ \operatorname{tr} X = 0$
$u(m)$	$X^\dagger = -X$
$su(m)$	$X^\dagger = -X,\ \operatorname{tr} X = 0$
$so(P, m-P)$	$X^T L + LX = 0,\ \operatorname{tr} X = 0$
$u(P, m-P)$	$X^\dagger L + LX = 0$
$su(P, m-P)$	$X^\dagger L + LX = 0,\ \operatorname{tr} X = 0$
$sp(2m)$	$X^T J + JX = 0.$

Recall that

$$
L := \begin{pmatrix} -I_P & 0 \\ 0 & I_{m-P} \end{pmatrix}, \qquad J := \begin{pmatrix} 0 & I_m \\ -I_m & 0 \end{pmatrix}.
$$

5.5   Important Concepts and Examples

In the following section we give a list of some of the important concepts in Lie algebras. Examples are also given to illustrate these concepts.

Let L be a Lie algebra.

Definition 5.3 L' *is a* **Lie subalgebra** *of* L *if* L' *is closed under commutation, that is, if* $[X_1', X_2'] \in L'$ *whenever* X_1' *and* X_2' *belong to* L'. *We write* $[L', L'] \subset L'$.

Example: The Euclidean Lie algebra $e(3)$ has the Lie subalgebras $\{R_1, R_2, R_3\}$ and $\{P_1, P_2, P_3\}$ known as $so(3)$ and $t(3)$ which generate rotations and translations respectively. In terms of infinitesimal generators we have

$$R_1 = x_1 \frac{\partial}{\partial x_2} - x_2 \frac{\partial}{\partial x_1}, \quad R_2 = x_1 \frac{\partial}{\partial x_3} - x_3 \frac{\partial}{\partial x_1}, \quad R_3 = x_2 \frac{\partial}{\partial x_3} - x_3 \frac{\partial}{\partial x_2}$$

and

$$P_1 = \frac{\partial}{\partial x_1}, \quad P_2 = \frac{\partial}{\partial x_2}, \quad P_3 = \frac{\partial}{\partial x_3}. \qquad\qquad \clubsuit$$

Definition 5.4 *Two Lie algebras* L *and* L' *are* **isomorphic** *if a vector space isomorphism* $\phi : L \to L'$ *exists such that*

$$\phi[X_1, X_2] = [\phi(X_1), \phi(X_2)]$$

for all $X_1, X_2 \in L$.

Example: Let $\{\, A_{ij} \,:\, i, j = 1, \ldots, m \,\}$ be the basis of a Lie algebra with commutation relation

$$[A_{ij}, A_{kl}] = \delta_{kj} A_{il} - \delta_{li} A_{jk}$$

where δ_{ij} is the Kronecker delta. The Lie algebra with the basis

$$\{\, x_i \frac{\partial}{\partial x_j} \,:\, i, j = 1, \ldots, m \,\}$$

as well as the Lie algebra with Bose operator basis

$$\{\, b_i^\dagger b_j \,:\, i, j = 1, \ldots, m \,\}$$

are isomorphic to the Lie algebra given above. $\qquad\qquad \clubsuit$

Definition 5.5 *A Lie algebra* L *is the* **direct sum** *of two Lie algebras* L' *and* L'' *if the sum of* L' *and* L'' *is a vector space and if* $[L', L''] = 0$. *We write*

$$L = A \oplus B$$

Note that the algebras L' and L'' are Lie subalgebras of L.

Example: $t(3)$ is the direct sum of the three Lie subalgebras $\{P_1\}$, $\{P_2\}$ and $\{P_3\}$. ♣

Definition 5.6 *A Lie algebra L is the* **semi-direct sum** *of two Lie subalgebras L' and L'' if the sum of L' and L'' is a vector space and $[L', L''] \subset L'$. We write*

$$L = L' \oplus_s L''.$$

Example: The Lie algebra $e(3)$ is the semi-direct sum of $t(3)$ and $so(3)$ so that

$$e(3) = t(3) \oplus_s so(3).$$ ♣

Definition 5.7 *A Lie subalgebra L' is an* **ideal** *of L if $[L', L] \subset L'$, that is, if $[X_1', X_2] \in L'$ whenever $X_1' \in L'$ and $X_2 \in L$.*

Example: In the Lie algebra, given by the commutation relations

$$[x_1, x_2] = x_3, \qquad [x_2, x_3] = x_1, \qquad [x_3, x_1] = x_2$$

the only ideals are $L' = \{0\}$ (0 vector) and $L' = L$. ♣

Note that if $L = L' \oplus_s L''$ then L' is an ideal of L. In a direct sum,

$$L = L_1' \oplus L_2' \oplus \cdots \oplus L_n',$$

each of the summands L_i' is an ideal of L.

Example: Consider $e(3)$. It follows that $t(3)$ is an ideal of $e(3)$, but $so(3)$ is not. ♣

Definition 5.8 *The* **centre** *of L is the largest ideal L' such that $[L, L'] = 0$. It is unique.*

Example: The centre of $e(3)$ is zero, while $\{P_3\}$ is the centre of the subalgebra L' of $e(3)$ consisting of R_3, P_1, P_2, P_3. ♣

Just as there is a notion of quotient group in the theory of groups there is a notion of **quotient algebra** in the theory of Lie algebras. If H is a subgroup of G we define an equivalence relation on G by

$$a \equiv b \,(\text{mod } H) \ \text{ if } \ a^{-1}b \in H.$$

The equivalence classes under this relation are called the left cosets of H and are denoted by aH. Similarly, we may define a second equivalence relation by

$$a \equiv b (\text{mod } H) \ \text{ if } \ ab^{-1} \in H.$$

The equivalence classes in this case are the right cosets of H, denoted by Ha. We say H is normal if $aH = Ha$ for all $a \in G$. In that case the cosets of H in G form a group, with the group operation defined by $(aH)(bH) := abH$. The fact that H is normal is used to prove that the operation is well defined. This group is called the quotient group and is denoted by g/H.

Now suppose L' is a subalgebra of the Lie algebra L. For any $X_1 \in L$ define $X_1 + L'$ to be the equivalence class of X_1 under the equivalence relation $X_1 \equiv X_2 (\text{mod } L')$ if $X_1 - X_2 \in L'$. In general these equivalence classes do not form a Lie algebra, but they do if L' is an ideal. In that case we define a Lie bracket on the classes by

$$[X_1 + L', X_2 + L'] = [X_1, X_2] + L'.$$

This bracket is well defined because L' is an ideal. The set of equivalence classes thus forms a new Lie algebra called the **quotient Lie algebra**. The quotient algebra is denoted by L/L'.

Example: $e(3)/t(3)$ is a Lie algebra which is isomorphic to $so(3)$.   ♣

The analogous group theoretical fact is that the quotient of the Euclidean group by the translations is the rotation group. In fact, ideals of Lie algebras always correspond to normal subgroups of the corresponding Lie group.

Definition 5.9 *The set of commutators* $[L, L]$ *is an ideal of L, called* $L^{(1)}$. *Similarly* $L^{(2)} = [L^{(1)}, L^{(1)}]$ *is an ideal of* $L^{(1)}$. *We define*

$$L^{(n+1)} := [L^{(n)}, L^{(n)}].$$

If this sequence terminates in zero, we say L is **solvable**.

Examples: Consider $e(3)$. It follows that $e(3)^{(1)} = e(3)$, so $e(3)^{(n)} = e(3)$. Hence $e(3)$ is not solvable. On the other hand the subalgebra $L' = \{R_3, P_1, P_2, P_3\}$ introduced above is solvable because $L'^{(2)}$ is zero. The set of all $n \times n$ upper triangular matrices

$$\begin{pmatrix} a_{11} & a_{12} & \cdots & a_{1n} \\ 0 & a_{22} & \cdots & a_{2n} \\ \vdots & & \ddots & \vdots \\ 0 & \cdots & 0 & a_{nn} \end{pmatrix}$$

is another solvable Lie algebra. Conversely, Lie proved that every complex solvable matrix algebra is isomorphic to a subalgebra of triangular matrices.   ♣

Definition 5.10 *Consider a different sequence of ideals given by*

$$L_{(1)} = [L, L], \quad L_{(2)} = [L, L_{(1)}], \quad \cdots, \quad L_{(n+1)} = [L, L_{(n)}].$$

This is a nested sequence with

$$L_{(n+1)} \subseteq L_{(n)} \subseteq \cdots \subseteq L_{(1)} = L^{(1)} \subseteq L.$$

We say L is **nilpotent** *if this sequence terminates in zero.*

Example: An important nilpotent algebra in quantum mechanics is the **Heisenberg algebra** $H = \{P, Q, I\}$, where the operators $Q = x$ and $P = \partial/\partial x$ are acting on smooth functions defined on the real line. The operator Q is multiplication by x, thus

$$(Qf)(x) := xf(x).$$

The commutation relations are

$$[P, Q] = I$$

where I is the identity operator. ♣

Note that $L_{(n)} \supseteq L^{(n)}$, so that nilpotency implies solvability.

Examples: The solvable algebra (2) is not nilpotent. Another example of a solvable algebra which is not nilpotent is the algebra $L' = \{R_3, P_1, P_2, P_3\}$. It follows that $L'_{(1)} = \{P_1, P_2\} = L'_{(n)}$ for all $n \geq 1$. It can be shown that the matrices

$$\begin{pmatrix} \lambda & a_{12} & \cdots & a_{1n} \\ 0 & \lambda & & \vdots \\ \vdots & & \ddots & \\ 0 & & & \lambda \end{pmatrix}$$

form a nilpotent algebra. ♣

Definition 5.11 *The* **radical** *of a Lie algebra is the maximal solvable ideal. It is unique and contains all other solvable ideals.*

Example: $t(3)$ is the radical of $e(3)$. ♣

Definition 5.12 *A Lie algebra L is* **simple** *if it contains no ideals other than L and* $\{0\}$*; it is* **semi-simple** *if it contains no abelian ideals other than* $\{0\}$*.*

Examples: The Lie algebra $su(2)$ with basis $\{x_1, x_2, x_3\}$ and commutation relations

$$[x_1, x_2] = x_3, \qquad [x_2, x_3] = x_1, \qquad [x_3, x_1] = x_2$$

is simple and thus also a semi-simple Lie algebra.   ♣

Levi's decomposition states that every Lie algebra is the semi-direct sum of its radical and a semi-simple Lie algebra. The semi-simple Lie algebras constitute an important class of Lie algebras and play a fundamental role in geometry and physics.

Example: $e(3) = t(3) \oplus_s so(3)$ is the Levi decomposition of $e(3)$.   ♣

If L is a Lie algebra and $X \in L$, the operator $\mathrm{ad}\, X$ that maps Y to $[X, Y]$ is a linear transformation of L onto itself, i.e.,

$$(\mathrm{ad}\, X)Y := [X, Y].$$

It is easily verified that $X \to \mathrm{ad}\, X$ is a representation of the Lie algebra L with L itself considered as the vector space of the representation. We have to check that

$$\mathrm{ad}\, [X, Y](Z) = [\mathrm{ad}\, X,\ \mathrm{ad}\, Y](Z)$$

where $X, Y, Z \in L$. We have

$$\mathrm{ad}[X, Y](Z) = [[X, Y], Z]$$

so that

$$
\begin{aligned}
[\mathrm{ad}\, X, \mathrm{ad}\, Y](Z) &= (\mathrm{ad}\, X(\mathrm{ad}\, Y) - \mathrm{ad}\, Y(\mathrm{ad}\, X))(Z) \\
&= \mathrm{ad}\, X(\mathrm{ad}\, Y)(Z) - \mathrm{ad}\, Y(\mathrm{ad}\, X)(Z) \\
&= \mathrm{ad}\, X([Y, Z]) - \mathrm{ad}\, Y([X, Z]) \\
&= [X, [Y, Z]] - [Y, [X, Z]] \\
&= [[X, Y], Z].
\end{aligned}
$$

The last step follows from the Jacobi identity.

Definition 5.13 *The representation* $\mathrm{ad}\, X$, *called the* **adjoint representation**, *provides a matrix representation of the Lie algebra.*

Example: Consider $sl(2, \mathbf{R})$ with basis $\{ X, Y, H \}$ where

$$X = \begin{pmatrix} 0 & 1 \\ 0 & 0 \end{pmatrix}, \qquad Y = \begin{pmatrix} 0 & 0 \\ 1 & 0 \end{pmatrix}, \qquad H = \begin{pmatrix} 1 & 0 \\ 0 & -1 \end{pmatrix}.$$

From

$$[X, H] = -2X, \qquad [X, Y] = H, \qquad [Y, H] = 2Y$$

we find

$$\begin{aligned}
\operatorname{ad} X(X) &= 0, & \operatorname{ad} H(X) &= 2X, & \operatorname{ad} Y(X) &= -H, \\
\operatorname{ad} X(Y) &= H, & \operatorname{ad} H(Y) &= -2Y, & \operatorname{ad} Y(Y) &= 0, \\
\operatorname{ad} X(H) &= -2X, & \operatorname{ad} H(H) &= 0, & \operatorname{ad} Y(H) &= 2Y.
\end{aligned}$$

From

$$(X, H, Y)\operatorname{ad} X := (\operatorname{ad} X(X),\ \operatorname{ad} X(H),\ \operatorname{ad} X(Y)) = (0, -2X, H)$$

it follows that

$$\operatorname{ad}(X) = \begin{pmatrix} 0 & -2 & 0 \\ 0 & 0 & 1 \\ 0 & 0 & 0 \end{pmatrix}.$$

Analogously

$$\operatorname{ad}(H) = \begin{pmatrix} 2 & 0 & 0 \\ 0 & 0 & 0 \\ 0 & 0 & -2 \end{pmatrix}, \qquad \operatorname{ad}(Y) = \begin{pmatrix} 0 & 0 & 0 \\ -1 & 0 & 0 \\ 0 & 2 & 0 \end{pmatrix}. \qquad \clubsuit$$

If $\{\, X_i\ :\ i = 1, \ldots, m \,\}$ is a basis for the Lie algebra L then

$$(\operatorname{ad} X_i) X_j = \sum_{k=1}^{m} C_{ij}^k X_k.$$

Therefore the matrix associated with the transformation $\operatorname{ad} X_i$ is

$$(M_i)_{jk} = C_{ik}^j.$$

Note the transposition of the indices j and k.

Example: The adjoint representation of $so(3)$ is given by

$$(M_i)_{jk} = C_{ik}^j = \varepsilon_{ikj},$$

so the matrices $X_1, X_2,$ and X_3 of $so(3)$ (first example of this chapter) are in fact also the matrices of the adjoint representation. $\qquad \clubsuit$

Definition 5.14 *The* **Killing form** *of a Lie algebra is the symmetric bilinear form*

$$K(X, Y) = tr(adX\, adY)$$

where tr denotes the trace.

If ρ is an automorphism of L then

$$K(\rho(X), \rho(Y)) = K(X, Y).$$

Moreover, K has the property

$$K([X, Y], Z) = K([Z, X], Y) = -K(Y, [X, Z]).$$

Definition 5.15 *If (X_i) form a basis for L then*

$$g_{ij} = K(X_i, X_j)$$

is called the **metric tensor** *for L. In terms of the structure constants,*

$$g_{ij} = \sum_{r,s=1}^{m} C_{is}^r C_{jr}^s.$$

Example: Consider $sl(2, \mathbf{R})$ with basis

$$X_1 = \begin{pmatrix} 0 & 1 \\ 0 & 0 \end{pmatrix}, \qquad X_2 = \begin{pmatrix} 1 & 0 \\ 0 & -1 \end{pmatrix}, \qquad X_3 = \begin{pmatrix} 0 & 0 \\ 1 & 0 \end{pmatrix}.$$

We calculate the metric tensor g_{ij} with $i, j = 1, 2, 3$. The adjoint representation for this Lie algebra is given in a previous example. We find

$$g_{12} = \mathrm{tr}(\mathrm{ad}X_1 \, \mathrm{ad}X_2) = \mathrm{tr} \begin{pmatrix} 0 & -2 & 0 \\ 0 & 0 & 1 \\ 0 & 0 & 0 \end{pmatrix} \begin{pmatrix} 2 & 0 & 0 \\ 0 & 0 & 0 \\ 0 & 0 & -2 \end{pmatrix} = \mathrm{tr} \begin{pmatrix} 0 & 0 & 0 \\ 0 & 0 & -2 \\ 0 & 0 & 0 \end{pmatrix} = 0.$$

Analogously we obtain

$$g_{11} = 0, \quad g_{13} = 4, \quad g_{21} = 0, \quad g_{22} = 8, \quad g_{23} = 0, \quad g_{33} = 0, \quad g_{31} = 4, \quad g_{32} = 0$$

so that

$$(g_{ij}) = \begin{pmatrix} 0 & 0 & 4 \\ 0 & 8 & 0 \\ 4 & 0 & 0 \end{pmatrix}. \qquad\qquad \clubsuit$$

Note that a Lie algebra is semi-simple if and only if the matrix (g_{ij}) is nonsingular, i.e., $\det((g_{ij})) \neq 0$. In the example given above $\det((g_{ij})) = -128$ which indicates that the Lie algebra is semi-simple.

The **Casimir-operators** play an important role in the applications of Lie algebras to quantum mechanics and elementary particle physics. For a given semi-simple Lie algebra L with basis $\{X_j \; : \; j = 1, \ldots, m\}$, the Casimir-operator is a quantity (which is not an element of the Lie algebra) that commutes with each element of the Lie algebra.

Definition 5.16 *The Casimir-operator C of a given Lie algebra L is given by*

$$C := \sum_{i=1}^{m} \sum_{j=1}^{m} (g^{ij}) X_i X_j$$

where (g^{ij}) is the inverse matrix of (g_{ij}) and $\{X_j \; : \; j = 1, \ldots, m\}$ is a basis of the Lie algebra.

Note that the Casimir-operator is independent of the choice of the basis.

Example: Consider the Lie algebra $so(3)$

$$L_{x_1} := -i \left(x_2 \frac{\partial}{\partial x_3} - x_3 \frac{\partial}{\partial x_2} \right)$$

$$L_{x_2} := -i \left(x_3 \frac{\partial}{\partial x_1} - x_1 \frac{\partial}{\partial x_3} \right)$$

$$L_{x_3} := -i \left(x_1 \frac{\partial}{\partial x_2} - x_2 \frac{\partial}{\partial x_1} \right)$$

with the commutation relations

$$[L_{x_1}, L_{x_2}] = iL_{x_3}, \qquad [L_{x_2}, L_{x_3}] = iL_{x_1}, \qquad [L_{x_3}, L_{x_1}] = iL_{x_2}.$$

The non-zero structure constants are

$$C_{12}^3 = i, \qquad C_{23}^1 = i, \qquad C_{31}^2 = i.$$

It follows that

$$g_{11} = \sum_{j=1}^{3} \sum_{k=1}^{3} C_{1j}^k C_{1k}^j = \sum_{j=1}^{3} (C_{1j}^2 C_{12}^j + C_{1j}^3 C_{13}^j) = C_{13}^2 C_{12}^3 + C_{12}^3 C_{13}^2 = 2.$$

Analogously $g_{22} = g_{33} = 2$. The remaining elements of the matrix (g_{jk}) are zero. It follows that

$$C = \sum_{j=1}^{3} \sum_{k=1}^{3} g^{jk} L_j L_k = g^{11} L_1^2 + g^{22} L_2^2 + g^{33} L_3^2 = \frac{1}{2} (L_1^2 + L_2^2 + L_3^2). \qquad \clubsuit$$

where

$$\sum_j g^{ij} g_{jk} = \sum_j g_{kj} g^{ji} = \delta_{ik}. \qquad \clubsuit$$

Remark: For every semisimple Lie algebra of rank r there exist r independent Casimir operators. Using the structure constants C_{ij}^k they can be found as

$$C_n = \sum_{\ldots i_l, k_l \ldots} C_{i_1 k_1}^{k_2} C_{i_2 k_2}^{k_3} \cdots C_{i_n k_n}^{k_1} J^{i_1} J^{i_2} \ldots J^{i_n}$$

where n takes all positive integers. Thus we arbitrarily often obtain every Casimir operator, or linear combinations of them. We have to choose r independent values for n (as small as possible). Here J_i are a basis of the semisimple Lie algebra and $J^i = \sum_j g^{ij} J_j$.

Another important tool in the study of Lie algebra is that of the **Cartan-Weyl basis** for a semi-simple Lie algebra. One can introduce new basis elements H_j, E_α such that

$$[H_i, H_j] = 0$$
$$[H_i, E_\alpha] = \alpha_i E_\alpha$$
$$[E_\alpha, E_\beta] = C_{\alpha\beta}^{\alpha+\beta} E_{\alpha+\beta}$$
$$[E_\alpha, E_{-\alpha}] = \sum_i \alpha_i H_i.$$

Example: Consider a family of operators c_j, $c_j^\dagger$ $(j = 1, \ldots, m)$ on a finite-dimensional vector space satisfying the anticommutator relations

$$[c_j, c_k]_+ = [c_j^\dagger, c_k^\dagger]_+ = 0, \quad [c_j, c_k^\dagger]_+ = \delta_{jk} I$$

where I is the identity operator. The operators $\{c_j, c_k^\dagger\}$ are called **Fermi operators**. For the **Sakata model** of elementary particles we need to consider the Fermi operators $c_p^\dagger$, c_p which denote the creation and annihilation of the proton; $c_n^\dagger$, c_n denote the creation and annihilation of the neutron and $c_\Lambda^\dagger$, c_Λ denote the creation and annihilation of the lambda particle. The nine operators

$$c_p^\dagger c_p, \ c_n^\dagger c_n, \ c_\Lambda^\dagger c_\Lambda, \ c_p^\dagger c_n, \ c_p^\dagger c_\Lambda, \ c_n^\dagger c_p, \ c_\Lambda^\dagger c_p, \ c_n^\dagger c_\Lambda, \ c_\Lambda^\dagger c_n$$

form a Lie algebra. The centre is given by the operator

$$c_p^\dagger c_p + c_n^\dagger c_n + c_\Lambda^\dagger c_\Lambda$$

and the zero operator. A Cartan-Weyl basis is given by the following eight operators

$$H_1 = \frac{1}{2}(c_p^\dagger c_p - c_n^\dagger c_n), \qquad H_2 = \frac{1}{3}(c_p^\dagger c_p + c_n^\dagger c_n - 2c_\Lambda^\dagger c_\Lambda)$$

$$E_1 = c_p^\dagger c_n, \quad E_2 = c_n^\dagger c_p, \quad E_3 = c_p^\dagger c_\Lambda, \quad E_4 = c_n^\dagger c_\Lambda, \quad E_5 = c_\Lambda^\dagger c_n, \quad E_6 = c_\Lambda^\dagger c_p.$$

♣

5.6 Computer Algebra Applications

In the classical case the *angular momentum* is given by

$$\mathbf{L} := \mathbf{x} \times \mathbf{p}$$

where $\times$ denotes the cross product. The components of $\mathbf{L}$ are given by

$$L_{x_1} := x_2 p_3 - x_3 p_2, \qquad L_{x_2} := x_3 p_1 - x_1 p_3, \qquad L_{x_3} := x_1 p_2 - x_2 p_1.$$

Introducing the quantization

$$p_1 \to -i\hbar\frac{\partial}{\partial x_1}, \qquad p_2 \to -i\hbar\frac{\partial}{\partial x_2}, \qquad p_3 \to -i\hbar\frac{\partial}{\partial x_3}$$

yields

$$\hat{L}_{x_1} := \frac{\hbar}{i}\left(x_2\frac{\partial}{\partial x_3} - x_3\frac{\partial}{\partial x_2}\right)$$

$$\hat{L}_{x_2} := \frac{\hbar}{i}\left(x_3\frac{\partial}{\partial x_1} - x_1\frac{\partial}{\partial x_3}\right)$$

$$\hat{L}_{x_3} := \frac{\hbar}{i}\left(x_1\frac{\partial}{\partial x_2} - x_2\frac{\partial}{\partial x_1}\right).$$

The angular momentum operators $\hat{L}_{x_1}$, $\hat{L}_{x_2}$ and $\hat{L}_{x_3}$ form a basis of a Lie algebra under the commutator. The commutators are given by

$$[\hat{L}_{x_1}, \hat{L}_{x_2}] = i\hbar\hat{L}_{x_3}, \qquad [\hat{L}_{x_3}, \hat{L}_{x_1}] = i\hbar\hat{L}_{x_2}, \qquad [\hat{L}_{x_2}, \hat{L}_{x_3}] = i\hbar\hat{L}_{x_1}.$$

where $[\,,\,]$ denotes the commutator.

In the program we evaluate the commutators. We consider the angular momentum operators as vector fields.

```
%lxlylz.red;

operator LX, LY, LZ, R12, x;
depend LX(j), x(1), x(2), x(3);
depend LY(j), x(1), x(2), x(3);
depend LZ(j), x(1), x(2), x(3);
depend R12(j), x(1), x(2), x(3);
LX(1) := 0;
LX(2) := i*hb*x(3);
LX(3) := -i*hb*x(2);
LY(1) := -i*hb*x(3);
LY(2) := 0;
LY(3) := i*hb*x(1);
```

```
LZ(1) := i*hb*x(2);
LZ(2) := -i*hb*x(1);
LZ(3) := 0;

%Commutator of LX and LY;
for k := 1:3 do
R12(k) := for j := 1:3 sum
LX(j)*df(LY(k),x(j)) - LY(j)*df(LX(k),x(j));

for k := 1:3 do
write R12(k);
```

The output is

```
- X(2)*HB**2$
X(1)*HB**2$
0$
```

In our second program we implement the commutation relations for Fermi and Bose operators. Then we give some applications.

```
%Bose Fermi system;
%Program name: bf2.red;

%The operators c(j) are the Fermi annihilation operators
%with quantum number j;
%The operators d(j) are the Fermi creation operators
%with quantum number j;
%vs is the vacuum state for Fermi operators;
%ds is the dual state of vs;

operator c, d, vs, ds;
noncom c, d, vs, ds;

for all j let c(j)**2=0;
for all j let d(j)**2=0;
for all j let c(j)*d(j)=-d(j)*c(j)+1;
for all j,k such that j neq k let c(j)*d(k)=-d(k)*c(j);
for all j,k such that j leq k let c(j)*c(k)=-c(k)*c(j);
for all j,k such that j leq k let d(j)*d(k)=-d(k)*d(j);

for all j let c(j)*vs(0)=0;
```

```
for all j let ds(0)*d(j)=0;

let ds(0)*vs(0)=1;

%The operators b are the Bose annihilation operators;
%The opertors bd are the Bose creation operators;
%bd(j) Bose creation operator with quantum number j;
%b(j) Bose annihilation operator with quantum number j;
%no is the vacuum state for Bose operators;
%mo is the dual state of no;

operator b, bd, no, mo;
noncom b, bd, no, mo;

for all j let b(j)*bd(j) = bd(j)*b(j) + 1;
for all j,k such that j neq k let
b(j)*bd(k) = bd(k)*b(j);

for all j,k such that j leq k let b(j)*b(k) = b(k)*b(j);
for all j,k such that j leq k let bd(j)*bd(k) = bd(k)*bd(j);

for all j let b(j)*no(0) = 0;
for all j let mo(0)*bd(j) = 0;
let mo(0)*no(0) = 1;

%We give now the commutation relation between
%the Bose and Fermi operators
%and an order for the Fermi and Bose vacuum states

for all j, k let c(j)*b(k) = b(k)*c(j);
for all j, k let d(j)*b(k) = b(k)*d(j);
for all j, k let c(j)*bd(k) = bd(k)*c(j);
for all j, k let d(j)*bd(k) = bd(k)*d(j);

for all j let c(j)*no(0) = no(0)*c(j);
for all j let d(j)*no(0) = no(0)*d(j);

for all k let mo(0)*d(k) = d(k)*mo(0);
for all k let mo(0)*c(k) = c(k)*mo(0);

let vs(0)*no(0) = no(0)*vs(0);
let mo(0)*ds(0) = ds(0)*mo(0);

%Example1:
res1 := mo(0)*ds(0)*no(0)*vs(0);
```

```
%Example2:
H1 := w*bd(0)*b(0) + J*d(0)*c(0) + gam*(bd(0) + b(0))*(d(0) + c(0));
H1*no(0)*vs(0);

%Example3:
H2 := w*bd(0)*b(0) + J*d(0)*c(0) + gam*(bd(0)*c(0) + b(0)*d(0));
NH := bd(0)*b(0) + d(0)*c(0); % Number operator
res2 := NH*H2 - H2*NH;

%Example4:
Q := (b(0) - lam*(b(0) + bd(0))**2)*d(0);
res3 := Q*Q;
QT := c(0)*(bd(0) - lam*(bd(0) + b(0))**2);
H3 := Q*QT + QT*Q;

%Example5:
H4 := om1*bd(1)*b(1) + om2*bd(2)*b(2) + om3*bd(3)*b(3) +
g*bd(1)*bd(2)*b(3) + gs*b(1)*b(2)*bd(3);
R0 := (1/3)*(bd(1)*b(1)+bd(2)*b(2)+2*bd(3)*b(3));
res4 := H4*R0 - R0*H1;
Y0 := (1/3)*(bd(1)*b(1) + bd(2)*b(2) - bd(3)*b(3));
res5 := H4*Y0 - Y0*H4;

%Example6:
H5 := w*bd(0)*b(0) + J*d(0)*c(0) + gam*(bd(0)*c(0) + b(0)*d(0));
p0 := no(0)*vs(0);   pd0 := ds(0)*mo(0);
p1 := no(0)*d(0)*vs(0); pd1 := ds(0)*c(0)*mo(0);
p2 := bd(0)*no(0)*vs(0); pd2 := ds(0)*mo(0)*b(0);
p3 := bd(0)*no(0)*d(0)*vs(0); pd3 := ds(0)*c(0)*mo(0)*b(0);
matrix A(4,4);
A(1,1) := pd0*H5*p0;  A(1,2) := pd0*H5*p1;
A(1,3) := pd0*H5*p2;  A(1,4) := pd0*H5*p3;
A(2,1) := pd1*H5*p0;  A(2,2) := pd1*H5*p1;
A(2,3) := pd1*H5*p2;  A(2,4) := pd1*H5*p3;
A(3,1) := pd2*H5*p0;  A(3,2) := pd2*H5*p1;
A(3,3) := pd2*H5*p2;  A(3,4) := pd2*H5*p3;
A(4,1) := pd3*H5*p0;  A(4,2) := pd3*H5*p1;
A(4,3) := pd3*H5*p2;  A(4,4) := pd3*H5*p3;
```

Chapter 6

Introductory Examples

In this chapter we study, as introductory examples, the following partial differential equations: the linear one-dimensional wave equation, the linear one-dimensional diffusion equation, system of equations for stationary flow and the linear Schrödinger equation in three space dimensions. We introduce the concept of invariance of the equation under the transformation group. Then we show how these transformation groups are generated by infinitesimal generators (the so-called vector fields or Lie symmetry vector fields) and how these vector fields form a Lie algebra under the commutator. The concept of gauge transformations for the coupling of the electromagnetic field with the Schrödinger equation is also discussed.

6.1   The One-Dimensional Linear Wave Equation

Consider the pseudo-orthogonal group $O(1,1)$ studied in chapter 3. Its dimension is one, so that it can be parametrized in the following form (see examples in chapter 3)

$$\begin{pmatrix} x_1' \\ x_2' \end{pmatrix} = \begin{pmatrix} \cosh \varepsilon & \sinh \varepsilon \\ \sinh \varepsilon & \cosh \varepsilon \end{pmatrix} \begin{pmatrix} x_1 \\ x_2 \end{pmatrix}$$

where $\varepsilon \in \mathbf{R}$ is the group parameter. This transformation is also known as the Lorentz boost. We find that $x_1^2 - x_2^2 = x_1'^2 - x_2'^2$. We now show that the one-dimensional wave equation

$$\frac{\partial^2 u}{\partial x_2^2} = \frac{\partial^2 u}{\partial x_1^2} \tag{1}$$

is invariant under the transformation

$$\begin{pmatrix} x_1' \\ x_2' \end{pmatrix} = \begin{pmatrix} \cosh \varepsilon & \sinh \varepsilon \\ \sinh \varepsilon & \cosh \varepsilon \end{pmatrix} \begin{pmatrix} x_1 \\ x_2 \end{pmatrix} \tag{2a}$$

$$u'(\mathbf{x}'(\mathbf{x})) = u(\mathbf{x}) \tag{2b}$$

83

where $\mathbf{x} = (x_1, x_2)$ and $\mathbf{x}' = (x_1', x_2')$. The transformation group is given by

$$
\begin{aligned}
x_1'(\mathbf{x}, u, \varepsilon) &= x_1 \cosh \varepsilon + x_2 \sinh \varepsilon \\
x_2'(\mathbf{x}, u, \varepsilon) &= x_1 \sinh \varepsilon + x_2 \cosh \varepsilon \\
u'(\mathbf{x}, u, \varepsilon) &= u.
\end{aligned}
$$

We write

$$
\mathbf{x}' = \boldsymbol{\varphi}(\mathbf{x}, u, \varepsilon), \qquad u' = \phi(\mathbf{x}, u, \varepsilon)
$$

where $\boldsymbol{\varphi} = (\varphi_1, \varphi_2)$. This transformation group corresponds to the following initial value problem

$$
\frac{d\mathbf{x}'}{d\varepsilon} = \boldsymbol{\xi}(\mathbf{x}', u'), \qquad \frac{du'}{d\varepsilon} = \boldsymbol{\eta}(\mathbf{x}', u')
$$

with initial conditions $\mathbf{x}' = \mathbf{x}$ and $u' = u$ for $\varepsilon = 0$. The infinitesimals are

$$
\boldsymbol{\xi}(\mathbf{x}, u) = \left.\frac{d\boldsymbol{\varphi}}{d\varepsilon}\right|_{\varepsilon=0}, \qquad \eta(\mathbf{x}, u) = \left.\frac{d\phi}{d\varepsilon}\right|_{\varepsilon=0}.
$$

The infinitesimal generator which is, in this context, also known as the Lie (point) symmetry vector field is given by $(m = 2)$

$$
Z = \sum_{j=1}^{m} \xi_j(\mathbf{x}, u) \frac{\partial}{\partial x_j} + \eta_i(\mathbf{x}, u) \frac{\partial}{\partial u}.
$$

To prove the invariance of the wave equation under the transformation (2) we consider

$$
\frac{\partial u'}{\partial x_2} = \frac{\partial u'}{\partial x_1'}\frac{\partial x_1'}{\partial x_2} + \frac{\partial u'}{\partial x_2'}\frac{\partial x_2'}{\partial x_2} = \frac{\partial u}{\partial x_2}. \tag{3}
$$

It follows that

$$
\frac{\partial^2 u'}{\partial x_2^2} = \left(\frac{\partial^2 u'}{\partial x_1'^2}\frac{\partial x_1'}{\partial x_2} + \frac{\partial^2 u'}{\partial x_1'\partial x_2'}\frac{\partial x_2'}{\partial x_2}\right)\frac{\partial x_1'}{\partial x_2} + \left(\frac{\partial^2 u'}{\partial x_1'\partial x_2'}\frac{\partial x_1'}{\partial x_2} + \frac{\partial^2 u'}{\partial x_2'^2}\frac{\partial x_2'}{\partial x_2}\right)\frac{\partial x_2'}{\partial x_2} = \frac{\partial^2 u}{\partial x_2^2} \tag{4}
$$

Analogously

$$
\frac{\partial^2 u'}{\partial x_1^2} = \left(\frac{\partial^2 u'}{\partial x_1'^2}\frac{\partial x_1'}{\partial x_1} + \frac{\partial^2 u'}{\partial x_1'\partial x_2'}\frac{\partial x_2'}{\partial x_1}\right)\frac{\partial x_1'}{\partial x_1} + \left(\frac{\partial^2 u'}{\partial x_1'\partial x_2'}\frac{\partial x_1'}{\partial x_1} + \frac{\partial^2 u'}{\partial x_2'^2}\frac{\partial x_2'}{\partial x_1}\right)\frac{\partial x_2'}{\partial x_1} = \frac{\partial^2 u}{\partial x_1^2}. \tag{5}
$$

From (2) we have

$$
\frac{\partial x_1'}{\partial x_2} = \sinh \varepsilon \tag{6a}
$$

$$
\frac{\partial x_1'}{\partial x_1} = \cosh \varepsilon \tag{6b}
$$

$$
\frac{\partial x_2'}{\partial x_2} = \cosh \varepsilon \tag{6c}
$$

$$
\frac{\partial x_2'}{\partial x_1} = \sinh \varepsilon. \tag{6d}
$$

Inserting (4), (5) and (6) into the wave equation we find

$$\frac{\partial^2 u'}{\partial x_2'^2} = \frac{\partial^2 u'}{\partial x_1'^2}.$$

Equation (1) is thus invariant under the transformation (2). The generator of the transformation group is given by the symmetry vector field

$$Z_1 = \xi_1(\mathbf{x}, u)\frac{\partial}{\partial x_1} + \xi_2(\mathbf{x}, u)\frac{\partial}{\partial x_2} + \eta(\mathbf{x}, u)\frac{\partial}{\partial u}$$

where the functions ξ_1, ξ_2 and η are given by

$$\xi_1(\mathbf{x}, u) = \left.\frac{d\varphi_1}{d\varepsilon}\right|_{\varepsilon=0}, \qquad \xi_2(\mathbf{x}, u) = \left.\frac{d\varphi_2}{d\varepsilon}\right|_{\varepsilon=0}, \qquad \eta(\mathbf{x}, u) = \left.\frac{d\phi}{d\varepsilon}\right|_{\varepsilon=0} = 0.$$

With

$$\begin{aligned}
\varphi_1(\mathbf{x}, u, \varepsilon) &= x_1\cosh\varepsilon + x_2\sinh\varepsilon \\
\varphi_2(\mathbf{x}, u, \varepsilon) &= x_1\sinh\varepsilon + x_2\cosh\varepsilon \\
\phi(\mathbf{x}, u, \varepsilon) &= u
\end{aligned}$$

it follows that

$$\xi_1(\mathbf{x}, u) = x_2, \qquad \xi_2(\mathbf{x}, u) = x_1, \qquad \eta(\mathbf{x}, u) = 0$$

so that the symmetry vector field

$$Z_1 = x_2\frac{\partial}{\partial x_1} + x_1\frac{\partial}{\partial x_2}$$

is the generator of the transformation group.

We also find that the scaling transformation

$$\begin{aligned}
x_1'(\mathbf{x}, \varepsilon) &= e^\varepsilon x_1 \\
x_2'(\mathbf{x}, \varepsilon) &= e^\varepsilon x_2 \\
u'(\mathbf{x}'(\mathbf{x}), \varepsilon) &= u(\mathbf{x})
\end{aligned}$$

leaves the wave equation invariant. The symmetry vector field for this scaling group is given by

$$Z_2 = x_1\frac{\partial}{\partial x_1} + x_2\frac{\partial}{\partial x_2}.$$

Since the independent variables x_1 and x_2 do not appear explicitly in the wave equation it is invariant under the translation groups

$$x_1'(\mathbf{x}, u, \varepsilon) = x_1 + \varepsilon, \qquad x_2'(\mathbf{x}, u, \varepsilon) = x_2, \qquad u'(\mathbf{x}, u, \varepsilon) = u$$

and

$$x_1'(\mathbf{x}, u, \varepsilon) = x_1, \qquad x_2'(\mathbf{x}, u, \varepsilon) = x_2 + \varepsilon, \qquad u'(\mathbf{x}, u, \varepsilon) = u \ .$$

These transformations correspond to the Lie symmetry vector fields

$$Z_3 = \frac{\partial}{\partial x_1}, \qquad Z_4 = \frac{\partial}{\partial x_2},$$

respectively. We can thus state that the wave equation admits the following Lie symmetry vector fields

$$\left\{ \frac{\partial}{\partial x_1}, \ \frac{\partial}{\partial x_2}, \ x_2\frac{\partial}{\partial x_1} + x_1\frac{\partial}{\partial x_2}, \ x_1\frac{\partial}{\partial x_1} + x_2\frac{\partial}{\partial x_2} \right\}.$$

These vector fields form a basis of a Lie algebra. The commutators are

$$\left[\frac{\partial}{\partial x_1}, \frac{\partial}{\partial x_2} \right] = 0$$

$$\left[\frac{\partial}{\partial x_1}, x_2\frac{\partial}{\partial x_1} + x_1\frac{\partial}{\partial x_2} \right] = \frac{\partial}{\partial x_2}$$

$$\left[\frac{\partial}{\partial x_2}, x_2\frac{\partial}{\partial x_1} + x_1\frac{\partial}{\partial x_2} \right] = \frac{\partial}{\partial x_1}$$

$$\left[x_1\frac{\partial}{\partial x_1} + x_2\frac{\partial}{\partial x_2}, \frac{\partial}{\partial x_1} \right] = -\frac{\partial}{\partial x_1}$$

$$\left[x_1\frac{\partial}{\partial x_1} + x_2\frac{\partial}{\partial x_2}, \frac{\partial}{\partial x_2} \right] = -\frac{\partial}{\partial x_2}$$

$$\left[x_2\frac{\partial}{\partial x_1} + x_1\frac{\partial}{\partial x_2}, x_1\frac{\partial}{\partial x_1} + x_2\frac{\partial}{\partial x_2} \right] = 0.$$

The commutator table is then given by

	Z_1	Z_2	Z_3	Z_4
Z_1	0	0	$-Z_4$	$-Z_3$
Z_2	0	0	$-Z_3$	$-Z_4$
Z_3	Z_4	Z_3	0	0
Z_4	Z_3	Z_4	0	0.

6.2 The One-Dimensional Diffusion Equation

The Lie point symmetry vector fields for the linear diffusion equation in one space dimension

$$\frac{\partial u}{\partial x_2} = \frac{\partial^2 u}{\partial x_1^2}$$

are given by

$$Z_1 = \frac{\partial}{\partial x_2}, \qquad Z_2 = \frac{\partial}{\partial x_1}, \qquad Z_3 = u\frac{\partial}{\partial u}$$

$$Z_4 = x_2\frac{\partial}{\partial x_1} - \frac{1}{2}x_1 u\frac{\partial}{\partial u}$$

$$Z_5 = x_1\frac{\partial}{\partial x_1} + 2x_2\frac{\partial}{\partial x_2}$$

$$Z_6 = x_2 x_1\frac{\partial}{\partial x_1} + x_2^2\frac{\partial}{\partial x_2} - \left(\frac{1}{4}x_1^2 + \frac{1}{2}x_2\right)u\frac{\partial}{\partial u}.$$

As an example we find the transformation group which corresponds to the generator Z_4. The autonomous system associated with the symmetry vector field Z_4 is

$$\frac{dx_2'}{d\varepsilon} = 0 \tag{7}$$

$$\frac{dx_1'}{d\varepsilon} = x_2' \tag{8}$$

$$\frac{du'}{d\varepsilon} = -\frac{1}{2}x_1' u' \tag{9}$$

with initial conditions

$$\mathbf{x}' = \mathbf{x} \quad \text{and} \quad u' = u$$

at $\varepsilon = 0$. The solution of this initial value problem will result in a one-parameter transformation group for the diffusion equation. Solving (7) we find

$$x_2'(\mathbf{x}, u, \varepsilon) = x_2.$$

Inserting this into (8) and integrating gives

$$x_1'(\mathbf{x}, u, \varepsilon) = x_2\varepsilon + x_1.$$

Inserting this into (9) and integrating gives

$$u'(\mathbf{x}, u, \varepsilon) = u\exp\left(-\frac{1}{2}\left(\frac{1}{2}x_2\varepsilon^2 + x_1\varepsilon\right)\right).$$

Consequently, the diffusion equation is invariant under the **Galilean space-time transformation**

$$x_2'(\mathbf{x}, \varepsilon) \;=\; x_2$$

$$x_1'(\mathbf{x}, \varepsilon) \;=\; x_2\varepsilon + x_1$$

$$u'(\mathbf{x}'(\mathbf{x}), \varepsilon) \;=\; u(\mathbf{x}) \exp\left(-\frac{1}{2}\left(\frac{1}{2}x_2\varepsilon^2 + x_1\varepsilon\right)\right) \;.$$

We can also find this transformation by considering the exponentiation of the generator Z_4 so that

$$\begin{pmatrix} x_1'(\mathbf{x}, \varepsilon) \\ x_2'(\mathbf{x}, \varepsilon) \\ u'(\mathbf{x}'(\mathbf{x}), \varepsilon) \end{pmatrix} = \exp(\varepsilon Z_4) \begin{pmatrix} x_1 \\ x_2 \\ u \end{pmatrix} \Bigg|_{u \to u(x_1, x_2)}$$

where

$$\exp(\varepsilon Z_4)x_1 \;=\; x_1 + \varepsilon \left(x_2\frac{\partial}{\partial x_1} - \frac{1}{2}x_1 u\frac{\partial}{\partial u}\right) x_1 + \frac{\varepsilon^2}{2!}\left(x_2\frac{\partial}{\partial x_1} - \frac{1}{2}x_1 u\frac{\partial}{\partial u}\right)^2 x_1 + \cdots$$

$$\;=\; x_1 + \varepsilon x_2$$

$$\exp(\varepsilon Z_4)x_2 \;=\; x_2$$

$$\exp(\varepsilon Z_4)u \;=\; u - \frac{\varepsilon}{2}x_1 u + \frac{\varepsilon^2}{2!}\left(-\frac{1}{2}x_2 u + \frac{1}{4}x_1^2 u\right) + \frac{\varepsilon^3}{3!}\left(\frac{3}{4}x_1 x_2 u - \frac{1}{8}x_1^3 u\right) + \cdots$$

$$\;=\; u \exp\left(-\frac{1}{2}\left(\frac{1}{2}x_2\varepsilon^2 + x_1\varepsilon\right)\right) \;.$$

This results in the transformation given above.

The Lie point symmetry vector fields $Z_1, \ldots Z_6$ form a basis of a non-abelian Lie algebra.

6.3 Stationary Flow

The following system of partial differential equations play an important role in hydrodynamics

$$u_1 \frac{\partial u_1}{\partial x_1} + u_2 \frac{\partial u_1}{\partial x_2} + \frac{1}{\rho} \frac{\partial p}{\partial x_1} = 0 \qquad (10)$$

$$u_1 \frac{\partial u_2}{\partial x_1} + u_2 \frac{\partial u_2}{\partial x_2} + \frac{1}{\rho} \frac{\partial p}{\partial x_2} = 0 \qquad (11)$$

$$\frac{\partial u_1}{\partial x_1} + \frac{\partial u_2}{\partial x_2} = 0. \qquad (12)$$

This system corresponds to the stationary flow, in the (x_1, x_2)-plane, for a nonresistant medium with constant density ρ. Here x_1 and x_2 are space coordinates, u_1 and u_2 are the velocity fields while p and ρ denote the pressure and density, respectively. Consider the transformation group $SO(2)$ where $x_1^2 + x_2^2 = x_1'^2 + x_2'^2$. The angle of rotation (group parameter) is ε. The transformation is given by

$$\begin{pmatrix} x_1'(\mathbf{x}, \varepsilon) \\ x_2'(\mathbf{x}, \varepsilon) \end{pmatrix} = \begin{pmatrix} \cos\varepsilon & \sin\varepsilon \\ -\sin\varepsilon & \cos\varepsilon \end{pmatrix} \begin{pmatrix} x_1 \\ x_2 \end{pmatrix}. \qquad (13)$$

The velocity vectors, $u_1(\mathbf{x})$ and $u_2(\mathbf{x})$ in the point (x_1, x_2), can be transformed to the velocity vectors $u_1'(\mathbf{x}')$ and $u_2'(\mathbf{x}')$ in the point (x_1', x_2') in the rotated coordinate system with the transformation

$$\begin{pmatrix} u_1'(\mathbf{x}'(\mathbf{x}), \varepsilon) \\ u_2'(\mathbf{x}'(\mathbf{x}), \varepsilon) \end{pmatrix} = \begin{pmatrix} \cos\varepsilon & \sin\varepsilon \\ -\sin\varepsilon & \cos\varepsilon \end{pmatrix} \begin{pmatrix} u_1(\mathbf{x}) \\ u_2(\mathbf{x}) \end{pmatrix}. \qquad (14)$$

The pressure is a scalar with

$$p'(\mathbf{x}'(\mathbf{x}), \varepsilon) = p(\mathbf{x}) . \qquad (15)$$

We now show that system (10), (11) and (12) is invariant under the transformation (13), (14) and (15).

Fist we find the transformation of the derivatives

$$\frac{\partial}{\partial x_1} = \cos\varepsilon \frac{\partial}{\partial x_1'} - \sin\varepsilon \frac{\partial}{\partial x_2'}$$

$$\frac{\partial}{\partial x_2} = \sin\varepsilon \frac{\partial}{\partial x_1'} + \cos\varepsilon \frac{\partial}{\partial x_2'}.$$

The inverse transformation of (14) is given by

$$\begin{pmatrix} u_1(\mathbf{x}) \\ u_2(\mathbf{x}) \end{pmatrix} = \begin{pmatrix} \cos\varepsilon & -\sin\varepsilon \\ \sin\varepsilon & \cos\varepsilon \end{pmatrix} \begin{pmatrix} u_1'(\mathbf{x}'(\mathbf{x}), \varepsilon) \\ u_2'(\mathbf{x}'(\mathbf{x}), \varepsilon) \end{pmatrix}.$$

We find for (12)

$$\frac{\partial u_1}{\partial x_1} + \frac{\partial u_2}{\partial x_2} = \frac{\partial u_1'}{\partial x_1'} + \frac{\partial u_2'}{\partial x_2'} \ .$$

Equation (10) transforms into

$$\left(u_1' \frac{\partial u_1'}{\partial x_1'} + u_2' \frac{\partial u_1'}{\partial x_2'} + \frac{1}{\rho}\frac{\partial p'}{\partial x_1'} \right) \cos \varepsilon - \left(u_1' \frac{\partial u_2'}{\partial x_1'} + u_2' \frac{\partial u_2'}{\partial x_2'} + \frac{1}{\rho}\frac{\partial p'}{\partial x_2'} \right) \sin \varepsilon = 0 \qquad (16)$$

and (13) becomes

$$\left(u_1' \frac{\partial u_2'}{\partial x_1'} + u_2' \frac{\partial u_2'}{\partial x_2'} + \frac{1}{\rho}\frac{\partial p'}{\partial x_2'} \right) \cos \varepsilon + \left(u_1' \frac{\partial u_1'}{\partial x_1'} + u_2' \frac{\partial u_1'}{\partial x_2'} + \frac{1}{\rho}\frac{\partial p'}{\partial x_1'} \right) \sin \varepsilon = 0 \ . \qquad (17)$$

ε can be eliminated from (16) and (17) by multiplying (16) by $\sin \varepsilon$ and (17) by $\cos \varepsilon$. Subtracting and adding these results one finds

$$u_1' \frac{\partial u_1'}{\partial x_1'} + u_2' \frac{\partial u_1'}{\partial x_2'} + \frac{1}{\rho}\frac{\partial p'}{\partial x_1'} = 0$$

$$u_1' \frac{\partial u_2'}{\partial x_1'} + u_2' \frac{\partial u_2'}{\partial x_2'} + \frac{1}{\rho}\frac{\partial p'}{\partial x_2'} = 0$$

respectively. System (10), (11), (12) is thus invariant under the transformation (13) and (14).

Let us now find the Lie symmetry vector field for this transformation group. We have

$$Z_1 = \xi_1(\mathbf{x}, \mathbf{u})\frac{\partial}{\partial x_1} + \xi_2(\mathbf{x}, \mathbf{u})\frac{\partial}{\partial x_2} + \eta_1(\mathbf{x}, \mathbf{u})\frac{\partial}{\partial u_1} + \eta_2(\mathbf{x}, \mathbf{u})\frac{\partial}{\partial u_2}.$$

The functions φ and ϕ are given by

$$\varphi_1(\mathbf{x}, \mathbf{u}, \varepsilon) = x_1 \cos \varepsilon + x_2 \sin \varepsilon$$
$$\varphi_2(\mathbf{x}, \mathbf{u}, \varepsilon) = -x_2 \sin \varepsilon + x_2 \cos \varepsilon$$
$$\phi_1(\mathbf{x}, \mathbf{u}, \varepsilon) = u_1 \cos \varepsilon + u_2 \sin \varepsilon$$
$$\phi_2(\mathbf{x}, \mathbf{u}, \varepsilon) = u_1 \sin \varepsilon + u_2 \cos \varepsilon$$

so that

$$\xi_1(\mathbf{x}, \mathbf{u}) = \left.\frac{d\varphi_1}{d\varepsilon}\right|_{\varepsilon=0} = x_2, \qquad \xi_2(\mathbf{x}, \mathbf{u}) = \left.\frac{d\varphi_2}{d\varepsilon}\right|_{\varepsilon=0} = -x_1$$

$$\eta_1(\mathbf{x}, \mathbf{u}) = \left.\frac{d\phi_1}{d\varepsilon}\right|_{\varepsilon=0} = u_2, \qquad \eta_2(\mathbf{x}, \mathbf{u}) = \left.\frac{d\phi_2}{d\varepsilon}\right|_{\varepsilon=0} = -u_1 \ .$$

Thus

$$Z_1 = x_2 \frac{\partial}{\partial x_1} - x_1 \frac{\partial}{\partial x_2} + u_2 \frac{\partial}{\partial u_1} - u_1 \frac{\partial}{\partial u_2} \ .$$

Since system (10), (11), (12) does not explicitly depend on x_1 and x_2, it is also invariant under the translation groups

$$x_1'(\mathbf{x}, u, \varepsilon) = x_1 + \varepsilon, \qquad x_2'(\mathbf{x}, u, \varepsilon) = x_2, \qquad u_1'(\mathbf{x}, u, \varepsilon) = u_1, \qquad u_2'(\mathbf{x}, u, \varepsilon) = u_2$$

and

$$x_1'(\mathbf{x}, u, \varepsilon) = x_1, \qquad x_2'(\mathbf{x}, u, \varepsilon) = x_2 + \varepsilon, \qquad u_1'(\mathbf{x}, u, \varepsilon) = u_1, \qquad u_2'(\mathbf{x}, u, \varepsilon) = u_2.$$

The symmetry vector fields for these transformation groups are

$$Z_2 = \frac{\partial}{\partial x_1}, \qquad Z_3 = \frac{\partial}{\partial x_2},$$

respectively. System (10), (11), (12) is also invariant under the scaling group

$$x_1'(\mathbf{x}, u, \varepsilon) = e^{-\varepsilon} x_1$$

$$x_2'(\mathbf{x}, u, \varepsilon) = e^{-\varepsilon} x_2$$

$$u_1'(\mathbf{x}, u, \varepsilon) = e^{\varepsilon} u_1$$

$$u_2'(\mathbf{x}, u, \varepsilon) = e^{\varepsilon} u_2.$$

This scaling group corresponds to the Lie symmetry vector field

$$Z_4 = -x_1 \frac{\partial}{\partial x_1} - x_2 \frac{\partial}{\partial x_2} + u_1 \frac{\partial}{\partial u_1} + u_2 \frac{\partial}{\partial u_2}.$$

The symmetry vector fields

$$\{ Z_1, \ Z_2, \ Z_3, \ Z_4 \}$$

form a bais of a four dimensional non-abelian Lie algebra.

6.4   Gauge Transformation

For a class of field equations, such as the Schrödinger equation, the Dirac equation and the Klein-Gordon equation, knowledge of the symmetry group plays an important role in connection with gauge theory.

Here we explain the coupling of the electromagnetic field with the Schrödinger equation in three space dimensions. The **Schrödinger equation** is given by

$$i\hbar\frac{\partial\psi(\mathbf{x})}{\partial x_4} = \hat{H}_0\psi(\mathbf{x}) \tag{18}$$

with

$$\hat{H}_0 := -\frac{\hbar^2}{2m}\triangle$$

where

$$\triangle := \sum_{j=1}^{3}\frac{\partial^2}{\partial x_j^2}.$$

The Schrödinger equation is invariant under the **global gauge transformation**

$$\psi'(\mathbf{x}'(\mathbf{x}),\varepsilon) = \exp\left(i\varepsilon\right)\psi(\mathbf{x}) \tag{19a}$$

$$x_l'(\mathbf{x},\varepsilon) = x_l \tag{19b}$$

where $l = 1,2,3,4$, $\mathbf{x} = (x_1,x_2,x_3,x_4)$ with $x_4 = t$. ε is a real dimensionless parameter, $\hbar$ is Planck's constant divided by 2π. The invariance of (18) under (19) means that

$$i\hbar\frac{\partial\psi'(\mathbf{x}')}{\partial x_4'} = -\frac{\hbar^2}{2m}\sum_{j=1}^{3}\frac{\partial^2\psi'(\mathbf{x}')}{\partial x_j'^2}.$$

We determine the symmetry vector field Z that generates the transformation (19).

Let

$$Z = \sum_{l=1}^{4}\xi_l(\mathbf{x},\psi)\frac{\partial}{\partial x_l} + \eta(\mathbf{x},\psi)\frac{\partial}{\partial\psi}$$

where

$$\varphi_l(\mathbf{x},\psi,\varepsilon) = x_l \qquad \phi(\mathbf{x},\psi,\varepsilon) = \exp(i\varepsilon)\psi$$

so that

$$\xi_l = \left.\frac{d\varphi_l}{d\varepsilon}\right|_{\varepsilon=0} = 0$$

$$\eta = \left.\frac{d\phi}{d\varepsilon}\right|_{\varepsilon=0} = i\psi.$$

Here $l = 1, 2, 3, 4$. The symmetry vector field is then given by

$$Z = i\psi \frac{\partial}{\partial \psi} \ . \tag{20}$$

One can rewrite the Schrödinger equation in its real and imaginary parts by

$$\psi(\mathbf{x}) = u_1(\mathbf{x}) + iu_2(\mathbf{x}) \ ,$$

where u_1 and u_2 are real fields. The Schrödinger equation then reads

$$\hbar \frac{\partial u_2}{\partial x_4} = -\hat{H}_0 u_1 \equiv \frac{\hbar^2}{2m} \triangle u_1 \tag{21a}$$

$$\hbar \frac{\partial u_1}{\partial x_4} = \hat{H}_0 u_2 \equiv -\frac{\hbar^2}{2m} \triangle u_2. \tag{21b}$$

We want to rewrite the symmetry vector field (20) to find the symmetry vector fields for (21).

Consider a (real or complex) differentiable function f that depends on u_1 and u_2. The total derivative is given by

$$df = \frac{\partial f}{\partial u_1} du_1 + \frac{\partial f}{\partial u_2} du_2 \ .$$

Note that the total derivative of the functions $\psi = u_1 + iu_2$ and $\psi^* = u_1 - iu_2$ (ψ^* is the conjugate complex of ψ) are given by

$$d\psi = du_1 + idu_2 \ , \qquad d\psi^* = du_1 - idu_2$$

so that

$$du_1 = \frac{1}{2}(d\psi + d\psi^*) \ , \qquad du_2 = \frac{1}{2i}(d\psi - d\psi^*) \ .$$

Thus we obtain

$$df = \frac{1}{2}\left(\frac{\partial f}{\partial u_1} - i\frac{\partial f}{\partial u_2}\right) d\psi + \frac{1}{2}\left(\frac{\partial f}{\partial u_1} + i\frac{\partial f}{\partial u_2}\right) d\psi^*$$

so that we can define

$$\frac{\partial}{\partial \psi} := \frac{1}{2}\left(\frac{\partial}{\partial u_1} - i\frac{\partial}{\partial u_2}\right)$$

$$\frac{\partial}{\partial \psi^*} := \frac{1}{2}\left(\frac{\partial}{\partial u_1} + i\frac{\partial}{\partial u_2}\right)$$

where

$$df = \frac{\partial f}{\partial \psi} d\psi + \frac{\partial f}{\partial \psi^*} d\psi^* \ .$$

The symmetry vector field (20) now becomes

$$Z = i(u_1 + iu_2)\frac{1}{2}\left(\frac{\partial}{\partial u_1} - i\frac{\partial}{\partial u_2}\right) = Z_R + iZ_I$$

with

$$Z_R = \frac{1}{2}\left(u_1\frac{\partial}{\partial u_2} - u_2\frac{\partial}{\partial u_1}\right)$$

$$Z_I = \frac{1}{2}\left(u_1\frac{\partial}{\partial u_1} + u_2\frac{\partial}{\partial u_2}\right)$$

where Z_R is the real and Z_I the imaginary part.

To summarize: The system of partial differential equations (21) admits the symmetry vector fields Z_R and Z_I, where $[Z_R, Z_I] = 0$.

We now describe the coupling of the electromagnetic field with the Schrödinger equation by considering the **local gauge transformation**

$$\psi'(\mathbf{x}'(\mathbf{x})) = \exp(i\varepsilon(\mathbf{x}))\,\psi(\mathbf{x}) \tag{22a}$$

$$x_j'(\mathbf{x}) = x_j \tag{22b}$$

where ε is a smooth function that depends on $\mathbf{x} = (x_1, x_2, x_3, x_4)$, and $j = 1, 2, 3, 4$. We find

$$\frac{\partial\psi'}{\partial x_4} = i\exp(i\varepsilon(\mathbf{x}))\frac{\partial\varepsilon}{\partial x_4}\psi + \exp(i\varepsilon(\mathbf{x}))\frac{\partial\psi}{\partial x_4}$$

$$\frac{\partial\psi'}{\partial x_j} = i\exp(i\varepsilon(\mathbf{x}))\frac{\partial\varepsilon}{\partial x_j}\psi + \exp(i\varepsilon(\mathbf{x}))\frac{\partial\psi}{\partial x_j}$$

$$\begin{aligned}\frac{\partial^2\psi'}{\partial x_j^2} &= -\exp(i\varepsilon(\mathbf{x}))\left(\frac{\partial\varepsilon}{\partial x_j}\right)^2\psi + i\exp(i\varepsilon(\mathbf{x}))\frac{\partial^2\varepsilon}{\partial x_j^2}\psi \\ &+ 2i\exp(i\varepsilon(\mathbf{x}))\frac{\partial\varepsilon}{\partial x_j}\frac{\partial\psi}{\partial x_j} + \exp(i\varepsilon(\mathbf{x}))\frac{\partial^2\psi}{\partial x_j^2}\end{aligned}$$

so that

$$\frac{\partial\psi}{\partial x_4} = \exp(-i\varepsilon(\mathbf{x}))\left(\frac{\partial\psi'}{\partial x_4'} - i\frac{\partial\varepsilon}{\partial x_4'}\psi'\right)$$

$$\frac{\partial^2\psi}{\partial x_j^2} = \exp(-i\varepsilon(\mathbf{x}))\left(\frac{\partial^2\psi'}{\partial x_j'^2} - \left(\frac{\partial\varepsilon}{\partial x_j'}\right)^2\psi' - i\frac{\partial^2\varepsilon}{\partial x_j'^2}\psi' - 2i\frac{\partial\varepsilon}{\partial x_j'}\frac{\partial\psi'}{\partial x_j'}\right)$$

where $j = 1, 2, 3$. Inserting these expressions into the Schrödinger equation we obtain

$$i\hbar \exp(-i\varepsilon(\mathbf{x})) \left(\frac{\partial \psi'}{\partial x'_4} - i\frac{\partial \varepsilon}{\partial x'_4}\psi' \right)$$

$$= -\frac{\hbar^2}{2m} \exp(-i\varepsilon(\mathbf{x})) \sum_{j=1}^{3} \left(\frac{\partial^2}{\partial x'^2_j} - i\frac{\partial^2 \varepsilon}{\partial x'^2_j} - 2i\frac{\partial \varepsilon}{\partial x'_j}\frac{\partial}{\partial x'_j} - \left(\frac{\partial \varepsilon}{\partial x'_j} \right)^2 \right) \psi'$$

$$= -\frac{\hbar^2}{2m} \exp(-i\varepsilon(\mathbf{x})) \sum_{j=1}^{3} \left(\frac{\partial}{\partial x'_j} - i\frac{\partial \varepsilon}{\partial x'_j} \right)^2 \psi'$$

so that the Schrödinger equation takes the form

$$i\hbar \frac{\partial \psi}{\partial x'_4} = -\frac{\hbar^2}{2m} \sum_{j=1}^{3} \left(\frac{\partial}{\partial x'_j} - i\frac{\partial \varepsilon}{\partial x'_j} \right)^2 \psi' - \hbar \frac{\partial \varepsilon}{\partial x'_4}\psi'.$$

We now make the following mapping

$$i\frac{\partial \varepsilon}{\partial x'_j} \longmapsto \frac{i}{\hbar}qA'_j \hbar \frac{\partial \varepsilon}{\partial x'_4} \longmapsto -qU'$$

$(j = 1, 2, 3)$ where $\mathbf{A}' = (A'_1, A'_2, A'_3)$ is the vector potential, U the scalar potential and q the charge. The Schrödinger equation for the coupled electromagnetic field now reads

$$i\hbar \frac{\partial \psi'}{\partial x'_4} = \left(-\frac{\hbar^2}{2m} \left(\nabla' - \frac{i}{\hbar}q\mathbf{A}' \right)^2 + qU' \right)\psi' \tag{23}$$

where $\nabla' = (\partial/\partial x'_1, \partial/\partial x'_2, \partial/\partial x'_3)$.

To find the transformation group that leaves the Schrödinger equation (23) invariant, we apply the local gauge transformation. We then obtain

$$i\hbar \frac{\partial \psi}{\partial x_4} = -\frac{\hbar^2}{2m} \sum_{j=1}^{3} \left(\frac{\partial}{\partial x_j} - i\frac{\partial \varepsilon}{\partial x_j} - i\frac{q}{\hbar}A'_j \right)^2 \psi + qU'\psi - \hbar \frac{\partial \varepsilon}{\partial x_4}\psi \tag{24}$$

where $(j = 1, 2, 3)$

$$U'(\mathbf{x}'(\mathbf{x})) = U(\mathbf{x}) + \frac{\hbar}{q}\frac{\partial \varepsilon}{\partial x_4}$$

$$A'_j(\mathbf{x}'(\mathbf{x})) = A_j(\mathbf{x}) - \frac{\hbar}{q}\frac{\partial \varepsilon}{\partial x_j}.$$

Then (24) becomes

$$i\hbar \frac{\partial \psi}{\partial x_4} = -\frac{\hbar^2}{2m} \sum_{j=1}^{3} \left(\frac{\partial}{\partial x_j} - i\frac{q}{\hbar}A_j \right)^2 \psi + qU\psi$$

or

$$i\hbar \frac{\partial \psi}{\partial x_4} = \left(-\frac{\hbar^2}{2m} \left(\nabla - i\frac{q}{\hbar}\mathbf{A} \right)^2 + qU \right)\psi, \tag{25}$$

where

$$\nabla := \left(\frac{\partial}{\partial x_1}, \frac{\partial}{\partial x_2}, \frac{\partial}{\partial x_3} \right)$$

and

$$\mathbf{A} := (A_1, A_2, A_3).$$

Thus we can now state that (25) is invariant under the transformation

$$x'_l(\mathbf{x}) = x_l$$

$$\psi'(\mathbf{x}'(\mathbf{x})) = \exp\left(i\varepsilon(\mathbf{x})\right) \psi(\mathbf{x})$$

$$U'(\mathbf{x}'(\mathbf{x})) = U(\mathbf{x}) + \frac{\hbar}{q} \frac{\partial \varepsilon}{\partial x_4}$$

$$A'_j(\mathbf{x}'(\mathbf{x})) = A_j(\mathbf{x}) - \frac{\hbar}{q} \frac{\partial \varepsilon}{\partial x_j}$$

($l = 1, 2, 3, 4$ and $j = 1, 2, 3$) which is called the **gauge transformation** for the electromagnetic field.

The approach described above can also be applied to the linear Dirac equation. The **linear Dirac equation** with nonvanishing rest mass m_0 is given by

$$\left(i\hbar \left(\gamma_0 \frac{\partial}{\partial x_0} + \gamma_1 \frac{\partial}{\partial x_1} + \gamma_2 \frac{\partial}{\partial x_2} + \gamma_3 \frac{\partial}{\partial x_3} \right) - m_0 c \right) \psi(\mathbf{x}) = 0 \qquad (26)$$

where

$$\mathbf{x} = (x_0, x_1, x_2, x_3), \qquad x_0 = ct$$

and the complex valued spinor ψ is

$$\psi(\mathbf{x}) := \left(\begin{array}{c} \psi_1(\mathbf{x}) \\ \psi_2(\mathbf{x}) \\ \psi_3(\mathbf{x}) \\ \psi_4(\mathbf{x}) \end{array} \right).$$

The **gamma matrices** are defined by

$$\gamma_0 \equiv \beta := \left(\begin{array}{cccc} 1 & 0 & 0 & 0 \\ 0 & 1 & 0 & 0 \\ 0 & 0 & -1 & 0 \\ 0 & 0 & 0 & -1 \end{array} \right), \qquad \gamma_1 := \left(\begin{array}{cccc} 0 & 0 & 0 & 1 \\ 0 & 0 & 1 & 0 \\ 0 & -1 & 0 & 0 \\ -1 & 0 & 0 & 0 \end{array} \right)$$

$$\gamma_2 := \left(\begin{array}{cccc} 0 & 0 & 0 & -i \\ 0 & 0 & i & 0 \\ 0 & i & 0 & 0 \\ -i & 0 & 0 & 0 \end{array} \right), \qquad \gamma_3 := \left(\begin{array}{cccc} 0 & 0 & 1 & 0 \\ 0 & 0 & 0 & -1 \\ -1 & 0 & 0 & 0 \\ 0 & 1 & 0 & 0 \end{array} \right).$$

The linear Dirac equation and the free wave equation

$$\left(\frac{\partial^2}{\partial x_0^2} - \frac{\partial^2}{\partial x_1^2} - \frac{\partial^2}{\partial x_2^2} - \frac{\partial^2}{\partial x_3^2}\right) A_\mu = 0$$

can be derived from Lagrangian densities

$$\mathcal{L}_D = c\bar{\psi}\left(i\hbar\gamma_\mu\frac{\partial}{\partial x_\mu} - m_0 c\right)\psi$$

$$\mathcal{L}_E = -\frac{1}{2\mu_0}\left(\frac{\partial A_\mu}{\partial x^\nu} - \frac{\partial A_\nu}{\partial x^\mu}\right)\frac{\partial A^\mu}{\partial x_\nu},$$

respectively. We used sum convention and $\bar{\psi} = \psi^\dagger \gamma_0$, $A_\mu = (A_0, A_1, A_2, A_3)$, $A^\mu = (A_0, -A_1, -A_2, -A_3)$, $x_\mu = (x_0, -x_1, -x_2, -x_3)$, $x^\mu = (x_0, x_1, x_2, x_3)$.

From gauge theory described above we obtain the Lagrangian density

$$\mathcal{L} = -\frac{1}{2\mu_0}\left(\frac{\partial A_\mu}{\partial x^\nu} - \frac{\partial A_\nu}{\partial x^\mu}\right)\frac{\partial A^\mu}{\partial x_\nu} + c\bar{\psi}\left(i\hbar\gamma_\mu\frac{\partial}{\partial x_\mu} - e\gamma_\mu A^\mu - m_0 c\right)\psi \qquad (27)$$

where

$$\bar{\psi} = \psi^\dagger \gamma_0 \equiv (\psi_1^*, \psi_2^*, -\psi_3^*, -\psi_4^*)$$

and

$$\gamma_\mu A^\mu = \gamma_0 A^0 + \gamma_1 A^1 + \gamma_2 A^2 + \gamma_3 A^3 = \gamma_0 A_0 - \gamma_1 A_1 - \gamma_2 A_2 - \gamma_3 A_3.$$

Here we make use of the sum convention. From this Lagrangian density and the **Euler-Lagrange equation**

$$\frac{\partial\mathcal{L}}{\partial\phi_r} - \frac{\partial}{\partial x^\mu}\frac{\partial\mathcal{L}}{\partial(\partial\phi_r/\partial x^\mu)} = 0$$

we obtain the **Maxwell-Dirac equation**

$$\left(\frac{\partial^2}{\partial x_0^2} - \frac{\partial^2}{\partial x_1^2} - \frac{\partial^2}{\partial x_2^2} - \frac{\partial^2}{\partial x_3^2}\right) A_\mu = \mu_0 ec\bar{\psi}\gamma_\mu\psi$$

$$\left(i\hbar\left(\gamma_0\frac{\partial}{\partial x_0} + \sum_{j=1}^{3}\gamma_j\frac{\partial}{\partial x_j}\right) - m_0 c\right)\psi = e\left(\sum_{j=1}^{3}\gamma_j A^j + \gamma_0 A_0\right)\psi$$

$$\frac{\partial A_0}{\partial x_0} + \frac{\partial A_1}{\partial x_1} + \frac{\partial A_2}{\partial x_2} + \frac{\partial A_3}{\partial x_3} = 0 \qquad \text{(Lorentz gauge condition)}$$

is the Dirac equation which is coupled with the vector potential A_μ where $A_0 = U/c$ and U denotes the scalar potential. Here $\mu = 0, 1, 2, 3$ and

$$\bar{\psi} = (\psi_1^*, \ \psi_2^*, \ -\psi_3^*, \ -\psi_4^*).$$

Here c is the speed of light, e the charge, μ_0 the permeability of free space, $\hbar = h/(2\pi)$ where h is Planck's constant and m_0 the particle rest mass. We set

$$\lambda = \frac{\hbar}{m_0 c}.$$

The Maxwell-Dirac equation can be written as the following coupled system of thirteen partial differential equations

$$\left(\frac{\partial^2}{\partial x_0^2} - \frac{\partial^2}{\partial x_1^2} - \frac{\partial^2}{\partial x_2^2} - \frac{\partial^2}{\partial x_3^2}\right) A_0 = (\mu_0 ec)(u_1^2 + v_1^2 + u_2^2 + v_2^2 + u_3^2 + v_3^2 + u_4^2 + v_4^2)$$

$$\left(\frac{\partial^2}{\partial x_0^2} - \frac{\partial^2}{\partial x_1^2} - \frac{\partial^2}{\partial x_2^2} - \frac{\partial^2}{\partial x_3^2}\right) A_1 = (2\mu_0 ec)(u_1 u_4 + v_1 v_4 + u_2 u_3 + v_2 v_3)$$

$$\left(\frac{\partial^2}{\partial x_0^2} - \frac{\partial^2}{\partial x_1^2} - \frac{\partial^2}{\partial x_2^2} - \frac{\partial^2}{\partial x_3^2}\right) A_2 = (2\mu_0 ec)(u_1 v_4 - u_4 v_1 + u_3 v_2 - u_2 v_3)$$

$$\left(\frac{\partial^2}{\partial x_0^2} - \frac{\partial^2}{\partial x_1^2} - \frac{\partial^2}{\partial x_2^2} - \frac{\partial^2}{\partial x_3^2}\right) A_3 = (2\mu_0 ec)(u_1 u_3 + v_3 v_1 - u_2 u_4 - v_2 v_4)$$

$$\lambda\frac{\partial u_1}{\partial x_0} + \lambda\frac{\partial u_3}{\partial x_3} + \lambda\frac{\partial u_4}{\partial x_1} + \lambda\frac{\partial v_4}{\partial x_2} - v_1 = \frac{e}{m_0 c}(A_0 v_1 - A_3 v_3 - A_1 v_4 + A_2 u_4)$$

$$\lambda\frac{\partial v_1}{\partial x_0} - \lambda\frac{\partial v_3}{\partial x_3} - \lambda\frac{\partial v_4}{\partial x_1} + \lambda\frac{\partial u_4}{\partial x_2} - u_1 = \frac{e}{m_0 c}(A_0 u_1 - A_3 u_3 - A_1 u_4 - A_2 v_4)$$

$$\lambda\frac{\partial u_2}{\partial x_0} + \lambda\frac{\partial u_3}{\partial x_1} - \lambda\frac{\partial v_3}{\partial x_2} - \lambda\frac{\partial u_4}{\partial x_3} - v_2 = \frac{e}{m_0 c}(A_0 v_2 - A_1 v_3 - A_2 u_3 + A_3 v_4)$$

$$-\lambda\frac{\partial v_2}{\partial x_0} - \lambda\frac{\partial v_3}{\partial x_1} - \lambda\frac{\partial u_3}{\partial x_2} + \lambda\frac{\partial v_4}{\partial x_3} - u_2 = \frac{e}{m_0 c}(A_0 u_2 - A_1 u_3 + A_2 v_3 + A_3 u_4)$$

$$-\lambda\frac{\partial u_1}{\partial x_3} - \lambda\frac{\partial u_2}{\partial x_1} - \lambda\frac{\partial v_2}{\partial x_2} - \lambda\frac{\partial u_3}{\partial x_0} - v_3 = \frac{e}{m_0 c}(A_3 v_1 + A_1 v_2 - A_2 u_2 - A_0 v_3)$$

$$\lambda\frac{\partial v_1}{\partial x_3} + \lambda\frac{\partial v_2}{\partial x_1} - \lambda\frac{\partial u_2}{\partial x_2} + \lambda\frac{\partial v_3}{\partial x_0} - u_3 = \frac{e}{m_0 c}(A_3 u_1 + A_1 u_2 + A_2 v_2 - A_0 u_3)$$

$$-\lambda\frac{\partial u_1}{\partial x_1} + \lambda\frac{\partial v_1}{\partial x_2} + \lambda\frac{\partial u_2}{\partial x_3} - \lambda\frac{\partial u_4}{\partial x_0} - v_4 = \frac{e}{m_0 c}(A_1 v_1 + A_2 u_1 - A_3 v_2 - A_0 v_4)$$

$$\lambda\frac{\partial v_1}{\partial x_1} + \lambda\frac{\partial u_1}{\partial x_2} - \lambda\frac{\partial v_2}{\partial x_3} + \lambda\frac{\partial v_4}{\partial x_0} - u_4 = \frac{e}{m_0 c}(A_1 u_1 - A_2 v_1 - A_3 u_2 - A_0 u_4)$$

$$\frac{\partial A_0}{\partial x_0} + \frac{\partial A_1}{\partial x_1} + \frac{\partial A_2}{\partial x_2} + \frac{\partial A_3}{\partial x_3} = 0.$$

6.5 Computer Algebra Applications

Let

$$i\hbar\frac{\partial\psi(\mathbf{x},t)}{\partial t} = -\frac{\hbar^2}{2m}\Delta\psi(\mathbf{x},t)$$

be the Schrödinger equation of the free particle in three-space dimensions. The equation is invariant under the transformation

$$\psi'(\mathbf{x}'(\mathbf{x},t),t'(\mathbf{x},t)) = \exp(i\epsilon)\psi(\mathbf{x},t)$$

$$x_j'(\mathbf{x},t) = x_j, \qquad t'(\mathbf{x},t) = t$$

where $j = 1,2,3$ and ϵ is independent of $\mathbf{x}$ and t. This transformation is called a global gauge transformation. Let

$$i\hbar\frac{\partial\psi}{\partial t} = -\frac{\hbar^2}{2m}\sum_{j=1}^{3}\left(\frac{\partial}{\partial x_j} - i\frac{q}{\hbar}A_j\right)^2\psi + qU\psi$$

where $\mathbf{A} = (A_1, A_2, A_3)$ is the vector potential and U the scalar potential. We show that this equation is invariant under the transformation

$$x_j'(\mathbf{x},t) = x_j, \qquad t'(\mathbf{x},t) = t$$

$$\psi'(\mathbf{x}'(\mathbf{x},t),t'(\mathbf{x},t)) = \exp(i\epsilon(\mathbf{x},t))\psi(\mathbf{x},t)$$

$$U'(\mathbf{x}'(\mathbf{x},t),t'(\mathbf{x},t)) = U(\mathbf{x},t) - \frac{\hbar}{q}\frac{\partial\epsilon(\mathbf{x},t)}{\partial t}$$

$$A_j'(\mathbf{x}'(\mathbf{x},t),t'(\mathbf{x},t)) = A_j(\mathbf{x},t) + \frac{\hbar}{q}\frac{\partial\epsilon(\mathbf{x},t)}{\partial x_j},$$

where $j = 1,2,3$ and q is the charge. This transformation is called a local gauge transformation, where ϵ depends on $\mathbf{x}$ and t. We show in the program that

$$i\hbar\frac{\partial\psi'}{\partial t'} = -\frac{\hbar^2}{2m}\sum_{j=1}^{3}\left(\frac{\partial}{\partial x_j'} - i\frac{q}{\hbar}A_j'\right)^2\psi' + qU'\psi'.$$

```
%gauge.red;

depend GP, x1, x2, x3, t;
depend G, x1, x2, x3, t;
depend psip, x1, x2, x3, t;
depend psi, x1, x2, x3, t;
depend AP1, x1, x2, x3, t;
depend AP2, x1, x2, x3, t;
depend AP3, x1, x2, x3, t;
depend A1, x1, x2, x3, t;
depend A2, x1, x2, x3, t;
depend A3, x1, x2, x3, t;
depend UP, x1, x2, x3, t;
depend U, x1, x2, x3, t;
depend ep, x1, x2, x3, t;
GP := i*hb*df(psip,t) +
hb*hb/(2*m)*(df(psip,x1,2)+df(psip,x2,2)+df(psip,x3,2)) +
(-i*q*hb)/(2*m)*(AP1*df(psip,x1)+AP2*df(psip,x2)+AP3*df(psip,x3))+
(-i*q*hb)/(2*m)*(df(psip*AP1,x1)+df(psip*AP2,x2)+df(psip*AP3,x3))-
q^2/(2*m)*(AP1*AP1*psip+AP2*AP2*psip+AP3*AP3*psip)-q*UP*psip;
```

```
G := sub({psip=psi*exp(i*ep),AP1=A1+hb/q*df(ep,x1),
AP2=A2+hb/q*df(ep,x2),AP3=A3+hb/q*df(ep,x3),UP=U-hb/q*df(ep,t)},GP);

on div;
G1 := G/EXP(i*ep);
G2 := sub({psi=psip,A1=AP1,A2=AP2,A3=AP3,U=UP},G1);
RES := GP - G2;
```

The output is

```
g1 :=  - 1/2*df(a1,x1)*i*m**(-1)*q*psi*hb -
1/2*df(a2,x2)*i*m**(-1)*q*psi*hb + df(psi,t)*i*hb +
1/2*df(psi,x1,2)*m**(-1)*hb**2 - df(psi,x1)*i*m**(-1)*q*a1*hb +
1/2*df(psi,x2,2)*m**(-1)*hb**2 - df(psi,x2)*i*m**(-1)*q*a2*hb +
1/2*df(psi,x3,2)*m**(-1)*hb**2 - df(psi,x3)*i*m**(-1)*q*a3*hb -
1/2*df(a3,x3)*i*m**(-1)*q*psi*hb - 1/2*m**(-1)*q**2*a1**2*psi -
1/2*m**(-1)*q**2*a2**2*psi - 1/2*m**(-1)*q**2*psi*a3**2 -
q*u*psi$

res := 0$
```

Chapter 7

Differential Forms and Tensor Fields

7.1 Vector Fields and Tangent Bundles

We begin the discussion by defining the tangent bundle on a differentiable manifold. For more details on differentiable manifolds we refer to appendix A.

Suppose C is a smooth curve on a manifold M with dimension m, parametrized by

$$\varphi : I \to M$$

where I is an open interval including $\{0\}$. In local coordinates $\mathbf{x} = (x_1, x_2, \ldots, x_m)$, C is given by the smooth function $\varphi(\varepsilon) = (\varphi_1(\varepsilon), \varphi_2(\varepsilon), \ldots, \varphi_m(\varepsilon))$ of the real variable $\varepsilon \in I$. At each point $\mathbf{x} = \varphi(\varepsilon)$ of C the curve has a **tangent vector**, namely the derivative

$$\dot{\varphi}(\varepsilon) \equiv \frac{d\varphi}{d\varepsilon} \equiv (\dot{\varphi}_1(\varepsilon), \ldots, \dot{\varphi}_m(\varepsilon))$$

where $\dot{\varphi}_j(\varepsilon) \equiv d\varphi_j/d\varepsilon$, $j = 1, \ldots, m$. The collection of all tangent vectors to all possible curves passing through a given point $\mathbf{x}$ in M is called a **tangent space** to M at $\mathbf{x}$ and is denoted by $TM|_{\mathbf{x}}$. If M is an m-dimensional manifold, then $TM|_{\mathbf{x}}$ is an m-dimensional vector space, with

$$\left\{ \frac{\partial}{\partial x_1}, \frac{\partial}{\partial x_2}, \ldots, \frac{\partial}{\partial x_m} \right\}$$

providing a basis for $TM|_{\mathbf{x}}$ in the given local coordinates.

Definition 7.1 *The collection of all tangent spaces corresponding to all points $\mathbf{x}$ in M is called the **tangent bundle** of M, denoted by*

$$TM = \bigcup_{\mathbf{x} \in M} TM|_{\mathbf{x}}.$$

The tangent vector $\dot{\varphi}(\varepsilon) \in TM|_{\varphi(\varepsilon)}$ will vary smoothly from point to point so that the tangent bundle TM is a smooth manifold of dimension $2m$.

Example: Consider $M = \mathbf{R}^m$. We can thus identify the tangent space $T\mathbf{R}^m|_{\mathbf{x}}$ at any $\mathbf{x} \in \mathbf{R}$ with $\mathbf{R}^m$ itself. This is true since the tangent vector $\dot{\boldsymbol{\varphi}}(\varepsilon)$ to a smooth curve $\boldsymbol{\varphi}(\varepsilon)$ can be realized as an actual vector in $\mathbf{R}^m$, namely $(\dot{\varphi}_1(\varepsilon), \ldots, \dot{\varphi}_m(\varepsilon))$. Another way of looking at this identification is that we are identifying the basis vector $\partial/\partial x_i$ of $T\mathbf{R}^m|_{\mathbf{x}}$ with the standard basis vector $\boldsymbol{e}_i$ of $\mathbf{R}^m$. The tangent bundle of $\mathbf{R}^m$ is thus a Cartesian product

$$T\mathbf{R}^m \simeq \mathbf{R}^m \times \mathbf{R}^m.$$

If S is a smooth surface in $\mathbf{R}^3$, then the tangent space $TS|_{\mathbf{x}}$ can be identified with the usual geometric tangent plane to S at each point $\mathbf{x} \in S$. This again uses the identification $T\mathbf{R}^3|_{\mathbf{x}}$ and so $TS|_{\mathbf{x}}$ is a plane in $\mathbf{R}^3$.   ♣

In local coordinates, $\mathbf{x} = (x_1, \ldots, x_m)$, a vector field $Z \in TM|_{\mathbf{x}}$ has the form

$$Z = \xi_1(\mathbf{x})\frac{\partial}{\partial x_1} + \cdots + \xi_m(\mathbf{x})\frac{\partial}{\partial x_m}$$

where each ξ_i is a smooth function of $\mathbf{x}$. An **integral curve** of a vector field Z is a smooth parametrized curve $\mathbf{x} = \boldsymbol{\varphi}(\varepsilon)$ whose tangent vector at any point coincides with the value of Z at the same point

$$\dot{\boldsymbol{\varphi}}(\varepsilon) = Z|_{\varphi(\varepsilon)}$$

for all ε. In local coordinates

$$\mathbf{x} = \boldsymbol{\varphi}(\varepsilon) = (\varphi_1(\varepsilon), \ldots, \varphi_m(\varepsilon))$$

must be a solution of the autonomous system of ordinary differential equations

$$\frac{dx_j}{d\varepsilon} = \xi_j(\mathbf{x}) \qquad j = 1, 2, \ldots, m.$$

For ξ_i smooth the standard existence and uniqueness theorems for systems of ordinary differential systems guarantee that there is a unique solution to the above autonomous system for each set of initial data $\boldsymbol{\varphi}(0) = \mathbf{x}_0$. This implies the existence of a unique maximal integral curve $\boldsymbol{\varphi} : I \to M$ passing through a given point $\mathbf{x}_0 = \varphi(0) \in M$, where maximal means that it is not contained in any other integral curve. For the vector field Z, the parametrized maximal integral curve passing through $\mathbf{x}$ in M is denoted by $\varphi(\mathbf{x}, \varepsilon)$. This is known as the action on $\mathbf{x}$ or the flow of the vector field vector Z, and is discussed in detail in chapters 4 and 5.

7.2 Differential Forms and Cotangent Bundles

We begin with the basic definition of differential k-forms.

Definition 7.2 *Let M be a smooth manifold and $TM|_{\mathbf{x}}$ its tangent space at $\mathbf{x}$. The space $\bigwedge_k T^*M|_{\mathbf{x}}$ of **differential k-forms** at $\mathbf{x}$ is the set of all k-linear alternating functions*

$$\omega : TM|_{\mathbf{x}} \times \cdots \times TM|_{\mathbf{x}} \to \mathbf{R}.$$

Specifically, if we denote the evaluation of ω on the tangent vectors $Z_1, \ldots, Z_k \in TM|_{\mathbf{x}}$ by $\langle \omega; Z_1 \ldots, Z_k \rangle$, the basic requirements are that for all tangent vectors at $\mathbf{x}$,

$$\langle \omega; Z_1, \ldots, cZ_i + c'Z_i', \ldots, Z_k \rangle = c\langle \omega; Z_1, \ldots, Z_i, \ldots, Z_k \rangle + c'\langle \omega; Z_1, \ldots, Z_i', \ldots, Z_k \rangle$$

for $c, c' \in \mathbf{R}$, $1 \leq i \leq k$, and

$$\langle \omega; Z_{i_1}, \ldots, Z_{i_k} \rangle = \epsilon_{i_1 \ldots i_k} \langle \omega; Z_1, \ldots, Z_k \rangle$$

where $\epsilon_{i_1 \cdots i_k}$ is the total antisymmetric tensor ($\epsilon_{12 \ldots k} = +1$). The space $\bigwedge_k T^*M|_{\mathbf{x}}$ is a vector space under the operation of addition and scalar multiplication.

Definition 7.3 *The **cotangent vector space**, denoted by $T^*M|_{\mathbf{x}} \equiv \bigwedge_1 T^*M|_{\mathbf{x}}$, is the space of 1-forms that is dual to the tangent vector space $TM|_{\mathbf{x}}$ at $\mathbf{x}$. In particular the 0-form is just a smooth real valued function $f \in C^\infty(M)$.*

Definition 7.4 *A smooth **differential k-form** ω on M is a collection of smoothly varying k-linear alternating maps $\omega|_{\mathbf{x}} \in \bigwedge_k T^*M|_{\mathbf{x}}$ for each $\mathbf{x} \in M$, where we require that for all smooth vector fields $Z_1, \ldots, Z_k$*

$$\langle \omega; Z_1, \ldots, Z_k \rangle(\mathbf{x}) \equiv \langle \omega|_{\mathbf{x}}; Z_1|_{\mathbf{x}}, \ldots, Z_k|_{\mathbf{x}} \rangle$$

be a smooth, real-valued function of $\mathbf{x}$.

Definition 7.5 *The **cotangent bundle** is the union of all cotangent vector spaces*

$$T^*M = \bigcup_{\mathbf{x} \in M} T^*M|_{\mathbf{x}}.$$

If $\omega \in T^*M|_{\mathbf{x}}$, then a point in T^*M is $\langle \omega; Z \rangle$.

If $(x_1, \ldots, x_m)$ are local coordinates on M, then the basis of the cotangent vector space $T^*M|_{\mathbf{x}}$ is given by

$$\{\, dx_1, \ldots, dx_m \,\}$$

which is dual to the basis of the tangent vector space $TM|_{\mathbf{x}}$, given in the same local coordinate system by

$$\left\{ \frac{\partial}{\partial x_1}, \ldots, \frac{\partial}{\partial x_m} \right\}.$$

Thus

$$\langle dx_i; \frac{\partial}{\partial x_j} \rangle = \delta_{ij}$$

for all i, j, where δ_{ij} is the Kronecker delta:

$$\delta_{ij} = \begin{cases} 1 & \text{for } i = j \\ 0 & \text{for } i \neq j \end{cases}.$$

A differential one-form ω has the local coordinate expression

$$\omega = h_1(\mathbf{x})dx_1 + \cdots + h_m(\mathbf{x})dx_m$$

where each coefficient function $h_j \in C^\infty(M)$ and $\mathbf{x} = (x_1, \ldots, x_m)$. Thus, for any vector field

$$Z = \sum_{i=1}^{m} \xi_i(\mathbf{x}) \frac{\partial}{\partial x_i}$$

it follows that

$$\langle \omega; Z \rangle = \sum_{i=1}^{m} h_i(\mathbf{x}) \xi_i(\mathbf{x})$$

is a smooth function. Consider the differential form

$$df(\mathbf{x}) := \sum_{i=1}^{m} \frac{\partial f}{\partial x_i} dx_i .$$

Thus we find

$$\langle df(\mathbf{x}); Z \rangle = \langle \sum_{i=1}^{m} \frac{\partial f}{\partial x_i} dx_i; \sum_{i=1}^{m} \xi_i(\mathbf{x}) \frac{\partial}{\partial x_i} \rangle = \sum_{i=1}^{m} \xi_i(\mathbf{x}) \frac{\partial f}{\partial x_i} = Z(f).$$

For a k-form we refer to k as the rank of the form. Among differential forms there is an algebraic operation $\wedge$, called the **exterior product** (also known as the **wedge product** or **Grassman product**) for making higher-rank forms out of ones with lower rank. Given a collection of differential 1-forms $\omega_1, \ldots, \omega_k$, we can form a **differential k-form**

$$\omega_1 \wedge \cdots \wedge \omega_k$$

as follows

$$\langle \omega_1 \wedge \cdots \wedge \omega_k; Z_1, \ldots, Z_k \rangle := \det(\langle \omega_i; Z_j \rangle)$$

the right-hand side being the determinant of a $k \times k$ matrix with indicated (i, j) entry. Note that the exterior product is both multilinear and alternating.

Example: Consider the one-forms ω and σ. The 2-form $\omega \wedge \sigma$ is then defined as follows

$$\langle \omega \wedge \sigma; Z_1, Z_2 \rangle := \det \begin{pmatrix} \langle \omega; Z_1 \rangle & \langle \omega; Z_2 \rangle \\ \langle \sigma; Z_1 \rangle & \langle \sigma; Z_2 \rangle \end{pmatrix}$$

where Z_1 and Z_2 are arbitrary vectors from $TM|_{\mathbf{x}}$. This is a real valued mapping on $TM|_{\mathbf{x}} \times TM|_{\mathbf{x}}$, it is linear in each argument, and changes sign when Z_1 and Z_2 are interchanged. If ω and σ are one-forms, then

$$\omega \wedge \sigma = -\sigma \wedge \omega.$$

It follows that

$$\omega \wedge \omega = 0$$

for differential one-forms. ♣

In local coordinates, $\bigwedge_k T^* M|_{\mathbf{x}}$ is spanned by the basis k-form

$$dx_I \equiv dx_{i_1} \wedge \cdots \wedge dx_{i_k},$$

where I ranges over all strictly increasing multi-indexes $1 \leq i_1 < i_2 < \cdots < i_k \leq m$. Thus $\bigwedge_k T^* M|_{\mathbf{x}}$ has dimension

$$\binom{m}{k}.$$

In particular, $\bigwedge_k T^* M|_{\mathbf{x}} \simeq \{0\}$ if $k > m$. Any smooth differential k-form on M has the local coordinate expression

$$\omega = \sum_I \alpha_I(\mathbf{x}) dx_I$$

where, for each strictly increasing multi-index I, the coefficient α_I is a smooth, real-valued function. If

$$
\begin{aligned}
\omega &= \omega_1 \wedge \cdots \wedge \omega_k \\
\sigma &= \sigma_1 \wedge \cdots \wedge \sigma_l
\end{aligned}
$$

their wedge product is the $(k + l)$-form

$$\omega \wedge \sigma = \omega_1 \wedge \cdots \wedge \omega_k \wedge \sigma_1 \wedge \cdots \wedge \sigma_l$$

with the definition extending bilinear to more general types of forms

$$
\begin{aligned}
(c\omega + c'\omega') \wedge \sigma &= c(\omega \wedge \sigma) + c'(\omega' \wedge \sigma) \\
\omega \wedge (c\sigma + c'\sigma') &= c(\omega \wedge \sigma) + c'(\omega \wedge \sigma')
\end{aligned}
$$

for c, $c' \in \mathbf{R}$. It can be shown that the wedge product is associative

$$\omega \wedge (\sigma \wedge \theta) = (\omega \wedge \sigma) \wedge \theta$$

and anti-commutative,

$$\omega \wedge \sigma = (-1)^{kl} \sigma \wedge \omega$$

where ω is a k-form and σ is an l-form.

Example: A basis for two-forms on $\mathbf{R}^3$ is given by

$$\{ \ dx_1 \wedge dx_2, \ dx_2 \wedge dx_3, \ dx_3 \wedge dx_1 \ \}.$$

A general two-form in $\mathbf{R}^3$ can then be written as

$$h_1(\mathbf{x})dx_1 \wedge dx_2 + h_2(\mathbf{x})dx_2 \wedge dx_3 + h_3(\mathbf{x})dx_3 \wedge dx_1$$

where $h_i \in C^\infty(\mathbf{R}^3)$ $(i = 1, 2, 3)$ and $\mathbf{x} = (x_1, x_2, x_3)$. The most general three form in $\mathbf{R}^3$ is

$$h(\mathbf{x})dx_1 \wedge dx_2 \wedge dx_3$$

where $h(\mathbf{x}) \in C^\infty(\mathbf{R}^3)$ and the basis is

$$\{ \ dx_1 \wedge dx_2 \wedge dx_3 \ \}. \qquad \qquad \clubsuit$$

Example: Let $M = \mathbf{R}^2$. In Cartesian coordinates in the $x_1 x_2-$plane $dx_1 \wedge dx_2$ should be regarded as the (oriented) area element. In polar coordinates

$$x_1(r, \theta) = r \cos \theta, \qquad x_2(r, \theta) = r \sin \theta$$

we have

$$dx_1 = -r \sin \theta d\theta + \cos \theta dr$$

$$dx_2 = r \cos \theta d\theta + \sin \theta dr$$

and

$$
\begin{aligned}
dx_1 \wedge dx_2 &= (-r \sin \theta d\theta + \cos \theta dr) \wedge (r \cos \theta d\theta + \sin \theta dr) \\
&= r \cos^2 \theta dr \wedge d\theta - r \sin^2 \theta d\theta \wedge dr \\
&= r dr \wedge d\theta
\end{aligned}
$$

since

$$dr \wedge dr = 0, \qquad d\theta \wedge d\theta = 0$$

and

$$d\theta \wedge dr = -dr \wedge d\theta. \qquad \qquad \clubsuit$$

7.3 The Exterior Derivative

We have already defined the exterior derivative d on $\bigwedge_0 T^*M|_{\mathbf{x}}$, namely

$$(df)(Z) = Zf.$$

The operator d can also be extended to all k-forms. We consider the smooth differential k-form in local coordinates

$$\omega = \sum_I \alpha_I(\mathbf{x}) dx_I$$

on a manifold M where I is the multi-index $I = (i_1, \ldots, i_k)$ and α_I is a smooth real valued function. Its **exterior derivative** is the $(k+1)$-form

$$d\omega = \sum_I d\alpha_I \wedge dx_I = \sum_{I,j} \frac{\partial \alpha_I}{\partial x_j} dx_j \wedge dx_I.$$

The operator d is a linear map that takes k-forms to $(k+1)$-forms

$$d : \bigwedge_k T^*M|_{\mathbf{x}} \to \bigwedge_{k+1} T^*M|_{\mathbf{x}}.$$

This map is subject to the following requirements:

1. If $f \in \bigwedge_0 T^*M|_{\mathbf{x}}$, $df \in \bigwedge_1 T^*M|_{\mathbf{x}}$, with values $df(Z) = Zf$ (which is nothing other than our original definition of df).

2. Linearity: If $\omega \in \bigwedge_k T^*M|_{\mathbf{x}}$, $\sigma \in \bigwedge_k T^*M|_{\mathbf{x}}$, then

$$d(c_1\omega + c_2\sigma) = c_1 d\omega + c_2 d\sigma$$

   where c_1, c_2 are constants.

3. Anti-derivation: If $\omega \in \bigwedge_k T^*M|_{\mathbf{x}}$, $\sigma \in \bigwedge_s T^*M|_{\mathbf{x}}$, then

$$d(\omega \wedge \sigma) = (d\omega) \wedge \sigma + (-1)^k \omega \wedge (d\sigma).$$

4. Closure (Lemma of Poincaré): If $\omega \in \bigwedge_k T^*M|_{\mathbf{x}}$, then

$$d(d\omega) \equiv 0.$$

Example: The closure properties of d translates into the familiar calculus identities:

$$\nabla \times (\nabla f) = 0, \qquad \nabla \cdot (\nabla \times \mathbf{f}) = 0.$$

In particular if $M = \mathbf{R}^3$ and $f_i \in C^\infty(\mathbf{R}^3)$ ($i = 1, 2, 3$) with $\mathbf{x} = (x_1, x_2, x_3)$, then for the differential of the one-form

$$f_1(\mathbf{x}) dx_1 + f_2(\mathbf{x}) dx_2 + f_3(\mathbf{x}) dx_3$$

we obtain

$$d(f_1 dx_1 + f_2 dx_2 + f_3 dx_3) = df_1 \wedge dx_1 + df_2 \wedge dx_2 + df_3 \wedge dx_3$$

$$= \frac{\partial f_1}{\partial x_2} dx_2 \wedge dx_1 + \frac{\partial f_1}{\partial x_3} dx_3 \wedge dx_1 + \frac{\partial f_2}{\partial x_1} dx_1 \wedge dx_2$$

$$+ \frac{\partial f_2}{\partial x_3} dx_3 \wedge dx_2 + \frac{\partial f_3}{\partial x_1} dx_1 \wedge dx_3 + \frac{\partial f_3}{\partial x_2} dx_2 \wedge dx_3$$

$$= \left(\frac{\partial f_2}{\partial x_1} - \frac{\partial f_1}{\partial x_2} \right) dx_1 \wedge dx_2 + \left(\frac{\partial f_3}{\partial x_2} - \frac{\partial f_2}{\partial x_3} \right) dx_2 \wedge dx_3 + \left(\frac{\partial f_1}{\partial x_3} - \frac{\partial f_3}{\partial x_1} \right) dx_3 \wedge dx_1$$

which can be identified with the curl, $\nabla \times \mathbf{f}$, where $\mathbf{f} = (f_1, f_2, f_3)$. Similarly, if $g_i \in C^\infty(\mathbf{R}^3)$ $(i = 1, 2, 3)$ the differential of the two-form

$$g_1(\mathbf{x}) dx_2 \wedge dx_3 + g_2(\mathbf{x}) dx_3 \wedge dx_1 + g_3(\mathbf{x}) dx_1 \wedge dx_2$$

results in the following

$$d(g_1 dx_2 \wedge dx_3 + g_2 dx_3 \wedge dx_1 + g_3 dx_1 \wedge dx_2) = \left(\frac{\partial g_1}{\partial x_1} + \frac{\partial g_2}{\partial x_2} + \frac{\partial g_3}{\partial x_3} \right) dx_1 \wedge dx_2 \wedge dx_3$$

which can be identified with $\nabla \cdot \mathbf{g}$, where $\mathbf{g} = (g_1, g_2, g_3)$. Consider

$$g_1 = \frac{\partial f_3}{\partial x_2} - \frac{\partial f_2}{\partial x_3}, \qquad g_2 = \frac{\partial f_1}{\partial x_3} - \frac{\partial f_3}{\partial x_1}, \qquad g_3 = \frac{\partial f_2}{\partial x_1} - \frac{\partial f_1}{\partial x_2}$$

so that

$$d(d(f_1 dx_1 + f_2 dx_2 + f_3 dx_3)) = \left(\frac{\partial g_1}{\partial x_1} + \frac{\partial g_2}{\partial x_2} + \frac{\partial g_3}{\partial x_3} \right) dx_1 \wedge dx_2 \wedge dx_3 = 0$$

which is equivalent to

$$\nabla \cdot (\nabla \times \mathbf{f}) = 0.$$

Note also that df can be identified with ∇f, so that

$$ddf = 0$$

as in

$$\nabla \times (\nabla f) = 0$$

where $f \in C^\infty(\mathbf{R}^3)$.   ♣

7.4 Pull-Back Maps

We consider the smooth map $F : M \to N$ between the manifolds M and N. There is an induced linear map F^*, called the **pull-back** or **codifferential** of F, which takes differential k-forms on N back to differential k-forms on M,

$$F^* : \bigwedge_k T^*N|_{F(x)} \to \bigwedge_k T^*M|_x \, .$$

If $f \in \bigwedge_0 T^*N|_{F(\mathbf{x})}$, then F^*f is the 0-form on M defined by

$$(F^*f)(\mathbf{x}) := f(F(\mathbf{x}))$$

so that $F^*f = f \circ F$, where $\circ$ denotes the composition of functions. If $\mathbf{x} = (x_1, \ldots, x_m)$ are local coordinates on M and $\mathbf{y} = (y_1, \ldots, y_n)$ local coordinates on N, then

$$F^*(dy_i) = d(F^*y_i) = \sum_{j=1}^m \frac{\partial y_i}{\partial x_j} dx_j \, ,$$

where $\mathbf{y} = F(\mathbf{x})$ gives the action of F^* on the basis one-forms. In general

$$F^*\left(\sum_I \alpha_I(\mathbf{y}) dy_I\right) = \sum_{I,J} \alpha_I(F(\mathbf{x})) \frac{\partial y_I}{\partial x_J} dx_J$$

where $\partial y_I / \partial x_J$ stands for the Jacobian determinant $\det(\partial y_{i_k}/\partial x_{j_\nu})$ corresponding to the increasing multi-indexes $I = (i_1, \ldots, i_n)$, $J = (j_1, \ldots, j_m)$, so that, if $\mathbf{y} = F(\mathbf{x})$ determines a change of coordinates on M, then the above relation provides the corresponding change of coordinates for differential k-forms on M. The codifferential of F has the following properties

$$F^*(\omega + \sigma) = F^*\omega + F^*\sigma$$
$$F^*(\omega \wedge \sigma) = (F^*\omega) \wedge (F^*\sigma)$$
$$F^*(d\omega) = d(F^*\omega)$$
$$F^*(df) = d(f \circ F)$$
$$F^*(f\omega) = (f \circ F)F^*\omega$$
$$(F \circ G)^*\omega = (G^* \circ F^*)\omega$$

where $f \in \bigwedge_0 T^*N|_{F(\mathbf{x})}$, $\omega, \sigma \in \bigwedge_k T^*N|_\mathbf{x}$ and

$$F : M \to N, \qquad G : Q \to M.$$

Example: If $G : \mathbf{R}^2 \to \mathbf{R}^2$ is given by $y_1(x_1, x_2) = x_1^2 - ax_2$, $y_2(x_1, x_2) = x_2^3$ (a is a constant) it follows that

$$G^*dy_1 = \sum_{j=1}^2 \frac{\partial y_1}{\partial x_j} dx_j = 2x_1 dx_1 - a dx_2, \qquad G^*dy_2 = 3x_2^2 dx_2.$$

Also

$$G^*(f(y_1, y_2) dy_1) = (G^*f)(G^*dy_1) = f(y_1(x_1, x_2), y_2(x_1, x_2))(2x_1 dx_1 - a dx_2). \quad \clubsuit$$

7.5 The Interior Product or Contraction

Let ω be a differential k-form and Z a smooth vector field. Then we can form a $(k-1)$-form $Z\lrcorner\omega$, called the **interior product** (also called the **contraction**) of Z with ω, defined so that

$$\langle Z\lrcorner\omega; Z_1, \ldots, Z_{k-1}\rangle = \langle \omega; Z, Z_1, \ldots, Z_{k-1}\rangle$$

for every set of vector fields $Z_1, \ldots, Z_{k-1}$. If $Z \in TM|_\mathbf{x}$ and $f \in C^\infty(M)$ is any 0-form, then we set

$$Z\lrcorner f := 0.$$

It follows from the linearity properties of k-forms that

$$(f_1 Z\lrcorner + f_2 Y\lrcorner)(f_3\omega + f_4\sigma) = f_1 f_3(Z\lrcorner\omega) + f_1 f_4(Z\lrcorner\sigma) + f_2 f_3(Y\lrcorner\omega) + f_2 f_4(Y\lrcorner\sigma)$$

where f_i ($i = 1, 2, 3, 4$) are smooth functions or 0-forms and $Z, Y \in TM|_\mathbf{x}$. In addition one can show that

$$Z\lrcorner(Z\lrcorner\omega) = (Z\lrcorner Z)\lrcorner\omega = 0$$
$$Z\lrcorner(\omega \wedge \sigma) = (Z\lrcorner\omega) \wedge \sigma + (-1)^k \omega \wedge (Z\lrcorner\sigma)$$
$$Z\lrcorner Y\lrcorner\omega = -Y\lrcorner Z\lrcorner\omega$$

and

$$\langle \omega; Z_1, \ldots, Z_k\rangle = Z_k\lrcorner Z_{k-1}\lrcorner \cdots \lrcorner Z_1\lrcorner\omega$$

where $Z, Z_1, \ldots, Z_k, Y \in TM|_\mathbf{x}$, ω is a differential k-form and σ a differential l-form. In the case $k = 1$, $Z\lrcorner\omega$ is a 0-form (i.e, a function)

$$Z\lrcorner\omega = \langle Z; \omega\rangle.$$

Obviously

$$\frac{\partial}{\partial x_i}\lrcorner dx_j = \delta_{ij}.$$

Example: Let

$$Z = \sum_{j=1}^m \xi_j \frac{\partial}{\partial x_j}$$

and

$$\omega = dx_i \wedge dx_j$$

with $i \neq j$. Then we find

$$Z\lrcorner(dx_i \wedge dx_j) = (Z\lrcorner dx_i) \wedge dx_j - dx_i \wedge (Z\lrcorner dx_j) = \xi_i dx_j - \xi_j dx_i.$$

Consider the k-form

$$\omega = dx_1 \wedge \cdots \wedge dx_k.$$

Then

$$Z\lrcorner\omega = \sum_{i=1}^k (-1)^{i+1}\xi_i(\mathbf{x})\, dx_1 \wedge \cdots \wedge \widehat{dx_i} \wedge \cdots \wedge dx_k.$$

The circumflex indicates omission.   ♣

7.6 Riemannian Manifolds

In this section we are concerned with the Riemannian and pseudo-Riemannian manifolds which have a metric tensor field structure. In order to understand these concepts we define tensor bundles and tensor fields. Let f be any r-linear function

$$f : TM|_{\mathbf{x}} \times \cdots \times TM|_{\mathbf{x}} \to \mathbf{R}$$

and g be any s-linear function

$$g : T^*M|_{\mathbf{x}} \times \cdots \times T^*M|_{\mathbf{x}} \to \mathbf{R}.$$

Definition 7.6 *With the above given f and g the **tensor product** of f and g is the map*

$$(f \otimes g)(Z_1, \ldots, Z_r, \omega_1, \ldots, \omega_s) = f(Z_1, \ldots, Z_s)g(\omega_1, \ldots, \omega_r).$$

Definition 7.7 *A **tensor bundle** of type (r, s) over M is given by*

$$T_s^r M := \bigcup_{\mathbf{x} \in M} T_s^r M|_{\mathbf{x}}$$

where $T_s^r M|_{\mathbf{x}}$ is given by

$$T_s^r M|_{\mathbf{x}} := TM|_{\mathbf{x}} \otimes \cdots \otimes TM|_{\mathbf{x}} \otimes T^*M|_{\mathbf{x}} \otimes \cdots \otimes T^*M|_{\mathbf{x}}.$$

Example: For $(r, s) = (0, 0)$ the tensor fields

$$T_0^0 M = M \times \mathbf{R}$$

are real valued functions on M. For $(r, s) = (1, 0)$ the tensor fields

$$T_0^1 M = TM$$

are $C^\infty(M)$ vector fields on M and for $(r, s) = (0, 1)$ the tensor fields

$$T_1^0 M = T^*M$$

are covariant vector fields or differential 1-forms on M. ♣

The local representation of tensor fields leads to the classical notation of tensors. For the local coordinates $x_1, \ldots, x_m$ with basis

$$\{\partial/\partial x_i\}_{1 \leq i \leq m}$$

of $TM|_{\mathbf{x}}$, the basis

$$\{dx_i\}_{1 \leq i \leq m}$$

yields a basis for $T_s^r M|_{\mathbf{x}}$ by r-fold tensor product

$$\frac{\partial}{\partial x_{i_1}} \otimes \cdots \otimes \frac{\partial}{\partial x_{i_r}}$$

and s-fold tensor product

$$dx_{j_1} \otimes \cdots \otimes dx_{j_s}.$$

Thus

$$T = \sum_{i_1,\ldots,i_r=1}^{m} \sum_{j_1,\ldots,j_s=1}^{m} t_{j_1,\ldots,j_s}^{i_1,\ldots,i_r}(\mathbf{x}) \frac{\partial}{\partial x_{i_1}} \otimes \cdots \otimes \frac{\partial}{\partial x_{i_r}} \otimes dx_{j_1} \otimes \cdots \otimes dx_{j_s},$$

so that the local field of type (r,s), $T_s^r M|_{\mathbf{x}}$, has local basis

$$\left\{ \frac{\partial}{\partial x_{i_1}} \otimes \cdots \otimes \frac{\partial}{\partial x_{i_r}} \otimes dx_{j_1} \otimes \cdots \otimes dx_{j_s} \right\}_{1 \le i_k \le m, \ 1 \le j_k \le m}$$

Covariant tensor fields of order s are given by

$$T = \sum_{j_1,\ldots,j_s=1}^{m} t_{j_1,\ldots,j_s}(\mathbf{x}) dx_{j_1} \otimes \cdots \otimes dx_{j_s}$$

and **contravariant tensor fields** of order r are given by

$$T = \sum_{i_1,\ldots,i_r=1}^{m} t^{i_1,\ldots,i_r}(\mathbf{x}) \frac{\partial}{\partial x_{i_1}} \otimes \cdots \otimes \frac{\partial}{\partial x_{i_r}}.$$

Example: The **Riemann curvature tensor**

$$R(\mathbf{x}) := \sum_{i,j,k,l=1}^{m} R_{jkl}^i(\mathbf{x}) dx_j \otimes dx_k \otimes dx_l \otimes \frac{\partial}{\partial x_i}$$

is a tensor field of type $(1,3)$ on M.   ♣

Definition 7.8 *An m-dimensional manifold M is said to be* **orientable** *if and only if there is a volume-form on M that is an m-form*

$$\Omega \in \bigwedge_m T^* M|_{\mathbf{x}}$$

such that

$$\Omega|_{\mathbf{x}} \ne 0$$

for all $\mathbf{x} \in M$.

Definition 7.9 *A* **Riemannian manifold** *is a differentiable manifold M with dimension m on which there is given, in any local coordinate system, a* **metric tensor field** *which is a covariant tensor field of type $(0,2)$, denoted by*

$$g = \sum_{i=1}^{m} \sum_{j=1}^{m} g_{ij}(\mathbf{x}) dx_i \otimes dx_j.$$

The function $g_{ij}(\mathbf{x})$ *of* $\mathbf{x} \in M$ *determines a Riemannian metric on* M. *The volume form* Ω *on* M *determined by this Riemannian metric is then given in local coordinates* $(x_1, \ldots, x_m)$ *by*

$$\Omega := \sqrt{|\det(g_{ij})|} dx_1 \wedge \cdots \wedge dx_m$$

and is called the **Riemannian volume form**. *If the determinant of the matrix* (g_{ij}) *is negative, the manifold is called* **pseudo-Riemannian**.

Example: Let $M = \mathbf{R}^3$. For the **Euclidean metric** tensor field

$$g = dx_1 \otimes dx_1 + dx_2 \otimes dx_2 + dx_3 \otimes dx_3$$

we have

$$\det(g_{ij}) = \begin{vmatrix} 1 & 0 & 0 \\ 0 & 1 & 0 \\ 0 & 0 & 1 \end{vmatrix} = 1$$

so that M is a Riemannian manifold. ♣

Example: Let $M = \mathbf{R}^4$. For the **Minkowski metric** in relativity theory,

$$g = dx_1 \otimes dx_1 + dx_2 \otimes dx_2 + dx_3 \otimes dx_3 - dx_4 \otimes dx_4,$$

we have

$$\det(g_{ij}) = \begin{vmatrix} 1 & 0 & 0 & 0 \\ 0 & 1 & 0 & 0 \\ 0 & 0 & 1 & 0 \\ 0 & 0 & 0 & -1 \end{vmatrix} = -1$$

so that M is a pseudo-Riemannian manifold. ♣

7.7   Hodge Star Operator

For Riemannian or pseudo-Riemannian manifolds the **Hodge star operator** $*$ (duality operation) is defined as a linear transformation between the spaces $\bigwedge_k(T^*M)$ and $\bigwedge_{m-k}(T^*M)$, i.e.

$$* : \bigwedge_k(T^*M) \to \bigwedge_{m-k}(T^*M)$$

where $k = 0, 1, \ldots, m$. The operator has the $*$-linearity which can be expressed as

$$*(f\omega + g\sigma) = f(*\omega) + g(*\sigma)$$

for all $f, g \in \bigwedge_0(T^*M)$ and all $\omega, \sigma \in \bigwedge_k(T^*M)$. The $*$ operator applied to a p-form ($p \leq m$) defined on an arbitrary Riemannian (or pseudo-Riemannian) manifold with metric tensor field g is given by

$$*(dx_{i_1} \wedge dx_{i_2} \wedge \cdots \wedge dx_{i_p}) := \sum_{j_1 \cdots j_m = 1}^{m} g^{i_1 j_1} \cdots g^{i_p j_p} \frac{1}{(m-p)!} \frac{g}{\sqrt{|g|}} \epsilon_{j_1 \cdots j_m} dx_{j_{p+1}} \wedge \cdots \wedge dx_{j_m}$$

where $\epsilon_{j_1 \cdots j_m}$ is the totally antisymmetric tensor ($\epsilon_{12\ldots m} = +1$), $g \equiv \det(g_{ij})$ and

$$\sum_{j=1}^{m} g^{ij} g_{jk} = \delta_{ik} \quad \text{(Kronecker symbol)}.$$

The double duality (composition of $*$ with itself) for $\omega \in \bigwedge_k T^*M$ is given by

$$* * (\omega) = (-1)^{k(m-k)} \omega.$$

The inverse of $*$ is

$$(*)^{-1} = (-1)^{k(m-k)} * .$$

Moreover,

$$\omega \wedge (*\sigma) = \sigma \wedge (*\omega)$$

for $\omega, \sigma \in \bigwedge_k T^*M$.

Example: Consider the Euclidean space $(\mathbf{R}^3, g)$ where the metric tensor field g is given by

$$g = dx_1 \otimes dx_1 + dx_2 \otimes dx_2 + dx_3 \otimes dx_3.$$

Let $f \in C^\infty(\mathbf{R}^3)$. Since f is a 0-form, we have $k = 0$ and $m = 3$. Applying the above we find

$$
\begin{aligned}
*f &= f(*1) = f \sum_{j_1, j_2, j_3 = 1}^{3} \frac{1}{3!} \epsilon_{j_1 j_2 j_3} dx_{j_1} \wedge dx_{j_2} \wedge dx_{j_3} \\
&= f \left(\frac{1}{3!} \epsilon_{123} dx_1 \wedge dx_2 \wedge dx_3 + \frac{1}{3!} \epsilon_{213} dx_2 \wedge dx_1 \wedge dx_3 + \frac{1}{3!} \epsilon_{312} dx_3 \wedge dx_1 \wedge dx_2 \right. \\
&\quad + \left. \frac{1}{3!} \epsilon_{132} dx_1 \wedge dx_3 \wedge dx_2 + \frac{1}{3!} \epsilon_{231} dx_2 \wedge dx_3 \wedge dx_1 + \frac{1}{3!} \epsilon_{321} dx_3 \wedge dx_2 \wedge dx_1 \right) \\
&= f \, dx_1 \wedge dx_2 \wedge dx_3.
\end{aligned}
$$

Analogously

$$*(f dx_1 \wedge dx_2 \wedge dx_3) = f * (dx_1 \wedge dx_2 \wedge dx_3) = f.$$ ♣

Example: Consider the Euclidean space $(\mathbf{R}^3, g)$ with the 1-form ω given by

$$\omega = a_1(\mathbf{x})dx_1 + a_2(\mathbf{x})dx_2 + a_3(\mathbf{x})dx_3$$

where $a_i \in C^\infty(\mathbf{R}^3)$ and $\mathbf{x} = (x_1, x_2, x_3)$. Now,

$$d\omega = \left(\frac{\partial a_2}{\partial x_1} - \frac{\partial a_1}{\partial x_2}\right) dx_1 \wedge dx_2 + \left(\frac{\partial a_1}{\partial x_3} - \frac{\partial a_3}{\partial x_1}\right) dx_3 \wedge dx_1 + \left(\frac{\partial a_3}{\partial x_2} - \frac{\partial a_2}{\partial x_3}\right) dx_2 \wedge dx_3.$$

To find $*d\omega$ we need to calculate

$$*(dx_1 \wedge dx_2) = \sum_{j_1, j_2, j_3 = 1}^{3} g^{1 j_1} g^{2 j_2} \frac{1}{(3-2)!} \epsilon_{j_1 j_2 j_3} dx_3 = g^{11} g^{22} \epsilon_{123} dx_3 = dx_3.$$

Analogously

$$*(dx_3 \wedge dx_1) = dx_2, \qquad *(dx_2 \wedge dx_3) = dx_1$$

so that

$$*d\omega = \left(\frac{\partial a_3}{\partial x_2} - \frac{\partial a_2}{\partial x_3}\right) dx_1 + \left(\frac{\partial a_1}{\partial x_3} - \frac{\partial a_3}{\partial x_1}\right) dx_2 + \left(\frac{\partial a_2}{\partial x_1} - \frac{\partial a_1}{\partial x_2}\right) dx_3.$$

We note that $*d$ is the curl-operator, operating on a 1-form. By calculating $*d(*\omega)$ we find

$$*d * (a_1 dx_1 + a_2 dx_2 + a_3 dx_3) = \frac{\partial a_1}{\partial x_1} + \frac{\partial a_2}{\partial x_2} + \frac{\partial a_3}{\partial x_3}$$

since

$$*dx_1 = dx_2 \wedge dx_3, \quad *dx_2 = dx_3 \wedge dx_1, \quad *dx_3 = dx_1 \wedge dx_2$$

so that $*d*$ is the divergence operating on a 1-form. ♣

Example: In the Euclidean space $(\mathbf{R}^3, g)$, at each point $\mathbf{x} \in \mathbf{R}^3$, (dx_1, dx_2, dx_3) is an orientable orthogonal basis of the cotangent space. In spherical coordinates (r, θ, φ) the elements

$$(dr, r d\theta, r \sin \theta d\varphi)$$

form another orthogonal basis of the cotangent space. The metric tensor field is then given by

$$g = dr \otimes dr + r^2 d\theta \otimes d\theta + r^2 \sin^2 \theta d\varphi \otimes d\varphi.$$

The volume element Ω is

$$\Omega = r^2 \sin\theta \; dr \wedge d\theta \wedge d\varphi$$

since

$$(g_{ij}) = \begin{pmatrix} 1 & 0 & 0 \\ 0 & r^2 & 0 \\ 0 & 0 & r^2 \sin^2\theta \end{pmatrix}$$

where

$$\det(g_{ij}) = \sqrt{r^4 \sin^2\theta} = r^2 \sin\theta.$$

Consider now a function $f \in C^\infty(\mathbf{R}^3)$. It follows that

$$df(r,\theta,\varphi) = \frac{\partial f}{\partial r}dr + \frac{\partial f}{\partial \theta}d\theta + \frac{\partial f}{\partial \varphi}d\varphi$$

and

$$*df(r,\theta,\varphi) = r^2 \sin\theta \frac{\partial f}{\partial r}d\theta \wedge d\varphi + \sin\theta \frac{\partial f}{\partial \theta}d\varphi \wedge dr + \frac{1}{\sin\theta}\frac{\partial f}{\partial \varphi}dr \wedge d\theta. \qquad \clubsuit$$

Example: We consider $M = \mathbf{R}^4$ with the Minkowski metric tensor field (pseudo-Riemannian manifold)

$$g = dx_1 \otimes dx_1 + dx_2 \otimes dx_2 + dx_3 \otimes dx_3 - dx_4 \otimes dx_4$$

and the one-form

$$du = \sum_{j=1}^{4} \frac{\partial u}{\partial x_j}dx_j.$$

We have

$$*du = \sum_{j=1}^{4} \frac{\partial u}{\partial x_j}(*dx_j)$$

with

$$
\begin{aligned}
*dx_1 &= -dx_2 \wedge dx_3 \wedge dx_4 \\
*dx_2 &= dx_3 \wedge dx_4 \wedge dx_1 \\
*dx_3 &= -dx_4 \wedge dx_1 \wedge dx_2 \\
*dx_4 &= -dx_1 \wedge dx_2 \wedge dx_3
\end{aligned}
$$

so that

$$d(*du) = -\left(\frac{\partial^2 u}{\partial x_1^2} + \frac{\partial^2 u}{\partial x_2^2} + \frac{\partial^2 u}{\partial x_3^2} - \frac{\partial^2 u}{\partial x_4^2}\right)\Omega$$

where

$$\Omega = dx_1 \wedge dx_2 \wedge dx_3 \wedge dx_4$$

is the volume element in Minkowski space. From the condition $d(*du) = 0$ we obtain the linear wave equation

$$\left(\frac{\partial^2}{\partial x_1^2} + \frac{\partial^2}{\partial x_2^2} + \frac{\partial^2}{\partial x_3^2} - \frac{\partial^2}{\partial x_4^2}\right)u = 0$$

with the Minkowski metric as underlying metric tensor field. $\qquad \clubsuit$

7.8 Computer Algebra Applications

Let V be a finite dimensional vector space over $\mathbf{R}$. Let r be an integer ≥ 1. Let $V^{(r)}$ be the set of all r-tuples of elements of V, i.e. $V^{(r)} = V \times \ldots \times V$. An element of $V^{(r)}$ is therefore an r-tuple $(v_1, \ldots, v_r)$ with $v_i \in V$. Each component of the r-tuple is an element of V. Let U be a finite dimensional vector space over $\mathbf{R}$. By an r-multilinear map of V into U one means a map

$$f : V \times \ldots \times V \to U$$

of $V^{(r)}$ into U which is linear in each component. In other words, for each $i = 1, \ldots, r$ we have

$$f(v_1, \ldots, v_i + v_i', \ldots, v_r) = f(v_1, \ldots, v_r) + f(v_1, \ldots, v_i', \ldots, v_r)$$

$$f(v_1, \ldots, cv_i, \ldots, v_r) = cf(v_1, \ldots, v_r)$$

for all $v_i, v_i' \in V$ and $c \in \mathbf{R}$. We say that a multilinear map f is alternating if it satisfies the condition $f(v_1, \ldots, v_r) = 0$ whenever two adjacent components are equal, i.e. whenever there exists an index $j < r$ such that $v_j = v_{j+1}$. We see that a multilinear map satisfies conditions completely similar to the properties satisfied by determinants.

Let V be a finite dimensional vector space over $\mathbf{R}$, of dimension n. Let r be an integer $1 \leq r \leq n$. There exists a finite dimensional space over $\mathbf{R}$, denoted by $\bigwedge^r V$, and an r-multilinear alternating map $V^{(r)} \to \bigwedge^r V$, denoted by

$$(u_1, \ldots, u_r) \mapsto u_1 \wedge \ldots \wedge u_r,$$

satisfying the following properties: If U is a vector space over $\mathbf{R}$, and $g : V^{(r)} \to U$ is an r-multilinear alternating map, then there exists a unique linear map $g_* : \bigwedge \to U$ such that for all $u_1, \ldots, u_r \in V$ we have

$$g(u_1, \ldots, u_r) = g_*(u_1 \wedge \ldots \wedge u_r).$$

If $\{v_1, \ldots, v_n\}$ is a basis of V, then the set of elements

$$\{v_{i_1} \wedge \ldots \wedge v_{i_r} \qquad (1 \leq i_1 < \ldots < i_r \leq n)\}$$

is a basis of $\bigwedge^r V$.

Remark: The composition $\wedge$ is also called exterior product or Grassmann product. In the program $\wedge$ is written as $*$. We evaluate the determinant of a 2×2 matrix and of the 4×4 matrix

$$A = \begin{pmatrix} 1 & 2 & 5 & 2 \\ 0 & 1 & 2 & 3 \\ 1 & 0 & 1 & 0 \\ 0 & 3 & 0 & 7 \end{pmatrix}.$$

```
%grass1.red;

operator e;
noncom e;

for all j let e(j)**2 = 0;
let e(2)*e(1) = - e(1)*e(2);

result := (a11*e(1)+a21*e(2))*(a12*e(1)+a22*e(2));
```

The output is

```
result := e(1)*e(2)*(a11*a22 - a12*a21);
```

```
%grass2.red;

operator e;
noncom e;

for all j let e(j)*e(j) = 0;
for all j, k such that (j > k) let e(j)*e(k) = - e(k)*e(j);

result := (e(1)+e(3))*(2*e(1)+e(2)+3*e(4))*
(5*e(1)+2*e(2)+e(3))*(2*e(1)+3*e(2)+7*e(4));
```

The output is

```
result := 24*e(1)*e(2)*e(3)*e(4);
```

Thus the determinant is equal to 24.

Soliton equations can be derived from pseudospherical surfaces. Using computer algebra we show how the sine-Gordon equation can be derived. Extensions to other soliton equations are straightforward. Soliton equations have several remarkable properties in common: (i) the initial value problem can be solved exactly in terms of the inverse scattering methods, (ii) they have an infinite number of conservation laws, (iii) they have Bäcklund transformations, (iv) they pass the Painlevé test. Furthermore they describe pseudospherical surfaces, i.e. surfaces of constant negative Gaussian curvature. An example is the **sine-Gordon equation**

$$\frac{\partial^2 u}{\partial x_1 \partial x_2} = \sin(u).$$

Here we show how computer algebra can be used to find the sine-Gordon equation from the line element of the surface. Extension to other soliton equations is straightforward.

The metric tensor field is given by

$$g = dx_1 \otimes dx_1 + \cos(u(x_1, x_2))dx_1 \otimes dx_2 + \cos(u(x_1, x_2))dx_2 \otimes dx_1 + dx_2 \otimes dx_2$$

i.e. the line element is

$$\left(\frac{ds}{d\lambda}\right)^2 = \left(\frac{dx_1}{d\lambda}\right)^2 + 2\cos(u(x_1, x_2))\frac{dx_1}{d\lambda}\frac{dx_2}{d\lambda} + \left(\frac{dx_2}{d\lambda}\right)^2.$$

Here u is a smooth function of x_1 and x_2. First we have to calculate the Riemann curvature scalar R from g. Then the equation (1) follows when we impose the condition

$$R = -2.$$

We have

$$g_{11} = g_{22} = 1, \qquad g_{12} = g_{21} = \cos(u(x_1, x_2)).$$

The quantity g can be written in matrix form

$$g = \left(\begin{array}{cc} g_{11} & g_{12} \\ g_{21} & g_{22} \end{array}\right).$$

Then the inverse of g is given by

$$g^{-1} = \left(\begin{array}{cc} g^{11} & g^{12} \\ g^{21} & g^{22} \end{array}\right)$$

where

$$g^{11} = g^{22} = \frac{1}{\sin^2 u}, \qquad g^{12} = g^{21} = -\frac{\cos u}{\sin u}.$$

Next we have to calculate the Christoffel symbols. They are defined as

$$\Gamma^a_{mn} := \frac{1}{2}g^{ab}(g_{bm,n} + g_{bn,m} - g_{mn,b})$$

where the sum convention is used and

$$g_{bm,1} := \frac{\partial g_{bm}}{\partial x_1} \qquad g_{bm,2} := \frac{\partial g_{bm}}{\partial x_2}.$$

Next we have to calculate the Riemann curvature tensor which is given by

$$R^r_{msq} := \Gamma^r_{mq,s} - \Gamma^r_{ms,q} + \Gamma^r_{ns}\Gamma^n_{mq} - \Gamma^r_{nq}\Gamma^n_{ms}.$$

The Ricci tensor follows as

$$R_{mq} := R^a_{maq} = -R^a_{mqa}$$

i.e. the Ricci tensor is constructed by contraction. From R_{nq} we obtain R^m_q via

$$R^m_q = g^{mn}R_{nq}.$$

Finally the curvature scalar R is given by

$$R := R^m_m.$$

With the metric tensor field given above we find that

$$R = -\frac{2}{\sin u}\frac{\partial^2 u}{\partial x_1 \partial x_2}.$$

If $R = -2$, then we obtain the sine-Gordon equation.

We give an implementation of this algorithm using REDUCE. We apply the concept of operators. Operators are the most general objects available in REDUCE. They are usually parametrized, and can be parametrized in a completely general way. Only the operator identifier is declared in an operator declaration. The number of parameters is not declared. Operators can be used to represent just about anything, although usually they represent mathematical operators or functions.

In our case u is declared as an operator and depends on x. x itself is also declared as an operator and depends on 1 and 2. The operator df() denotes differentiation. Since terms of the form $\cos^2(x)$ and $\sin^2(x)$ result from our calculation we have to include the identity $\sin^2(u) + \cos^2(u) \equiv 1$ in order to simplify expressions. Thus the implementation is as follows:

```
% tensor.red
matrix g(2,2);
matrix g1(2,2);  % inverse of g;
array gamma(2,2,2); array R(2,2,2,2); array Ricci(2,2);

operator u, x;

g(1,1)  := 1;  g(2,2)  := 1;
g(1,2)  := cos(u(x(1),x(2))); g(2,1)  := cos(u(x(1),x(2)));
```

```
g1 := g^(-1); % calculating the inverse
for a := 1:2 do
for m := 1:2 do
for n := 1:2 do
gamma(a,m,n) := (1/2)*(for b := 1:2 sum
g1(a,b)*(df(g(b,m),x(n)) + df(g(b,n),x(m)) - df(g(m,n),x(b))));

for a := 1:2 do
for m := 1:2 do
for n := 1:2 do
write "gamma(",a,",",m,",",n,") = ", gamma(a,m,n);

for b := 1:2 do
for m := 1:2 do
for s := 1:2 do
for q := 1:2 do
R(b,m,s,q) := df(gamma(b,m,q),x(s)) - df(gamma(b,m,s),x(q))
+ (for n := 1:2 sum gamma(b,n,s)*gamma(n,m,q))
- (for n := 1:2 sum gamma(b,n,q)*gamma(n,m,s));

cos(u(x(1),x(2)))**2 := 1 - sin(u(x(1),x(2)))**2;

for m := 1:2 do
for q := 1:2 do
Ricci(m,q) := for s := 1:2 sum R(s,m,s,q);

for m := 1:2 do
for q := 1:2 do
write "Ricci(",m,",",q,") = ", Ricci(m,q);

array Ricci1(2,2);
for m := 1:2 do
for q := 1:2 do
Ricci1(m,q) := (for b := 1:2 sum g1(m,b)*Ricci(q,b));

CS := for m := 1:2 sum Ricci1(m,m);
```

The calculation of the curvature scalar from a metric tensor field (for example Goedel metric, Schwarzschild metric) is one of the oldest applications of computer algebra. Here we showed that it can be extended to find soliton equations. By modifying the metric tensor field we can obtain other soliton equations.

Let E^3 be the three dimensional Euclidean space. The metric tensor field is given by

$$g = dx_1 \otimes dx_1 + dx_2 \otimes dx_2 + dx_3 \otimes dx_3.$$

The two dimensional unit sphere

$$S^2 := \{(x_1, x_2, x_3) \; : \; x_1^2 + x_2^2 + x_3^2 = 1\}$$

is embedded into E^3. We evaluate the metric tensor field for the unit sphere. A parameter representation is given by

$$x_1(u, v) = \cos(u)\sin(v), \quad x_2(u, v) = \sin(u)\sin(v), \quad x_3(u, v) = \cos(v)$$

where $0 < v < \pi$. Since

$$dx_1 = -\sin(u)\sin(v)du + \cos(u)\cos(v)dv$$

$$dx_2 = \cos(u)\sin(v)du + \sin(u)\cos(v)dv$$

$$dx_3 = -\sin(v)dv$$

we obtain for the metric tensor field of the unit sphere

$$\tilde{g} = \sin^2(v)du \otimes du + dv \otimes dv.$$

In the program the tensor product $\otimes$ is denoted by $*$

```
%Program name: metrics.our;

operator DU, DV;
noncom DU, DV;

depend x1, u, v;
depend x2, u, v;

x1 := cos(u)*sin(v);
x2 := sin(u)*sin(v);
x3 := cos(v);

DX1 := df(x1,u)*DU + df(x1,v)*DV;
DX2 := df(x2,u)*DU + df(x2,v)*DV;
DX3 := df(x3,u)*DU + df(x3,v)*DV;

GT := DX1*DX1 + DX2*DX2 + DX3*DX3;

for all q let (cos(q))^2 = 1 - (sin(q))^2;

GT;
```

Chapter 8

Lie Derivative and Invariance

8.1 Introduction

The concept of the Lie derivative of functions, vector fields, differential forms and tensor fields with respect to a vector field plays an important role in many domains in physics. Applications have been made to classical mechanics, hydrodynamics, optics, quantum mechanics, theory of relativity, statistical mechanics and supergravity. The Lie derivative also plays a leading role in the study of symmetries of differential equations. This will be discussed in detail in the next chapter. In the present chapter we give a survey on the geometrical setting of the Lie derivative, (Choquet-Bruhat *et al* [17], von Westenholz [141], Olver [87], Steeb [116], [128]). For the applications of the differential forms and the Lie derivative we show how some physical laws can be derived within this approach.

8.2 Definitions

In this section we discuss the **Lie derivative** of functions, vector fields, differential forms and tensor fields on a differentiable manifold M. The Lie derivative of a certain geometrical object on a manifold M gives the infinitesimal change of these objects under the flow $\exp(\varepsilon Z)$ induced by Z, where Z is a vector field on M. By a **geometrical object** we mean that σ can be a smooth function, a smooth vector field, a smooth differential form or a smooth metric tensor field.

Let us first define the Lie derivative of a C^∞-function which is a 0-form on M.

Definition 8.1 *Let Z be a vector field on M and $\{\varphi(\varepsilon)\}$ the local 1-parameter group of transformations generated by Z. This vector field induces a differential operator L_Z on the C^∞-function f referred to as the Lie derivative of f with respect to Z, defined by*

$$(L_Z f)(\mathbf{x}) := \lim_{\varepsilon \to 0} \frac{1}{\varepsilon}[f(\varphi(\mathbf{x}, \varepsilon)) - f(\mathbf{x})] = \lim_{\varepsilon \to 0} \frac{1}{\varepsilon}[(\varphi(\varepsilon)^* f)(\mathbf{x}) - f(\mathbf{x})] = \frac{d}{d\varepsilon} f(\varphi(\mathbf{x}, \varepsilon))\Big|_{\varepsilon=0}$$

where $\varphi^*(\varepsilon)$ is the pull back map.

For differential forms we define

Definition 8.2 *If ω is any k-form on M, and Z any vector field on M, the Lie derivative $L_Z\omega$ of ω with respect to Z is the k-form*

$$(L_Z\omega)(\mathbf{x}) := \lim_{\varepsilon \to 0} \frac{1}{\varepsilon}[(\varphi(\varepsilon)^*\omega)(\mathbf{x}) - \omega(\mathbf{x})]$$

where $\{\varphi(\varepsilon)\}$ is the local 1-parameter group generated by Z.

It can be shown that the Lie derivative of k-forms with respect to a vector field Z can be written in terms of the exterior derivative and interior product

$$L_Z\omega := Z\lrcorner d\omega + d(Z\lrcorner\omega).$$

We now give the properties of the Lie derivative. Let Z and Y be any two vector fields on M:

1. If $f_1, f_2 \in \bigwedge_0 T^*M$, then

$$L_Z(f_1 f_2) = f_1(L_Z f_2) + f_2(L_Z f_1)$$

   where

$$L_Z f_i = Z\lrcorner df_i + d(Z\lrcorner f_i) = Zf_i \qquad i = 1, 2.$$

   Obviously with $f = c$ where c is a constant it follows that $L_Z c = 0$.

2. The operators d and L_Z commute, i.e.,

$$L_Z d\omega = d(L_Z\omega)$$

   for any $\omega \in \bigwedge T^*M$. This can be seen by direct calculation:

$$L_Z d\omega = Z\lrcorner d(d\omega) + d(Z\lrcorner d\omega) = d(Z\lrcorner d\omega)$$

   and

$$d(L_Z\omega) = d(Z\lrcorner d\omega) + d(d(Z\lrcorner\omega)) = d(Z\lrcorner d\omega).$$

3. If $\omega, \sigma \in \bigwedge_k T^*M$ and c_1, c_2 are constants, then

$$L_Z(c_1\omega + c_2\sigma) = c_1 L_Z\omega + c_2 L_Z\sigma,$$

   and

$$L_{(c_1 Z + c_2 Y)}\omega = c_1 L_Z\omega + c_2 L_Y\omega.$$

4. If $\omega \in \bigwedge_k T^*M$ and $\theta \in \bigwedge_r T^*M$, then the product rule holds, i.e.,

$$L_Z(\omega \wedge \theta) = (L_Z\omega) \wedge \theta + \omega \wedge (L_Z\theta).$$

5. If $f \in \bigwedge_0 T^*M$ and $\omega \in \bigwedge_k T^*M$, then

$$L_Z(f\omega) = (Zf)\omega + f(L_Z\omega)$$

and

$$L_{fZ}\omega = f(L_Z\omega) + df \wedge (Z\lrcorner\omega).$$

6. If $\omega \in \bigwedge_k T^*M$, then

$$L_{[Z,Y]}\omega = (L_ZL_Y - L_YL_Z)\omega = [L_Z, L_Y]\omega.$$

Let us now investigate how a Lie derivative of a vector field with respect to a vector field can be defined. Let Z and Y be any two vector fields on M. We define the Lie derivative of Y with Z by requiring that L_Z, when applied to $Y\lrcorner\omega$, be a derivation in the sense that

$$L_Z(Y\lrcorner\omega) = (L_ZY)\lrcorner\omega + Y\lrcorner L_Z\omega$$

for all $\omega \in \bigwedge T^*M$. We now consider a 0-form f where we put $\omega = df$. It follows that

$$L_Z(Y\lrcorner\omega) = L_Z(Y\lrcorner df) = L_Z(Yf) = ZYf$$

while

$$Y\lrcorner L_Z\omega = Y\lrcorner(L_Z df) = Y\lrcorner\{Z\lrcorner d(df) + d(Z\lrcorner df)\} = Y\lrcorner d(Zf) = YZf.$$

Substituting into the above yields

$$Z(Yf) = (L_ZY)\lrcorner df + Y(Zf).$$

Since

$$(L_ZY)\lrcorner df = (L_ZY)f = (ZY - YZ)f$$

it follows that

$$L_ZY = [Z, Y]$$

where $[\;,\;]$ is the commutator.

Example: Let us consider the vector fields Z and Y in local coordinates so that

$$Z = \sum_{j=1}^m \xi_j(\mathbf{x})\frac{\partial}{\partial x_j}, \qquad Y = \frac{\partial}{\partial x_i}$$

where $\mathbf{x} = (x_1, \ldots, x_m)$. Since

$$L_Z\frac{\partial}{\partial x_i} = \left[\sum_{j=1}^m \xi_j(\mathbf{x})\frac{\partial}{\partial x_j}, \frac{\partial}{\partial x_i}\right] = -\sum_{j=1}^m \left(\frac{\partial \xi_j(\mathbf{x})}{\partial x_i}\right)\frac{\partial}{\partial x_j}.$$

and

$$L_Y Z = \left[\frac{\partial}{\partial x_i}, \sum_{j=1}^m \xi_j(\mathbf{x})\frac{\partial}{\partial x_j}\right] = \sum_{j=1}^m \left(\frac{\partial \xi_j(\mathbf{x})}{\partial x_i}\right)\frac{\partial}{\partial x_j}$$

we obtain

$$L_Z Y + L_Y Z = 0. \qquad \clubsuit$$

Example: Let V, X be two vector fields defined on M. Let α be a one-form defined on M and let f be a function $f : M \to \mathbf{R}$. If

$$L_X \alpha = df \qquad \text{and} \qquad V \lrcorner \, d\alpha = 0$$

then

$$L_V(f - X \lrcorner \alpha) = 0 \qquad \text{and} \qquad L_X(f - X \lrcorner \alpha) = 0.$$

The vector field V describes a dynamical system and $V \lrcorner \, d\alpha = 0$ is the equation of motion, where α is the **Cartan form**

$$\alpha := \sum_{i=1}^{N} \sum_{j=1}^{3} p_{ij} dq_{ij} - H(\mathbf{p}, \mathbf{q}) dt.$$

The assumption $L_X \alpha = df$ is the invariance requirement of α under the vector field X. The equation $L_V(f - X \lrcorner \alpha) = 0$ tells that the function $f - X \lrcorner \alpha$ is a constant of motion. The Cartan form given above applies in non-relativistic classical mechanics. In relativistic classical mechanics we have

$$\alpha := \sum_{i=1}^{N} \sum_{j=1}^{3} p_{ij} dq_{ij} \quad \sum_{i=1}^{N} E_i dt_i - K(\mathbf{p}, \mathbf{E}, \mathbf{q}, t) d\lambda \qquad \clubsuit$$

Some physical quantities, for example the energy-momentum tensor, cannot be expressed as a differential form. It is a tensor field. We thus give the following definition for the Lie derivative of a tensor field of type (r, s) on M.

Definition 8.3 *Let Z be a vector field on M, $\{\varphi(\varepsilon)\}$ the 1-parameter group of transformation generated by Z and T a tensor field of type (r, s) on M. Then the Lie derivative of T with respect to Z is a tensor field of type (r, s) on M defined by*

$$(L_Z T)(\mathbf{x}) := \lim_{\varepsilon \to 0} \frac{1}{\varepsilon} [(\varphi(\varepsilon)^* T)(\mathbf{x}) - T(\mathbf{x})].$$

Let S and T be (r, s) tensor fields and Z and Y vector fields on M. Then

$$L_Z(T + S) = L_Z T + L_Z S$$
$$L_{Z+Y} S = L_Z S + L_Y S$$
$$L_Z(fT) = Z(f)T + f(L_Z T).$$

Let S be a (r, s) tensor field and T be a (u, v) tensor field. Then (product rule)

$$L_Z(T \otimes S) = (L_Z T) \otimes S + T \otimes (L_Z S).$$

Application of the Lie derivative will be discussed in the next section.

8.3 Invariance and Conformal Invariance

Let Z be a vector field on M and σ some geometrical object on M. Let us first give the definitions of conformal invariance and invariance. Then we give some examples to illustrate these definitions for various geometrical objects (i.e., functions, differential forms, vector fields and tensor fields).

Definition 8.4 *The geometrical object σ is called* **conformal invariant** *w.r.t. the vector field Z if*

$$L_Z \sigma = \rho \sigma$$

where $\rho \in C^\infty(M)$.

Definition 8.5 *The geometrical object σ is called* **invariant** *w.r.t. the vector field Z if*

$$L_Z \sigma = 0.$$

Geometrically the condition of invariance implies that the object σ does not change as it propagates down the trajectories of Z. For short, σ is sometimes called Z-invariant.

Let f be an invariant w.r.t the vector field

$$Z = \sum_{i=1}^{m} \xi_i(\mathbf{x}) \frac{\partial}{\partial x_i}$$

on $M = \mathbf{R}^m$ so that $L_Z f = 0$. It follows that f is a first integral of the autonomous system of differential equations

$$\frac{d\mathbf{x}}{d\varepsilon} = \boldsymbol{\xi}(\mathbf{x}),$$

where $\mathbf{x} = (x_1, \ldots, x_m)$ and $\boldsymbol{\xi} = (\xi_1, \ldots, \xi_m)$. This follows from straightforward calculations:

$$L_Z f(\mathbf{x}) \equiv Z f(\mathbf{x}) = \sum_{i=1}^{m} \xi_i(\mathbf{x}) \frac{\partial f}{\partial x_i} = 0.$$

For f to be a first integral we have

$$\frac{df}{d\varepsilon} = \sum_{i=1}^{m} \frac{\partial f(\mathbf{x})}{\partial x_i} \frac{dx_i}{d\varepsilon} = \sum_{i=1}^{m} \frac{\partial f(\mathbf{x})}{\partial x_i} \xi_i(\mathbf{x}) = 0.$$

An important observation is that if some invariant quantities are known, other invariants can be calculated. We have the following properties: Let $L_{Z_1}\sigma = 0$ and $L_{Z_2}\sigma = 0$. Then

$$L_{[Z_1, Z_2]}\sigma = 0.$$

Let Z be a vector field on M, and ω, θ invariant forms of Z. Then $Z \lrcorner \omega$, $d\omega$, and $\omega \wedge \theta$ are invariant forms of Z. This follows from the properties of the Lie derivative given in the previous section. For the conformal invariance the following can easily be shown. If ω is conformal invariant w.r.t Z then $Z \lrcorner \omega$ is also conformal invariant w.r.t the same Z: Applying the properties of the Lie derivative with $L_Z \omega = f\omega$ we find

$$L_Z(Z \lrcorner \omega) = [Z, Z] \lrcorner \omega + Z \lrcorner f\omega = f(Z \lrcorner \omega).$$

If ω and θ are conformal invariants w.r.t Z then $\omega \wedge \theta$ is also conformal invariant w.r.t the same Z: Applying the properties of the Lie derivative with $L_Z \omega = f\omega$ and $L_Z \theta = g\theta$ we find

$$L_Z(\omega \wedge \theta) = (L_Z \omega) \wedge \theta + \omega \wedge (L_Z \theta) = f\omega \wedge \theta + \omega \wedge g\theta = (f + g)(\omega \wedge \theta).$$

In general $d\omega$ is not a conformal invariant w.r.t. Z if ω is a conformal invariant w.r.t. Z. Consider $d(L_Z \omega) = L_Z(d\omega)$ and the relation $d(f\omega) = (df) \wedge \omega + f d\omega$. It follows that only if f is constant do we find that $L_Z d\omega = f d\omega$.

We now give some examples to illustrate the definitions given above.

Example: We find that the quadratic form

$$x_1^2 + x_2^2 + x_3^2 - x_4^2$$

is invariant under the vector fields associated with the Lorentz group, given by

$$Z_1 = x_4 \frac{\partial}{\partial x_1} + x_1 \frac{\partial}{\partial x_4}, \qquad Z_2 = x_4 \frac{\partial}{\partial x_2} + x_2 \frac{\partial}{\partial x_4}$$
$$Z_3 = x_4 \frac{\partial}{\partial x_3} + x_3 \frac{\partial}{\partial x_4}, \qquad Z_4 = x_2 \frac{\partial}{\partial x_1} - x_1 \frac{\partial}{\partial x_2}$$
$$Z_5 = x_3 \frac{\partial}{\partial x_2} - x_2 \frac{\partial}{\partial x_3}, \qquad Z_6 = x_1 \frac{\partial}{\partial x_3} - x_3 \frac{\partial}{\partial x_1}.$$

This means $L_{Z_i}(x_1^2 + x_2^2 + x_3^2 - x_4^2) = 0$, where $i = 1, \ldots, 6$.   ♣

Example: Let f, g be smooth functions. We find that the differential one-form

$$\omega = (x_1 f(r) - x_2 g(r)) dx_2 - (x_2 f(r) + x_1 g(r)) dx_1$$

is invariant w.r.t the vector field

$$Z = x_2 \frac{\partial}{\partial x_1} - x_1 \frac{\partial}{\partial x_2}$$

where $r^2 := x_1^2 + x_2^2$. Straightforward calculation yields

$$
\begin{aligned}
L_Z \omega &= L_Z(x_1 f(r) dx_2) - L_Z(x_2 g(r) dx_2) - L_Z(x_2 f(r) dx_1) - L_Z(x_1 g(r) dx_1) \\
&= (Z x_1 f(r)) dx_2 + x_1 f(r)(L_Z dx_2) - (Z x_2 g(r)) dx_2 + x_2 g(r)(L_Z dx_2)
\end{aligned}
$$

$$- \ (Z(x_2 f(r))dx_1 + x_2 f(r)(L_Z dx_1)) - (Z(x_1 g(r))dx_1 + x_1 g(r)(L_Z dx_1))$$

$$= \ (x_2 f(r)dx_2 - x_1 f(r)dx_1) - (-x_1 g(r)dx_2 - x_2 g(r)dx_1)$$

$$- \ (-x_1 f(r)dx_1 + x_2 f(r)dx_2) - (x_2 g(r)dx_1 + x_1 g(r)dx_2)$$

$$= \ 0.$$

♣

Example: We consider the forms

$$\omega = d\Phi - u_1 dx_2 - u_2 dx_1,$$
$$\sigma = du_1 \wedge dx_1 + du_2 \wedge dx_2 - m^2 \Phi dx_1 \wedge dx_2$$

and the vector field

$$Z = x_1 \frac{\partial}{\partial x_2} + x_2 \frac{\partial}{\partial x_1} - u_2 \frac{\partial}{\partial u_1} - u_1 \frac{\partial}{\partial u_2},$$

where m^2 is a constant. To obtain $L_Z \omega$ we make use of $L_Z \omega \equiv d(Z \lrcorner \omega) + Z \lrcorner (d\omega)$. Straightforward calculations yield

$$Z \lrcorner \omega = -u_1 x_1 - u_2 x_2$$
$$d(Z \lrcorner \omega) = -x_1 du_1 - u_1 dx_1 - x_2 du_2 - u_2 dx_2$$
$$d\omega = -du_1 \wedge dx_2 - du_2 \wedge dx_1$$
$$Z \lrcorner (d\omega) = Z \lrcorner (-du_1 \wedge dx_2 - du_2 \wedge dx_1).$$

We recall that $Z \lrcorner (\alpha \wedge \beta) = (Z \lrcorner \alpha) \wedge \beta + (-1)^r \alpha \wedge (Z \lrcorner \beta)$ where α is an r-form and β an s-form. It follows that

$$Z \lrcorner (d\omega) = (Z \lrcorner (-du_1)) \wedge dx_2 + du_1 \wedge (Z \lrcorner dx_2) - ((Z \lrcorner du_2) \wedge dx_1 - du_2 \wedge (Z \lrcorner dx_1))$$
$$= u_2 dx_2 + x_1 du_1 + u_1 dx_1 + x_2 du_2$$

so that

$$L_Z \omega = 0$$

which implies that $L_Z d\omega = 0$. In the same way it can also be shown that $L_Z \sigma = 0$. The example has physical meaning. It describes the one dimensional Klein-Gordon equation

$$\frac{\partial^2 \Phi}{\partial x_1^2} - \frac{\partial^2 \Phi}{\partial x_2^2} + m^2 \Phi = 0$$

in the so called geometrical approach of partial differential equations. Within this approach the partial differential equation is cast into an equivalent set of differential forms. In the present case we have put

$$\frac{\partial \Phi}{\partial x_2} = u_1, \qquad \frac{\partial \Phi}{\partial x_1} = u_2.$$

The vector field Z is associated with the Lorentz transformation. The conclusion is then that the differential equation is invariant under the Lorentz transformation described by the vector field Z. The invariance of differential equations under transformation groups is discussed in detail in the next chapter where we consider the jet bundle formalism. ♣

Let us now consider the invariance of vector fields. The trajectories of a vector field Z_2 will be invariant under a one-parameter group of transformation generated by a vector field Z_1, if

$$[Z_1, Z_2] = f Z_2.$$

Obviously, any vector parallel to Z_2 trivially satisfies this equation. The special case

$$[Z_1, Z_2] = 0$$

is an integrability condition. The geometrical meaning is that the (local) flows (φ_1, φ_2) associated with Z_1 and Z_2 commute

$$\varphi_1 \circ \varphi_2 = \varphi_2 \circ \varphi_1.$$

Example: Consider the vector fields

$$Z_1 = x_1 \frac{\partial}{\partial x_2} - x_2 \frac{\partial}{\partial x_1},$$
$$Z_2 = x_1 \frac{\partial}{\partial x_1} + x_2 \frac{\partial}{\partial x_2}.$$

Obviously $[Z_1, Z_2] = 0$. The vector fields can be cast into a matrix form,

$$\begin{pmatrix} Z_1 \\ Z_2 \end{pmatrix} = \begin{pmatrix} -x_2 & x_1 \\ x_1 & x_2 \end{pmatrix} \begin{pmatrix} \partial/\partial x_1 \\ \partial/\partial x_2 \end{pmatrix}.$$

The determinant of the 2×2 matrix gives the function

$$I(x_1, x_2) = x_1^2 + x_2^2$$

where I is an invariant w.r.t the vector field Z_1

$$L_{Z_1} I = 0$$

and conformal invariant w.r.t the vector field Z_2

$$L_{Z_2} I = 2I. \qquad\qquad ♣$$

The relation between the conformal invariance $L_{Z_1} \sigma = f \sigma$ and $[Z_1, Z_2] = f Z_2$ can be stated as the following

Theorem 8.1 *Let $M = \mathbf{R}^m$ (or an open subset of $\mathbf{R}^m$) and $\theta = dx_1 \wedge \ldots \wedge dx_m$. Let $Z_1 = \sum_{i=1}^m \xi_i \partial/\partial x_i$ and $Z_2 = \sum_{i=1}^m \eta_i \partial/\partial x_i$ be two vector fields such that $[Z_1, Z_2] = f Z_2$. Then*

$$L_{Z_1}(Z_2 \lrcorner \theta) = (f + div\ \boldsymbol{\xi})\ Z_2 \lrcorner \theta.$$

Proof:

$$
\begin{aligned}
L_{Z_1}(Z_2 \lrcorner \theta) &= (L_{Z_1} Z_2) \lrcorner \theta + Z_2 \lrcorner (L_{Z_1}\theta) \\
&= [Z_1, Z_2] \lrcorner \theta + Z_2 \lrcorner (L_{Z_1}\theta) \\[2mm]
&= (fZ_2) \lrcorner \theta + Z_2 \lrcorner \{d(Z_1 \lrcorner \theta) + Z_1 \lrcorner d\theta\} \\[2mm]
&= f(Z_2 \lrcorner \theta) + Z_2 \lrcorner (\mathrm{div}\boldsymbol{\xi})\theta \\[2mm]
&= (f + \mathrm{div}\ \boldsymbol{\xi})(Z_2 \lrcorner \theta).
\end{aligned}
$$

We now consider the definition for invariance and conformal invariance for the metric tensor field g on a Riemannian manifold. In particular we consider the metric tensor field on $\mathbf{R}^m$ given by

$$g = \frac{1}{2} \sum_{i,j=1}^m g_{ij}(\mathbf{x}) dx_i \otimes dx_j.$$

Definition 8.6 *The vector fields Z_j on $\mathbf{R}^m$ are called* **Killing vector fields** *if*

$$L_{Z_j} g = 0$$

for all $j = 1, \ldots, p$. These vector fields span a p-dimensional Lie algebra.

Definition 8.7 *The vector fields Z_j on $\mathbf{R}^m$ are called* **conformal vector fields** *if*

$$L_{Z_j} g = \rho g$$

for all $j = 1, \ldots, p$ with $\rho \in C^\infty(\mathbf{R}^m)$. These vector fields span a p-dimensional Lie algebra under the commutator.

Example: Let us consider the vector space $\mathbf{R}^3$ and the vector field

$$Z = \sum_{i=1}^3 \xi_i(\mathbf{x}) \frac{\partial}{\partial x_i}.$$

We also make the assumption that $g_{jk} = \delta_{jk}$ where δ_{jk} is the Kronecker symbol. We now calculate the Lie derivative. Then we give an interpretation of $L_Z g = 0$. Since

$$
\begin{aligned}
L_Z g &= L_Z(dx_1 \otimes dx_1 + dx_2 \otimes dx_2 + dx_3 \otimes dx_3) \\
&= L_Z(dx_1 \otimes dx_1) + L_Z(dx_2 \otimes dx_2) + L_Z(dx_3 \otimes dx_3),
\end{aligned}
$$

$$L_Z(dx_j \otimes dx_k) = (L_Z dx_j) \otimes dx_k + dx_j \otimes (L_Z dx_k)$$

and

$$L_Z dx_j = d(Z \lrcorner dx_j) = d\xi_j = \sum_{k=1}^{3} \frac{\partial \xi_j}{\partial x_k} dx_k,$$

we find

$$L_Z g = \frac{1}{2} \sum_{j,k=1}^{3} \left(\frac{\partial \xi_j}{\partial x_k} + \frac{\partial \xi_k}{\partial x_j} \right) dx_j \otimes dx_k.$$

Since $dx_j \otimes dx_k$ $(j, k = 1, 2, 3)$ are basic elements, the right-hand side of the above equation can also be written as a symmetric 3×3 matrix, namely

$$\frac{1}{2} \begin{pmatrix} 2\partial \xi_1/\partial x_1 & \partial \xi_1/\partial x_2 + \partial \xi_2/\partial x_1 & \partial \xi_1/\partial x_3 + \partial \xi_3/\partial x_1 \\ \partial \xi_1/\partial x_2 + \partial \xi_2/\partial x_1 & 2\partial \xi_2/\partial x_2 & \partial \xi_2/\partial x_3 + \partial \xi_3/\partial x_2 \\ \partial \xi_1/\partial x_3 + \partial \xi_3/\partial x_1 & \partial \xi_2/\partial x_3 + \partial \xi_3/\partial x_2 & 2\partial \xi_3/\partial x_3 \end{pmatrix}.$$

The physical meaning is as follows: Consider a deformable body B in $\mathbf{R}^3$, $B \subset \mathbf{R}^3$. A displacement with or without deformation of the body B is a diffeomorphism of $\mathbf{R}^3$ defined in a neighbourhood of B. All such diffeomorphisms form a local group generated by the so-called displacement vector field Z. Consequently, the strain tensor field can be considered as the Lie derivative of the metric tensor field g in $\mathbf{R}^3$ with respect to the vector field Z. The metric tensor field gives rise to the distance between two points, namely

$$(ds)^2 = \sum_{j,k=1}^{3} g_{jk} dx_j dx_k.$$

The strain tensor field measures the variation of the distance between two points under a displacement generated by Z. Then the equation $L_Z g = 0$ tells us that two points of the body B do not change during the displacement.

To summarise: The strain tensor is the Lie derivative of the metric tensor field with respect to the deformation (or exactly the displacement vector field).   ♣

Example: Here we consider the **Gödel metric tensor field** which is given by

$$g = a^2 [dx \otimes dx - \frac{1}{2} e^{2x} dy \otimes dy + dz \otimes dz - c^2 dt \otimes dt - ce^x dy \otimes dt - ce^x dt \otimes dy]$$

where a and c are constants. One can show that the vector fields

$$Z_1 = \frac{\partial}{\partial y}, \qquad Z_2 = \frac{\partial}{\partial z}, \qquad Z_3 = \frac{\partial}{\partial t}$$

$$Z_4 = \frac{\partial}{\partial x} - y \frac{\partial}{\partial y}$$

$$Z_5 = y \frac{\partial}{\partial x} + (e^{-2x} - \frac{1}{2} y^2) \frac{\partial}{\partial y} - \frac{2}{c} e^{-x} \frac{\partial}{\partial t}$$

are Killing vector fields of the Gödel metric tensor. Since the functions g_{jk} do not depend on y, z and t it is obvious that Z_1, Z_2 and Z_3 are Killing vector fields. Applying the product rule and

$$L_{\partial/\partial x}e^{2x}dt \;=\; 2e^{2x}dt \qquad L_{\partial/\partial x}e^{2x}dy = 2e^{2x}dy$$
$$L_{y\partial/\partial y}dy \;=\; dy \qquad L_{y\partial/\partial y}e^{x}dy = e^{x}dy$$

we find that

$$L_{Z_4}g = 0.$$

For Z_5 we apply the product rule and obtain

$$L_{Z_5}g = 0.$$

The Gödel metric tensor field contains closed timelike lines; that is, an observer can influence his own past. ♣

Example: We consider the Minkowski metric

$$g = dx_1 \otimes dx_1 + dx_2 \otimes dx_2 + dx_3 \otimes dx_3 - dx_4 \otimes dx_4.$$

The vector fields with the corresponding functions ρ in $L_Z g = \rho g$ are listed below.

$$T = \frac{\partial}{\partial x_4} \qquad\qquad\qquad \rho_T = 0$$

$$P_i = \frac{\partial}{\partial x_i} \qquad\qquad\qquad \rho_{P_i} = 0 \quad i = 1, 2, 3$$

$$R_{ij} = x_i\frac{\partial}{\partial x_j} - x_j\frac{\partial}{\partial x_i} \qquad\qquad \rho_{R_{ij}} = 0 \quad i \neq j,\ i,\ j = 1, 2, 3$$

$$L_i = x_i\frac{\partial}{\partial x_4} + x_4\frac{\partial}{\partial x_i} \qquad\qquad \rho_{L_i} = 0 \quad i = 1, 2, 3$$

$$S = x_1\frac{\partial}{\partial x_1} + x_2\frac{\partial}{\partial x_2} + x_3\frac{\partial}{\partial x_3} + x_4\frac{\partial}{\partial x_4} \qquad \rho_S = 2$$

$$I_4 = 2x_4\left(x_1\frac{\partial}{\partial x_1} + x_2\frac{\partial}{\partial x_2} + x_3\frac{\partial}{\partial x_3}\right)$$

$$+(x_1^2 + x_2^2 + x_3^2 + x_4^2)\frac{\partial}{\partial x_4} \qquad\qquad \rho_{I_4} = 4x_4$$

$$I_i = -2x_i\left(x_1\frac{\partial}{\partial x_1} + x_2\frac{\partial}{\partial x_2} + x_3\frac{\partial}{\partial x_3} + x_4\frac{\partial}{\partial x_4}\right)$$

$$+(x_1^2 + x_2^2 + x_3^2 - x_4^2)\frac{\partial}{\partial x_i} \qquad\qquad \rho_{I_i} = -4x_i, \quad i = 1, 2, 3.$$

Note that the vector fields T, P_i, R_{ij} and L_i are the Killing vector fields for the Minkowski metric. The physical interpretation of the given vector fields (Lie group generators) is the following: T generates time translation; P_i space translation; R_{ij} space rotation; L_i space-time rotations, and S the uniform dilatations ($\mathbf{x} \mapsto \varepsilon \mathbf{x}$, $\varepsilon > 0$). I_4 is the conjugation of T by the inversion in the unit hyperboloid $Q : (\mathbf{x}) \mapsto (\mathbf{x})/(x_1^2+x_2^2+x_3^2-x_4^2)$, and the I_i are the conjugations of the P_i by Q.   ♣

Finally we give some examples which illustrate the application to physics.

Example: In physics we study problems in space-time so that the vector field under investigation takes the form

$$Z = \sum_{i=1}^{3} \xi_i(\mathbf{x}, t) \frac{\partial}{\partial x_i} + \frac{\partial}{\partial t}.$$

The above vector field is usually called the Eulerian (velocity) field. We can apply the vector field, via the Lie derivative, to 0-forms, 1-forms, 2-forms, 3-forms and 4-forms. We require that $L_Z \omega = 0$ and identify both the vector field Z and the form ω with physical quantities. The invariance requirement will then result in physical laws. As useful abbreviations we introduce

$$\Omega = dx_1 \wedge dx_2 \wedge dx_3 \wedge dt$$

and

$$\omega = dx_1 \wedge dx_2 \wedge dx_3,$$

where Ω is the volume element in space-time and ω the spatial volume element. Consider first the vector field $f(x, t)Z$ and the 4-form Ω. It follows that

$$L_{fZ}\Omega = \left(\frac{\partial f}{\partial t} + \xi_1 \frac{\partial f}{\partial x_1} + \xi_2 \frac{\partial f}{\partial x_2} + \xi_3 \frac{\partial f}{\partial x_3} + f \operatorname{div} \boldsymbol{\xi} \right) \Omega$$

where $\boldsymbol{\xi} = (\xi_1, \xi_2, \xi_3)$. Demanding that $L_{fZ}\Omega = 0$ we obtain, after a little algebraic manipulation, the relation

$$\frac{\partial f}{\partial t} + \sum_{i=1}^{3} \frac{\partial (f \xi_i)}{\partial x_i} = 0,$$

since Ω is a basis element in $\mathbf{R}^4$. If we identify the quantity f with the mass density and Z with the velocity vector field, then the above equation describes nothing more than the well-known continuity equation in local (differential) form. Obviously we get the same result when we consider the field Z and the form $f\Omega$.   ♣

Example: The energy equation can readily be obtained by slightly modifying the example given above. Now we start with the vector field $f(\mathbf{x}, t)g(\mathbf{x}, t)Z_1 + Z_2$, where

$$Z_2 = a_1(\mathbf{x}, t) \frac{\partial}{\partial x_1} + a_2(\mathbf{x}, t) \frac{\partial}{\partial x_2} + a_3(\mathbf{x}, t) \frac{\partial}{\partial x_3}$$

and where $\mathbf{x} = (x_1, x_2, x_3)$. We require that

$$L_{(f_g Z_1 + Z_2)}\Omega = (fp)\Omega$$

so that we consider conformal invariance. By straightforward calculation we find that

$$\frac{\partial(fg)}{\partial t} + \sum_{i=1}^{3}\frac{\partial(fg\xi_i)}{\partial x_i} + \sum_{i=1}^{3}\frac{\partial a_i}{\partial x_i} = fp.$$

In the above equation we make the following identification: f is identified with the mass density, g with the total energy per unit mass, Z_2 with the heat flux density, and fp the internal heat generation rate per unit volume. Requiring that

$$L_{Z_1}(f\omega) = 0$$

we also obtain the continuity equation. However, an additional condition appears. Since

$$
\begin{aligned}
L_{Z_1} f\omega &= \left(\frac{\partial f}{\partial t} + \sum_{i=1}^{3}\frac{\partial(f\xi_i)}{\partial x_i}\right)\omega \\
&\quad + f\left(\frac{\partial\xi_1}{\partial t}dx_2 \wedge dx_3 \wedge dt + \frac{\partial\xi_2}{\partial t}dx_3 \wedge dx_1 \wedge dt + \frac{\partial\xi_3}{\partial t}dx_1 \wedge dx_2 \wedge dt\right)
\end{aligned}
$$

we obtain, besides the continuity equation, the conditions

$$\frac{\partial\xi_1}{\partial t} = \frac{\partial\xi_2}{\partial t} = \frac{\partial\xi_3}{\partial t} = 0,$$

which state that the flow is steady.   ♣

8.4   Computer Algebra Applications

We give an application of the Lie derivative and Killing vector field. In the REDUCE program we calculate

$$L_V g = \sum_{j=1}^{4} \sum_{k=1}^{4} \left[\sum_{l=1}^{4} \left(V_l \frac{\partial g_{jk}}{\partial x_l} + g_{lk} \frac{\partial V_l}{\partial x_j} + g_{jl} \frac{\partial V_l}{\partial x_k} \right) \right] dx_j \otimes dx_k.$$

for the Gödel metric tensor field.

```
%Lie derivative and Killing vector field;
%Program name: Lieder.our;

operator V, g, x;

depend V(j), x(1), x(2), x(3), x(4);
depend g(i,j), x(1), x(2), x(3), x(4);
operator LD;

%This is the Goedel metric;
g(1,1):=1; g(1,2):=0; g(1,3):=0; g(1,4):=0;
g(2,1):=0; g(2,2):=-exp(2*x(1))/2;
g(2,3):=0; g(2,4):=-exp(x(1));
g(3,1):=0; g(3,2):= 0; g(3,3):=1; g(3,4):=0;
g(4,1):=0; g(4,2):=-exp(x(1)); g(4,3):=0; g(4,4):=-1;

%This is a Killing vector field of the Goedel metric;
V(1) := x(2); V(2) := exp(- 2*x(1)) - x(2)*x(2)/2;
V(3) := 0; V(4) := -2*exp(- x(1));

for j := 1:4 do
for k := 1:4 do
begin
LD(j,k) := for l := 1:4 sum V(l)*df(g(j,k),x(1)) +
g(l,k)*df(V(l),x(j)) +
g(j,l)*df(V(l),x(k));
write LD(j,k);
end;
```

Chapter 9

Invariance of Differential Equations

9.1 Prolongation of Vector Fields

9.1.1 Introductory Example

Consider an ordinary first order differential equation of the form

$$F\left(x, u(x), \frac{du}{dx}\right) = 0. \tag{1}$$

The set

$$F(x, u, u_1) = 0$$

is assumed to be a smooth submanifold of $\mathbf{R}^3$. A solution of the differential equation is a curve $(x, u(x), u_1(x))$ on the surface $F = 0$ such that

$$u_1(x) = \frac{du}{dx}.$$

Let us assume that the differential equation is invariant under the group of rotations ($\varepsilon \in \mathbf{R}$)

$$x' = x \cos \varepsilon - u \sin \varepsilon, \qquad u' = x \sin \varepsilon + u \cos \varepsilon \tag{2}$$

where

$$\varphi(x, u, \varepsilon) = x \cos \varepsilon - u \sin \varepsilon, \qquad \phi(x, u, \varepsilon) = x \sin \varepsilon + u \cos \varepsilon \tag{3}$$

are acting in the (x, u)-plane. From (2) we obtain the transformation

$$x'(x) = x \cos \varepsilon - u(x) \sin \varepsilon, \qquad u'(x'(x)) = x \sin \varepsilon + u(x) \cos \varepsilon. \tag{4}$$

It follows that

$$\frac{du'}{dx} = \frac{du'}{dx'} \frac{dx'}{dx} = \sin \varepsilon + \frac{du}{dx} \cos \varepsilon.$$

Consequently

$$\frac{du'}{dx'} = \frac{\sin \varepsilon + (du/dx) \cos \varepsilon}{\cos \varepsilon - (du/dx) \sin \varepsilon}.$$

Thus the action induced on the (x, u, u_1)-space is given by

$$\begin{aligned}
\varphi(x, u, \varepsilon) &= x \cos \varepsilon - u \sin \varepsilon \\
\phi(x, u, \varepsilon) &= x \sin \varepsilon + u \cos \varepsilon \\
\phi^{(1)}(x, u, u_1, \varepsilon) &= \frac{\sin \varepsilon + u_1 \cos \varepsilon}{\cos \varepsilon - u_1 \sin \varepsilon}
\end{aligned}$$

which form a one-parameter transformation group on $\mathbf{R}^3$

$$x' = \varphi(x, u, \varepsilon) \tag{5a}$$

$$u' = \phi(x, u, \varepsilon) \tag{5b}$$

$$u_1' = \phi^{(1)}(x, u, u_1, \varepsilon). \tag{5c}$$

Thus the group action (φ, ϕ) on $\mathbf{R}^2$ has induced a group action $(\varphi, \phi, \phi^{(1)})$ on $\mathbf{R}^3$ called the **first prolongation**.

The symmetry vector field (infinitesimal generator) of the transformation group (2) is given by

$$Z = -u \frac{\partial}{\partial x} + x \frac{\partial}{\partial u}.$$

Let us now find the vector field for the prolonged transformation group (5). We consider

$$\bar{Z}^{(1)} = \xi(x, u) \frac{\partial}{\partial x} + \eta(x, u) \frac{\partial}{\partial u} + \gamma(x, u, u_1) \frac{\partial}{\partial u_1}$$

with the infinitesimals

$$\xi(x, u) = \left. \frac{d\varphi}{d\varepsilon} \right|_{\varepsilon = 0} = -u$$

$$\eta(x, u) = \left. \frac{d\phi}{d\varepsilon} \right|_{\varepsilon = 0} = x$$

$$\gamma(x, u, u_1) = \left. \frac{d\phi^{(1)}}{d\varepsilon} \right|_{\varepsilon = 0} = 1 + u_1^2.$$

The vector field for the prolonged transformation is then given by

$$\bar{Z}^{(1)} = Z + (1 + u_1^2) \frac{\partial}{\partial u_1}.$$

♣

Definition 9.1 *The vector field $\bar{Z}^{(1)}$ is called the* **first prolonged vector field** *of Z.*

It should be clear that the condition of invariance on the function F, given in (1), w.r.t. the transformation (2) is obtained by calculating the Lie derivative of F w.r.t. the above prolonged vector field $Z^{(1)}$, i.e.,

$$L_{Z^{(1)}} F = 0$$

whereby $F \hat{=} 0$. Thus, the condition of invariance of (1) under the transformation (2), is given by

$$\left(-u \frac{\partial F}{\partial x} + x \frac{\partial F}{\partial u} + (1 + u_1^2) \frac{\partial F}{\partial u_1} \right) \bigg|_{F=0} = 0.$$

As an example, let

$$F \equiv u_1 - f(x, u)$$

so that the condition of invariance becomes

$$u \frac{\partial f}{\partial x} - x \frac{\partial f}{\partial u} + (1 + f^2) = 0$$

with general solution

$$f(x, u) = \tan \left(\arcsin \left(\frac{u}{\sqrt{u^2 + x^2}} \right) + g(u^2 + x^2) \right).$$

The conclusion is that the differential equation

$$\frac{du}{dx} = \tan \left(\arcsin \left(\frac{u}{\sqrt{u^2 + x^2}} \right) + g(u^2 + x^2) \right)$$

is invariant under the transformation (2), whereby g is an arbitrary function of one variable.

Remark: *Several authors use the notation $pr^{(r)} Z$ for the r-th prolongation $\bar{Z}^{(r)}$.*

This prolongation can, in general, be found for any Lie transformation group. Moreover, the above argument can be extended to include all derivatives of all orders of the initial function $u(x)$. What we are, however, interested in is to find the prolonged vector fields for r-th order partial differential equations with n dependent variables and m independent variables. To study this problem it is useful to consider the jet bundle formalism for differential equations.

9.1.2　Local Jet Bundle Formalism

Here the jet bundle formulation (Rogers and Shadwick [97]) for differential equations is discussed.

In general, the utility of jet bundles in this connection is due to the fact that they have local coordinates which adopt the roles of the field variables and their derivatives. In this context the derivatives may then be regarded as independent quantities.

Let us begin the discussion with some definitions.

Let M and N be smooth manifolds of dimension m and n respectively. In most cases we consider $M = \mathbf{R}^m$, $N = \mathbf{R}^n$ with local coordinates $\mathbf{x} = (x_1, \ldots, x_m)$ and $\mathbf{u} = (u_1, \ldots, u_n)$, respectively. Let $C^\infty(M, N)$ denote the set of smooth maps from M to N. Then, if

$$f \in C^\infty(M, N)$$

is defined at $\mathbf{x} \in M$, it follows that f is determined by n coordinate functions

$$u_j := f_j(\mathbf{x}). \tag{6}$$

Definition 9.2 *If f, $g \in C^\infty(M, N)$ are defined at $\mathbf{x} \in M$, then f and g are said to be r-equivalent at $\mathbf{x}$ if and only if f and g have the same Taylor expansion up to order r at $\mathbf{x}$, i.e.,*

$$f_j(\mathbf{x}) \;=\; g_j(\mathbf{x})$$

$$\frac{\partial^k f_j(\mathbf{x})}{\partial x_{i_1} \partial x_{i_2} \ldots \partial x_{i_k}} \;=\; \frac{\partial^k g_j(\mathbf{x})}{\partial x_{i_1} \partial x_{i_2} \ldots \partial x_{i_k}}$$

where

$$i_1, \ldots, i_r \in \{1, \ldots, m\}, \qquad j \in \{1, \ldots, n\}$$

$$1 \leq i_1 \leq i_2 \leq \cdots \leq i_r \leq m$$

$$k \in \{1, \ldots, r\}.$$

Definition 9.3 *The r-equivalent class of f at $\mathbf{x}$ is called the r-**jet** of f at $\mathbf{x}$ and is denoted by $j_{\mathbf{x}}^r f$.*

Definition 9.4 *The collection of all r-jets $j_{\mathbf{x}}^r f$ as $\mathbf{x}$ ranges over M and f ranges over $C^\infty(M, N)$ is called the r-**jet** bundle of maps from M to N and is denoted by $J^r(M, N)$. Thus*

$$J^r(M, N) := \bigcup_{\mathbf{x} \in M,\ f \in C^\infty(M, N)} j_{\mathbf{x}}^r f.$$

From the definitions above it follows that if $\mathbf{p}$ is a point in $J^r(M,N)$, then $\mathbf{p} = j_{\mathbf{x}}^r f$ for some $\mathbf{x} \in M$ and $f \in C^\infty(M,N)$. Consequently $\mathbf{p}$ is determined uniquely by the numbers

$$x_i, \ u_j, \ u_{j,i}, \ u_{j,i_1 i_2}, \ \ldots, \ u_{j,i_1 \ldots i_r} \tag{7}$$

where x_i, u_j ($i = 1, \ldots, m$ and $j = 1, \ldots, n$) are the coordinates of $\mathbf{x}$ and f, respectively and

$$u_{j,i_1 \ldots i_r} = \frac{\partial^r f_j(\mathbf{x})}{\partial x_{i_1} \partial x_{i_2} \ldots \partial x_{i_r}}.$$

The conditions on $i_1, \ldots, i_r$ are given in definition (9.2). Note that the latter quantities are, by definition, independent of the choice of f in the equivalence class $j_{\mathbf{x}}^r f$. Any collection of numbers $x_i, u_j, u_{j,i_1 \ldots i_k}$, $k = 1, \ldots, r$ with the $u_{j,i_1 \ldots i_k}$ symmetric in the indices $i_1, \ldots, i_k$ determines a point of $J^r(M,N)$.

Remark: *The r-jet bundle $J^r(M,N)$ with $M = \mathbf{R}^m$, $N = \mathbf{R}^n$ may, accordingly, be identified by*

$$\mathbf{R}^{\dim J^r}$$

with coordinates (7).

Example: Let $M = \mathbf{R}^2$, $N = \mathbf{R}$, and choose coordinates x_1, x_2 on M and u_1 on N. Then, if $f, g \in C^\infty(M,N)$ are given by

$$f(x_1, x_2) = x_1^2 + x_2$$
$$g(x_1, x_2) = x_2,$$

f and g belong to the same 1-jet bundle $J^1(M,N)$ at $(0,1)$ but not to the same 2-jet. The 2-jet bundle $J^2(M,N)$ may be identified with $\mathbf{R}^8$ with the coordinates

$$x_1, \ x_2, \ u_1, \ u_{1,1}, \ u_{1,2}, \ u_{1,11}, \ u_{1,12}, \ u_{1,22}. \qquad \clubsuit$$

If two maps are r-equivalent at $\mathbf{x} \in M$ they are also j-equivalent for all $j \leq r$. We can now state the following

Definition 9.5 *The map*

$$\pi_{r-l}^r : j_{\mathbf{x}}^r f \to j_{\mathbf{x}}^{r-l} f$$

is defined as the **canonical projection map** *from $J^r(M,N)$ to $J^{r-l}(M,N)$, where $l = 0, 1, \ldots, r-1$. Here $J^0(M,N)$ is identified with $M \times N$ and*

$$\pi_0^r : j_{\mathbf{x}}^r f \to (\mathbf{x}, f(\mathbf{x})).$$

Definition 9.6 *The two maps*

$$\alpha : J^r(M,N) \to M$$
$$\beta : J^r(M,N) \to N$$

defined by

$$\alpha : j_{\mathbf{x}}^r f \to \mathbf{x}$$
$$\beta : j_{\mathbf{x}}^r f \to f$$

are called the **source** *and* **target maps***, respectively.*

Definition 9.7 *A map*

$$h : M \to J^k(M, N)$$

which satisfies

$$\alpha \circ h = id_M$$

where id_M is the identity map on M, is called the **cross section** *of the source map α.*

Definition 9.8 *A cross section of a map $f \in C^\infty(M, N)$ which is denoted by $j^r f$ and defined by*

$$j^r f : \mathbf{x} \to j_{\mathbf{x}}^r f$$

*is known as the r-***jet extension** *of the map f. For $r = 0$, $j^r f$ is the graph of f.*

Example: Let $M = \mathbf{R}^2$ and $N = \mathbf{R}$ with the same coordinates as in the previous example. The 1-jet extension $j^1 f$ of $f \in C^\infty(\mathbf{R}^2, \mathbf{R})$ is defined by

$$j^1 f : (x_1, x_2) \to \left(x_1, x_2, f(x_1, x_2), \frac{\partial f(x_1, x_2)}{\partial x_1}, \frac{\partial f(x_1, x_2)}{\partial x_2} \right)$$

and the 2-jet extension $j^2 f$ is

$$j^2 f : (x_1, x_2)$$
$$\to \left(x_1, x_2, f(x_1, x_2), \frac{\partial f(x_1, x_2)}{\partial x_1}, \frac{\partial f(x_1, x_2)}{\partial x_2}, \frac{\partial^2 f(x_1, x_2)}{\partial x_1^2}, \frac{\partial^2 f(x_1, x_2)}{\partial x_1 \partial x_2}, \frac{\partial^2 f(x_1, x_2)}{\partial x_2^2} \right).$$

♣

Those cross sections of the source map α that are r-jet extensions of maps from M to N may be conveniently characterized in terms of differential forms on $J^r(M, N)$.

Let us first consider the case $r = 1$ and define differential 1-forms θ_j, $j = 1, \ldots, n$, on $J^1(M, N)$ by

$$\theta_j := du_j - \sum_{i=1}^m u_{j,i} dx_i.$$

Since u_j and $u_{j,i}$ are coordinates on $J^1(M, N)$, the 1-forms θ_j are not identically zero. There are, however, certain privileged submanifolds on $J^1(M, N)$ on which the θ_j vanish. Consider

$$g : M \to J^1(M, N)$$

as a cross section of α which is given in coordinates by

$$g : \mathbf{x} \to (x_i, g_j(\mathbf{x}), g_{j,i}(\mathbf{x})).$$

We now consider the 1-forms $g^*\theta_j$ on M where g^* is the pull-back map. We have

$$g^*\theta_j = dg_j - \sum_{i=1}^{m} g_{j,i} dx_i = \sum_{i=1}^{m} \left(\frac{\partial g_j}{\partial x_i} - g_{j,i} \right) dx_i,$$

so that

$$g^*\theta_j = 0,$$

if and only if

$$g_{j,i} = \frac{\partial g_j}{\partial x_i}$$

and this, in turn, holds if and only if

$$g = j^1(\beta \circ g)$$

where $\beta \circ g : M \to N$ and is given in coordinates by

$$\beta \circ g : \mathbf{x} \to g_j(\mathbf{x}).$$

Definition 9.9 *The one-forms θ_j are called* **contact forms** *and the set of all finite linear combinations of the differential one-form θ_j over the ring of C^∞ functions on $J^1(M, N)$ is called the* **first order contact module** *and is denoted by $\Omega^1(M, N)$. Thus*

$$\Omega^1(M, N) := \left\{ \sum_{j=1}^{n} f_j \theta_j \ : \ f_j \in C^\infty(J^1(M, N), \mathbf{R}) \right\}.$$

The set of differential one-forms $\{\theta_j\}$ is called the standard basis for $\Omega(M, N)$.

We can now generalize the above discussion to the case $k > 1$ by introducing the differential one-forms $\theta_j, \theta_{j,i_1 \ldots i_l}, l = 1, \ldots, r - 1$, according to

$$\begin{aligned}
\theta_j &:= du_j - \sum_{k=1}^{m} u_{j,k} dx_k \\
\theta_{j,i_1} &:= du_{j,i_1} - \sum_{k=1}^{m} u_{j,i_1 k} dx_k \\
&\vdots \\
\theta_{j,i_1 \ldots i_{r-1}} &:= du_{j,i_1 \ldots i_{r-1}} - \sum_{k=1}^{m} u_{j,i_1 \ldots i_{r-1} k} dx_k.
\end{aligned}$$

It follows that if $g : M \to J^r(M, N)$ is a cross section of the source map α, then

$$g^*\theta_{j,i_1 \ldots i_l} = 0$$

with $l = 1, \ldots, r - 1$, if and only if

$$g = j^r (\beta \circ g).$$

Thus, if g is given in coordinates by

$$g : \mathbf{x} \rightarrow (x_i, g_j(x_i), g_{j,i_1 \ldots i_p}(x_i))$$

with $p = 1, \ldots, r$, it follows that

$$g^* \theta_j = 0,$$

if and only if

$$g_{j,i} = \frac{\partial g_j}{\partial x_i},$$

and then iteratively that

$$g^* \theta_{j,i_1 \ldots i_l} = 0$$

if and only if

$$g_{j,i_1 \ldots i_{l+1}} = \frac{\partial^{l+1} g_j}{\partial x_{i_1} \ldots \partial x_{i_{l+1}}}$$

with $l = 1, \ldots, r - 1$.

Definition 9.10 *The collection*

$$\Omega^r(M, N) := \left\{ \sum_{j=1}^n f_j \theta_j + \sum_{j=1}^n \sum_{i_1, \ldots, i_p = 1}^m \left(f_{j,i_1 \ldots i_p} \theta_{j,i_1 \ldots i_p} \right) \right\}$$

is called the r-th order contact module, where

$$i_1 \leq i_2 \leq \ldots \leq i_p.$$

One may equally characterize k-jet extensions in terms of differential m-forms on $J^r(M, N)$ or in terms of differential p-forms with $1 \leq p \leq m$.

9.1.3 Prolongation of Vector Fields

We are now interested in the rth prolongation of a vector field Z on $M \times N$.

Definition 9.11 *We introduce the **total derivative operator** $D_i^{(r)}$ defined on $J^r(M, N)$ for $r \geq 1$ by ($i_1 \leq \ldots \leq i_{r-1}$)*

$$D_i^{(r)} := \frac{\partial}{\partial x_i} + \sum_{j=1}^{n} u_{j,i} \frac{\partial}{\partial u_j} + \cdots + \sum_{j=1}^{n} \sum_{i_1,\ldots,i_{r-1}=1}^{m} u_{j,ii_1\ldots i_{r-1}} \frac{\partial}{\partial u_{j,i_1\ldots i_{r-1}}}$$

The operator $D_i^{(r)}$ denotes the r-th prolonged total derivative operator.

$D_i^{(r)}$ acts as the total derivative d/dx_i on sections.

Example: If $(x_i, u_j, u_{j,i}, \ldots, u_{j,i_1\ldots i_r})$ is a section then

$$D_i f(x_i, u_j, u_{j,i}, \ldots, u_{j,i_1\ldots i_r}) = \frac{d}{dx_i} f(x_i, u_j, u_{j,i}, \ldots, u_{j,i_1\ldots i_r}). \qquad \clubsuit$$

Now, consider a general one parameter transformation group

$$x_i' = \varphi_i(\mathbf{x}, \mathbf{u}, \varepsilon) \qquad (8a)$$

$$u_j' = \phi_j(\mathbf{x}, \mathbf{u}, \varepsilon) \qquad (8b)$$

where $i = 1, \ldots, m$ and $j = 1, \ldots, n$. Let Z, defined on $M \times N$, be a vector field (infinitesimal generator) of this one parameter group given by

$$Z = \sum_{i=1}^{m} \xi_i(\mathbf{x}, \mathbf{u}) \frac{\partial}{\partial x_i} + \sum_{j=1}^{n} \eta_j(\mathbf{x}, \mathbf{u}) \frac{\partial}{\partial u_j} \qquad (9)$$

where

$$\left. \frac{d\varphi_i}{d\varepsilon} \right|_{\varepsilon=0} = \xi_i$$

$$\left. \frac{d\phi_j}{d\varepsilon} \right|_{\varepsilon=0} = \eta_j.$$

Definition 9.12 *Associated with the vector field Z is a vector field Z_V on $J^1(M, N)$ which is vertical over M and given by*

$$Z_V = \sum_{j=1}^{n} \left(\eta_j - \sum_{i=1}^{m} \xi_i u_{j,i} \right) \frac{\partial}{\partial u_j}. \qquad (10)$$

Z_V *is known as the **vertical vector field** of Z.*

Since

$$Z \lrcorner \theta_j = \left(\sum_{i=1}^{m} \xi_i \frac{\partial}{\partial x_i} + \sum_{j=1}^{n} \eta_j \frac{\partial}{\partial u_j} \right) \lrcorner \left(du_j - \sum_{i=1}^{m} u_{j,i} dx_i \right) = \eta_j - \sum_{i=1}^{m} \xi_i u_{j,i},$$

it follows that

$$Z_V \lrcorner \theta_j = (Z \lrcorner \theta_j) \frac{\partial}{\partial u_j} \lrcorner \theta_j = Z \lrcorner \theta_j.$$

Definition 9.13 *The r-th prolongation of the vector field Z, denoted by $\bar{Z}^{(r)}$, is defined by*

$$\bar{Z}^{(r)} := Z + \sum_{j=1}^{n} \sum_{i=1}^{m} \gamma_{j,i} \frac{\partial}{\partial u_{j,i}} + \cdots + \sum_{j=1}^{n} \sum_{i_1 \ldots i_r = 1} \gamma_{j,i_1 \ldots i_r} \frac{\partial}{\partial u_{j,i_1 \ldots i_r}}, \tag{11}$$

where

$$\gamma_{j,i} = D_i(\eta_j) - \sum_{k=1}^{m} u_{j,k} D_i(\xi_k)$$

$$\vdots \qquad \vdots \qquad \vdots$$

$$\gamma_{j,i_1 \ldots i_r} = D_{i_r}(\gamma_{j,i_1 \ldots i_{r-1}}) - \sum_{k=1}^{m} u_{j,i_1 \ldots i_{r-1}k} D_{i_r}(\xi_k).$$

Definition 9.14 *The prolongation of the vertical vector field Z_V, denoted by $\bar{Z}_V$, up to infinite order, is defined by*

$$\bar{Z}_V := \sum_{j=1}^{n} U_j \frac{\partial}{\partial u_j} + \sum_{j=1}^{n} \sum_{i=1}^{m} D_i(U_j) \frac{\partial}{\partial u_{j,i}} + \cdots + \sum_{j=1}^{n} \sum_{i_1, \ldots, i_r} D_{i_1 \ldots i_r}(U_j) \frac{\partial}{\partial u_{j,i_1, \ldots, i_r}} + \cdots \tag{12}$$

where $i_1 \leq \ldots \leq i_r$ and

$$U_j := \eta_j - \sum_{i=1}^{m} \xi_i u_{j,i}. \tag{13}$$

Now an action $\varphi = (\varphi_1, \ldots, \varphi_m, \phi_1, \ldots, \phi_n)$ on $M \times N$ induces, by means of the rth prolongation, an action $\varphi^{(r)} = (\varphi_1^{(r)}, \ldots, \varphi_m^{(r)}, \phi_1^{(r)}, \ldots, \phi_n^{(r)})$ on $J^r(M, N)$. In particular, if G is a Lie group of transformations acting on $M \times N$, then transformations sufficiently close to the identity will induce a transformation on $J^k(M, N)$.

Remark: *Since the group composition is preserved by the prolongation the prolonged vector field $\bar{Z}^{(k)}$ will form a representation of the Lie algebra L, i.e.,*

$$[\bar{Z}_1^{(k)}, \bar{Z}_2^{(k)}] = \overline{[Z_1, Z_2]}^{(k)}.$$

9.1.4   Partial Differential Equations on Jet Bundles

Consider a general system of partial differential equations of order r

$$F_\nu \left(x_i, u_j, \frac{\partial u_j}{\partial x_i}, \ldots, \frac{\partial^r u_j}{\partial x_{i_1} \partial x_{i_2} \ldots \partial x_{i_r}} \right) = 0 \tag{14}$$

where $\nu = 1, \ldots, q$ and $i_k = 1, \ldots, m$, $j = 1, \ldots, n$. Such a system determines a submanifold of the r-jet bundle $J^r(M, N)$, namely, the submanifold R^r given by the constraint equations

$$F_\nu(x_i, u_j, u_{j,i}, \ldots, u_{j,i_1 \ldots i_r}) = 0. \tag{15}$$

This submanifold, being a geometric object, provides a convenient setting whereby transformation properties of the system (14) may be analyzed.

Definition 9.15 *A solution of R^r is defined as a map $s \in C^\infty(M, N)$ where*

$$s : U \subset M \to N$$

such that

$$(j^k s)^* \subset R^r.$$

Thus, if s is a solution of R^r and if s is given in coordinates by $u_j = s_j(x_i)$, then it follows that

$$F_\nu \left(x_i, s_j, \frac{\partial s_j}{\partial x_i}, \ldots, \frac{\partial^r s_j}{\partial x_{i_1} \ldots \partial x_{i_r}} \right) = 0$$

and s is a solution of system (14).

Example: Let $M = \mathbf{R}^2$, $N = \mathbf{R}$, and choose coordinates $x_i, u, u_i, u_{i_1 i_2}, u_{i_1 i_2 i_3}$, on $J^3(M, N)$. The Korteweg de Vries equation

$$\frac{\partial u}{\partial x_2} + u \frac{\partial u}{\partial x_1} + \sigma \frac{\partial^3 u}{\partial x_1^3} = 0$$

(σ is a constant) determines the submanifold R^3 of $J^3(M, N)$ given by

$$u_2 + u u_1 + \sigma u_{111} = 0.$$

A solution of R^3 is a map $s \in C^\infty(M, N)$ such that, if $\mathbf{x} \in M$, then $j^3 s(\mathbf{x}) \in R^3$, that is,

$$\frac{\partial s(x_1, x_2)}{\partial x_2} + s(x_1, x_2) \frac{\partial s(x_1, x_2)}{\partial x_1} + \sigma \frac{\partial^3 s(x_1, x_2)}{\partial x_1^3} = 0. \qquad \clubsuit$$

We have thus replaced the system of partial differential equations by a submanifold and an appropriate jet bundle. This will be the setting of the investigation of symmetries and conservation laws for partial differential equations.

We now discuss the important notion of the prolongation of a system of differential equations. Thus we want to characterize a system of differential equations and its derivatives as a new system in terms of jets. Here we make use of the total differential operator $D^{(r)}$.

Definition 9.16 *The lth prolongation of R^r is the submanifold*

$$R^{r+l} \subset J^{r+l}(M, N)$$

defined by the constraint equations

$$
\begin{aligned}
F_\nu &= 0 \\
D_i^{(r+l)} F_\nu &= 0 \\
&\vdots \\
D_{i_1 \dots i_l}^{(r+l)} F_\nu &= 0
\end{aligned}
$$

($\nu = 1, \dots, q$) where repeated total derivatives

$$D_{i_1}^{(r)} \circ D_{i_2}^{(r)} \circ \cdots \circ D_{i_p}^{(r)}$$

are written as $D_{i_1 \dots i_p}^{(r)}$. It follows from the above that if s is a solution of R^{r+l}, then

$$F_\nu\left(x_i, s_j, \dots, \frac{\partial^r s_j}{\partial x_{i_1} \dots \partial x_{i_r}}\right) = 0$$

$$\frac{\partial}{\partial x_i} F_\nu\left(x_i, s_j, \dots, \frac{\partial^r s_j}{\partial x_{i_1} \dots \partial x_{i_r}}\right) = 0$$

$$\vdots$$

$$\frac{\partial^r}{\partial x_{i_1} \dots \partial x_{i_r}} F_\nu\left(x_i, s_j, \dots, \frac{\partial^r s_j}{\partial x_{i_1} \dots \partial x_{i_r}}\right) = 0.$$

Example: Consider again the Korteweg-de Vries equation. The total derivatives on $J^4(M, N)$ are given by

$$D_i^{(4)} = \frac{\partial}{\partial x_i} + u_i \frac{\partial}{\partial u} + \sum_{i_1=1}^{2} u_{i i_1} \frac{\partial}{\partial u_{i_1}} + \sum_{i_1,i_2=1}^{2} u_{i i_1 i_2} \frac{\partial}{\partial u_{i_1 i_2}} + \sum_{i_1,i_2,i_3=1}^{2} u_{i i_1 i_2 i_3} \frac{\partial}{\partial u_{i_1 i_2 i_3}}$$

where $i_1 \leq i_2$ and $i_1 \leq \dots \leq i_3$. Here $u_j = u_1 \equiv u$, where $i = 1, 2$. Consequently the first prolongation of the Korteweg de Vries equation is given by

$$D_i^{(4)} F = 0$$

where $F \equiv u_2 + u u_1 + \sigma u_{111} = 0$ and $i = 1, 2$. It follows that

$$
\begin{aligned}
D_1^{(4)}(u_2 + u u_1 + \sigma u_{111}) &= u_{12} + u_1^2 + u u_{11} + \sigma u_{1111} = 0 \\
D_2^{(4)}(u_2 + u u_1 + \sigma u_{111}) &= u_{22} + u_2 u_1 + u u_{12} + \sigma u_{1112} = 0.
\end{aligned}
$$

♣

9.2 Invariance of Differential Equations

In chapter 8 we introduced the concept of invariance of a geometrical object σ by $L_Z \sigma = 0$, where $L_Z(\cdot)$ is the Lie derivative and Z the vector field for some one parameter transformation group. This means that σ does not change as it propagates down the trajectories of Z. We are now concerned with the invariance of a differential equation, i.e. the submanifold R^r of the r-jet bundle $J^r(M, N)$ given by the constraint equations

$$F_\nu(x_i, u_j, u_{j,i}, \ldots, u_{j,i_1}, \ldots, u_{j,i_1 \ldots i_r}) = 0. \tag{16}$$

From the definition of invariance and the Lie derivative it should be clear that system (16) is invariant under the vector field Z_V if the following condition holds:

$$L_{\tilde{Z}_V} F_\nu \hat{=} 0 \tag{17}$$

i.e., the Lie derivative of (16) w.r.t. the vertical vector field Z_V, where $\hat{=}$ stands for the restriction to solutions of (16) and $\nu = 1, \ldots, q$.

Definition 9.17 *Vector fields which admit condition (17) for system (16) are called* **Lie point symmetry vector fields** *of system (16).*

As was shown in previous chapters, these infinitesimal vector fields generate local Lie groups. We can thus give the following

Definition 9.18 *A local transformation acting on an open subset of the independent and dependent variables of the system of differential equations, which is generated by the Lie point symmetry vector fields, form a local Lie group of transformations called the* **Lie symmetry group**.

A symmetry group G of a system of differential equations has the property that whenever $u_j = s_j(\mathbf{x})$ is a solution of that system and whenever $g \cdot s$ is defined for $g \in G$, then $u_j = g \cdot s_j(\mathbf{x})$ is also a solution of that system of differential equations. One can thus use the symmetry group of a system of differential equations to construct solutions from solutions. In some cases one can construct multi-parameter solutions from trivial solutions. This is the subject of section 8.4. In the current section we need to understand how to find Lie symmetry vector fields by the use of condition (17).

After applying the Lie derivative of the prolonged vertical vector field on the constraint equations $F_\nu = 0$ of the submanifold R^r, we obtain the expression

$$\sum_{j=1}^n U_j \frac{\partial F_\nu}{\partial u_j} + \sum_{j=1}^n \sum_{i=1}^m D_i(U_j) \frac{\partial F_\nu}{\partial u_{j,i}} + \ldots \sum_{j=1}^n \sum_{i_1, \ldots, i_r = 1}^m D_{i_1 \ldots i_r}(U_j) \frac{\partial F_\nu}{\partial u_{j,i_1 \ldots i_r}} \hat{=} 0 \tag{18}$$

where

$$
\begin{aligned}
D_i(U_j) &= D_i(\eta_j - \sum_{k=1}^m \xi_k u_{j,k}) \\
&= \frac{\partial \eta_j}{\partial x_i} + \sum_{k=1}^n u_{k,i} \frac{\partial \eta_j}{\partial u_k} - \sum_{k=1}^m \frac{\partial \xi_k}{\partial x_i} u_{j,k} - \sum_{l=1}^n \sum_{k=1}^m \frac{\partial \xi_k}{\partial u_l} u_{l,i} u_{j,k} - \sum_{i_1=1}^m \sum_{j=1}^n u_{j,i_1} \xi_{i_1}.
\end{aligned}
$$

After insertion of the constraint equations of the l-th prolonged submanifold R^{r+l} the so-called **determining equations** for $\xi_1, \ldots, \xi_m$ and $\eta_1, \ldots, \eta_n$ are found by equating the coefficients of the coordinates $u_{j,i}$, u_{j,ii_1}, $\ldots$, on the prolonged jet bundle, to zero so that equation (17) is satisfied. Solution of the determining equations provide the Lie point symmetry vector fields of (16).

Definition 9.19 *The general solution of the determining equations for a system (16) provide the Lie point symmetry vector fields which span a Lie algebra, called the* **fundamental Lie algebra** *of system (16).*

We now give several examples to illustrate the procedure for calculating Lie point symmetry vector fields.

Example: We calculate the Lie point symmetry vector fields for the **inviscid Burgers'** **equation**

$$\frac{\partial u}{\partial x_2} = u \frac{\partial u}{\partial x_1}.$$

Here $M = \mathbf{R}^2$ and $N = \mathbf{R}$. This partial differential equation determines a submanifold R^1 on the 1-jet bundle $J^1(M, N)$ given by the constraint equation

$$F(x_1, x_2, u, u_1, u_2) \equiv u_2 - u u_1 = 0.$$

The contact form on $J^1(M, N)$ is

$$\theta = du - u_1 dx_1 - u_2 dx_2.$$

The operators of total differentiation are

$$D_1 := \frac{\partial}{\partial x_1} + u_1 \frac{\partial}{\partial u} + u_{11} \frac{\partial}{\partial u_1} + u_{12} \frac{\partial}{\partial u_2}$$

$$D_2 := \frac{\partial}{\partial x_2} + u_2 \frac{\partial}{\partial u} + u_{21} \frac{\partial}{\partial u_1} + u_{22} \frac{\partial}{\partial u_2}.$$

Consider the following Lie point symmetry vector field

$$Z = \xi_1(x_1, x_2, u) \frac{\partial}{\partial x_1} + \xi_2(x_1, x_2, u) \frac{\partial}{\partial x_2} + \eta(x_1, x_2, u) \frac{\partial}{\partial u}. \tag{19}$$

The corresponding vertical vector field is given by

$$Z_V = (\eta - \xi_1 u_1 - \xi_2 u_2) \frac{\partial}{\partial u}$$

with the prolonged vertical vector field

$$\bar{Z}_V = U \frac{\partial}{\partial u} + D_1(U) \frac{\partial}{\partial u_1} + D_2(U) \frac{\partial}{\partial u_2}.$$

Here $U := \eta - \xi_1 u_1 - \xi_2 u_2$ and

$$D_1(U) = \frac{\partial \eta}{\partial x_1} - \frac{\partial \xi_1}{\partial x_1} u_1 - \frac{\partial \xi_2}{\partial x_1} u_2 + \frac{\partial \eta}{\partial u} u_1 - \frac{\partial \xi_1}{\partial u} u_1^2 - \frac{\partial \xi_2}{\partial u} u_1 u_2 - \xi_1 u_{11} - \xi_2 u_{12}$$

$$D_2(U) = \frac{\partial \eta}{\partial x_2} - \frac{\partial \xi_1}{\partial x_2} u_1 - \frac{\partial \xi_2}{\partial x_2} u_2 + \frac{\partial \eta}{\partial u} u_2 - \frac{\partial \xi_1}{\partial u} u_1 u_2 - \frac{\partial \xi_2}{\partial u} u_2^2 - \xi_1 u_{12} - \xi_2 u_{22}.$$

We now have to determine the infinitesimals ξ_1, ξ_2 and η. The prolongations of $F \equiv u_2 - u u_1 = 0$ are

$$D_1(u_2 - u u_1) = u_{12} - u_1^2 - u u_{11} = 0$$

$$D_2(u_2 - u u_1) = u_{22} - u_2 u_1 - u u_{12} = 0$$

so that we can consider the following relations

$$u_2 = u u_1, \qquad u_{12} = u_1^2 + u u_{11}, \qquad u_{22} = 2 u u_1^2 + u^2 u_{11}. \tag{20}$$

The invariant condition $L_{\tilde{z}_V} F \hat{=} 0$ gives

$$\left(U \frac{\partial}{\partial u} + D_1(U) \frac{\partial}{\partial u_1} + D_2(U) \frac{\partial}{\partial u_2} \right) (u_2 - u u_1) = U(-u_1) + D_1(U)(-u) + D_2(U)$$

so that

$$-u_1 \eta + \xi_2 u_1 u_2 - \frac{\partial \eta}{\partial x_1} u + \frac{\partial \xi_1}{\partial x_1} u u_1 + \frac{\partial \xi_2}{\partial x_1} u u_2 - \frac{\partial \eta}{\partial u} u u_1$$

$$+ \frac{\partial \xi_1}{\partial u} u u_1^2 + \frac{\partial \xi_2}{\partial u} u u_1 u_2 + \xi_1 u_1^2 + u u_{12} \xi_2 + \frac{\partial \eta}{\partial x_2} - \frac{\partial \xi_1}{\partial x_2} u_1 - \frac{\partial \xi_2}{\partial x_2} u_2$$

$$+ \frac{\partial \eta}{\partial u} u_2 - \frac{\partial \xi_1}{\partial u} u_1 u_2 - \frac{\partial \xi_2}{\partial u} u_2^2 - \xi_1 u_{12} - \xi_2 u_{22} + \xi_1 u u_{11} \hat{=} 0.$$

Inserting (20) into the above expression we obtain

$$-u_1 \eta - \frac{\partial \eta}{\partial x_1} u + u u_1 \frac{\partial \xi_1}{\partial x_1} + \frac{\partial \xi_2}{\partial x_1} u^2 u_1 + \frac{\partial \eta}{\partial x_2} - \frac{\partial \xi_1}{\partial x_2} u_1 - \frac{\partial \xi_2}{\partial x_2} u u_1 = 0.$$

To find the solutions for ξ_1, ξ_2 and η so that the above relation holds we recall that on the $J^1(M, N)$ jet bundle, the derivatives of u are independent coordinates, so that the coefficients of the coordinates u_1^0, u_1 must be equal to zero. Consequently

$$\frac{\partial \eta}{\partial x_2} - u \frac{\partial \eta}{\partial x_1} = 0 \tag{21}$$

$$\frac{\partial \xi_1}{\partial x_1} u - \frac{\partial \xi_1}{\partial x_2} + \frac{\partial \xi_2}{\partial x_1} u^2 - \frac{\partial \xi_2}{\partial x_2} u - \eta = 0. \tag{22}$$

The system of linear differential equations (21) and (22) provide the conditions on ξ_1, ξ_2 and η for finding symmetry vector fields of the form (19). We now consider special solutions for this linear system of partial differential equations.

Case 1: $\xi_1 = 1$, $\xi_2 = 0$, $\eta = 0$, so that

$$Z_1 = \frac{\partial}{\partial x_1}.$$

Case 2: $\xi_1 = 0$, $\xi_2 = 1$, $\eta = 0$, so that

$$Z_2 = \frac{\partial}{\partial x_2}.$$

Case 3: $\xi_1 = x_1$, $\xi_2 = x_2$, $\eta = 0$, so that

$$Z_3 = x_1 \frac{\partial}{\partial x_1} + x_2 \frac{\partial}{\partial x_2}.$$

Case 4: $\xi_1 = x_1$, $\xi_2 = 0$, $\eta = u$, so that

$$Z_4 = x_1 \frac{\partial}{\partial x_1} + u \frac{\partial}{\partial u}.$$

Case 5: $\xi_1 = x_2$, $\xi_2 = 0$, $\eta = -1$, so that

$$Z_5 = x_2 \frac{\partial}{\partial x_1} - \frac{\partial}{\partial u}.$$

Case 6: $\xi_1 = 0$, $\xi_2 = x_1$, $\eta = u^2$, so that

$$Z_6 = x_1 \frac{\partial}{\partial x_2} + u^2 \frac{\partial}{\partial u}.$$

Case 7: $\xi_1 = x_1 x_2$, $\xi_2 = x_2^2$, $\eta = -(x_1 + x_2 u)$, so that

$$Z_7 = x_1 x_2 \frac{\partial}{\partial x_1} + x_2^2 \frac{\partial}{\partial x_2} - (x_1 + x_2 u) \frac{\partial}{\partial u}.$$

Case 8: $\xi_1 = x_1^2$, $\xi_2 = x_1 x_2$, $\eta = (x_1 + x_2 u)u$, so that

$$Z_8 = x_1^2 \frac{\partial}{\partial x_1} + x_1 x_2 \frac{\partial}{\partial x_2} + (x_1 + x_2 u)u \frac{\partial}{\partial u}.$$

Case 9: $\xi_1 = u$, $\xi_2 = 0$, $\eta = 0$, where u satisfies the inviscid Burgers equation, so that

$$Z_9 = u \frac{\partial}{\partial x_1}.$$

Case 10: $\xi_1 = 0$, $\xi_2 = u$, $\eta = 0$, where u satisfies the inviscid Burgers equation, so that

$$Z_{10} = u \frac{\partial}{\partial x_2}.$$

The symmetry vector fields $\{Z_1, \ldots, Z_{10}\}$ form a ten-dimensional Lie algebra.   ♣

Let us now consider a set of Lie generators which were introduced in chapter 7 as Killing vector fields for the Minkowski metric. It was shown that the linear wave equation follows from $d(*du) = 0$, where $*$ is the Hodge star operator and d is the exterior derivative. These vector fields are known as the conformal vector fields. In this section we show how to extend the Killing vector fields for the Minkowski metric in order to obtain the conformal Lie symmetry vector fields for particular wave equations of the form

$$\left(\sum_{j=1}^{m-1} \frac{\partial^2}{\partial x_j^2} - \frac{\partial^2}{\partial x_m^2} \right) u + f(x_1, \ldots, x_m, u) = 0,$$

In the calculation of the Lie symmetry vector fields of wave equations, the following set of first order partial differential equations play a fundamental role:

$$\frac{\partial \xi_m}{\partial x_m} - \frac{\partial \xi_j}{\partial x_j} = 0$$

$$\frac{\partial \xi_j}{\partial x_k} + \frac{\partial \xi_k}{\partial x_j} = 0$$

$$\frac{\partial \xi_j}{\partial x_m} - \frac{\partial \xi_m}{\partial x_j} = 0$$

where $j \neq k = 1, \ldots, m - 1$. This system is known as the **Killing equations**. It is easy to show that the general solution of this set of over determined equations is given by

$$\xi_m u = 2x_\mu \left(b_m x_m - \sum_{j=1}^{m-1} b_j x_j \right) - b_\mu \left(x_m^2 - \sum_{j=1}^{m-1} x_j^2 \right) + \sum_{\nu=1}^{m} c_{\mu\nu} x_m u + d_\mu$$

where b_μ, $c_{\mu\nu}$ and d_μ are arbitrary real constants with the conditions

$$c_{mj} = c_{jm}, \qquad c_{ij} = -c_{ji}, \qquad c_{jj} = c_{mm},$$
$$i \neq j = 1, \ldots, m - 1, \qquad \mu = 1, \ldots, m.$$

This solution is known as the **Killing solution**. Let us now give some examples of conformal invariant wave equations.

Example: Let us calculate the Lie symmetries of the simplest wave equation in Minkowski space, namely

$$\left(\sum_{j=1}^{m-1} \frac{\partial^2}{\partial x_j^2} - \frac{\partial^2}{\partial x_m^2} \right) u = 0.$$

After applying the invariance condition, the following set of determining equations is obtained

$$\frac{\partial \xi_m}{\partial x_m} - \frac{\partial \xi_j}{\partial x_j} = 0, \qquad 2\frac{\partial^2 \eta}{\partial u^2} x_m - \left(\sum_{j=1}^{m-1} \frac{\partial^2}{\partial x_j^2} - \frac{\partial^2}{\partial x_m^2} \right) \xi_m = 0$$

$$\frac{\partial \xi_j}{\partial x_k} + \frac{\partial \xi_k}{\partial x_j} = 0, \qquad 2\frac{\partial^2 \eta}{\partial u^2} x_j + \left(\sum_{j=1}^{m-1} \frac{\partial^2}{\partial x_j^2} - \frac{\partial^2}{\partial x_m^2}\right)\xi_j = 0$$

$$\frac{\partial \xi_j}{\partial x_m} - \frac{\partial \xi_m}{\partial x_j} = 0, \qquad \frac{\partial^2 \eta}{\partial u^2} = 0$$

$$\left(\sum_{j=1}^{m-1} \frac{\partial^2}{\partial x_j^2} - \frac{\partial^2}{\partial x_m^2}\right)\eta = 0$$

where $j \neq k = 1, \ldots, m-1$. Note the coupling with the Killing equations. By using the Killing solutions, we are left with the problem of solving the determining equations for η. Two cases have to be considered.

Case 1: For $m \geq 3$ the determining equations are solved by the Killing solution and

$$\eta = \left((2-m)\left(b_m x_m - \sum_{j=1}^{m-1} b_j x_j\right) + \lambda\right)u + \alpha(x_1, \ldots, x_m),$$

where

$$\left(\sum_{j=1}^{m-1} \frac{\partial^2}{\partial x_j^2} - \frac{\partial^2}{\partial x_m^2}\right)\alpha = 0.$$

Here λ is an additional real constant. The basis elements of the Lie algebra are now obtained by choosing one of the arbitrary constants as one, and the rest as zero. This is done for all constants b_μ, $c_{\mu\nu}$, d_μ, and λ, where μ, $\nu = 1, \ldots, m$. The Lie symmetry vector fields are then given by

$$T = \frac{\partial}{\partial x_m}$$

$$P_i = \frac{\partial}{\partial x_j} \qquad\qquad\qquad\qquad i = 1, \ldots, m-1$$

$$R_{ij} = x_i \frac{\partial}{\partial x_j} - x_j \frac{\partial}{\partial x_i} \qquad\qquad\qquad i \neq j = 1, \ldots, m-1$$

$$L_i = x_i \frac{\partial}{\partial x_m} + x_m \frac{\partial}{\partial x_i} \qquad\qquad\qquad i = 1, \ldots, m-1$$

$$S_x = \sum_{\mu=1}^{m} x_\mu \frac{\partial}{\partial x_\mu}$$

$$I_k = -2x_k\left(\sum_{\mu=1}^{m} x_\mu \frac{\partial}{\partial x_\mu}\right) - \left(x_m^2 - \sum_{j=1}^{m-1} x_j^2\right)\frac{\partial}{\partial x_k} - (2-m)x_k u\frac{\partial}{\partial u} \quad k = 1, \ldots, m-1$$

$$I_m = 2x_m\left(\sum_{j=1}^{m-1} x_j \frac{\partial}{\partial x_j}\right) + \left(x_m^2 + \sum_{j=1}^{m-1} x_j^2\right)\frac{\partial}{\partial x_m} + (2-m)x_m u\frac{\partial}{\partial u}$$

These vector fields form the basis of an $(m+1)(m+2)/2$-dimensional Lie algebra (with $m \geq 3$), which is know as the **conformal Lie algebra**. The additional vector fields

$$S_u = u\frac{\partial}{\partial u}, \qquad Z_\infty = \alpha(x_1, \ldots, x_m)\frac{\partial}{\partial u},$$

where α is an arbitrary solution of

$$\left(\sum_{j=1}^{m-1} \frac{\partial^2}{\partial x_j^2} - \frac{\partial^2}{\partial x_m^2}\right)\alpha = 0$$

reflect the linearity of the equation.

Case 2: For $m = 2$, the wave equation admits the following infinite Lie symmetry vector field

$$Z_\infty = (f_1(x_1 + x_2) + f_2(x_1 - x_2))\frac{\partial}{\partial x_1} + (f_1(x_1 + x_2) - f_2(x_1 - x_2))\frac{\partial}{\partial x_2}$$

$$+ (f_3(x_1 + x_2) + f_4(x_1 - x_2) + \lambda u)\frac{\partial}{\partial u},$$

where f_j $(j = 1, \ldots, 4)$ are arbitrary smooth functions.

♣

Example: Let us consider the two dimensional wave equation

$$\left(\frac{\partial^2}{\partial x_1^2} - \frac{\partial^2}{\partial x_2^2}\right)u + g(x_1, x_2, u) = 0 \qquad (23)$$

where g is an arbitrary smooth function. It follows that this equation admits the Lie symmetry vector fields

$$L_1 = x_1\frac{\partial}{\partial x_2} + x_2\frac{\partial}{\partial x_1}$$

$$\tilde{I}_1 = (x_1^2 + x_2^2)\frac{\partial}{\partial x_1} + 2x_1 x_2\frac{\partial}{\partial x_2} + x_1\left(\frac{\lambda_1}{x_1^2 - x_2^2} + \lambda_2\right)\frac{\partial}{\partial u}$$

$$\tilde{I}_2 = 2x_1 x_2\frac{\partial}{\partial x_1} + (x_1^2 + x_2^2)\frac{\partial}{\partial x_2} + +x_2\left(\frac{\lambda_1}{x_1^2 - x_2^2} + \lambda_2\right)\frac{\partial}{\partial u},$$

if and only if,

$$g(x_1, x_2, u) = \frac{1}{(x_1^2 - x_2^2)^2}\,\tilde{g}\left(\frac{\lambda_1}{2(x_1^2 - x_2^2)} - \frac{\lambda_2}{2}\ln(x_1^2 - x_2^2) + u\right).$$

Here $\tilde{g}$ is an arbitrary smooth function with one argument, as given above. The vector fields form a three dimensional Lie algebra, with commutation relations

$$[L_1, \tilde{I}_1] = \tilde{I}_2, \qquad [L_1, \tilde{I}_2] = \tilde{I}_1, \qquad [\tilde{I}_1, \tilde{I}_2] = 0.$$

♣

The formalism described above can also be applied to ordinary differential equations.

Example: For the linear ordinary differential equation $d^3u/dx^3 = 0$ we obtain seven Lie symmetry vector fields

$$\left\{ \frac{\partial}{\partial x}, \; \frac{\partial}{\partial u}, \; x\frac{\partial}{\partial u}, \; x^2\frac{\partial}{\partial u}, \; x\frac{\partial}{\partial x}, \; u\frac{\partial}{\partial u}, \; x^2\frac{\partial}{\partial x} + 2xu\frac{\partial}{\partial u} \right\}. \qquad \clubsuit$$

Example: Consider the **Lorenz equations**

$$\frac{du_1}{dx} = \sigma(u_2 - u_1), \qquad \frac{du_2}{dx} = -u_1u_3 + ru_1 - u_2, \qquad \frac{du_3}{dx} = u_1u_2 - bu_3$$

where σ, r and b are three real positive parameters. By eliminating u_2 and u_3 from the above system, we obtain the third-order equation (we put $u_1 = u$)

$$u\frac{d^3u}{dx^3} - \left(\frac{du}{dx} - (\sigma + b + 1)u \right)\frac{d^2u}{dx^2} - (\sigma + 1)\left(\frac{du}{dx} \right)^2$$

$$+ \left(u^3 + b(\sigma + 1)u \right)\frac{du}{dx} + \sigma\left(u^4 + b(1 - r)u^2 \right) = 0.$$

The following form of the vertical symmetry vector field is considered

$$Z = \left(\gamma_0(x, u, \dot{u}) + \gamma_1(x, u, \dot{u})\,\ddot{u} \right)\frac{\partial}{\partial u}$$

where $\dot{u} \to du/dx$ and $\ddot{u} \to d^2u/dx^2$. The third order ordinary differential equation given above then admits the following symmetry vector fields:

Case I: $\sigma = 1/2$, $b = 1$, $r = 0$.

$$Z_1 = e^{5x/2}\left(\left(u\dot{u} + \frac{1}{2}u^2 \right)\ddot{u} - \dot{u}^3 + \left(\frac{1}{4}u^4 + \frac{1}{2}u^2 \right)\dot{u} + \frac{1}{8}(u^5 + u^3) \right)\frac{\partial}{\partial u}$$

$$Z_2 = e^{3x/2}\left(\left(\frac{\dot{u}}{u} + \frac{1}{2} \right)\ddot{u} + \frac{2}{3}\frac{\dot{u}^2}{u} + \left(\frac{1}{2}u^2 + \frac{5}{4} \right)\dot{u} + \frac{1}{4}(u^3 + u) \right)\frac{\partial}{\partial u}$$

$$Z_3 = e^{x/2}\left(\dot{u} + \frac{1}{2}u \right)\frac{\partial}{\partial u}$$

$$Z_4 = \dot{u}\frac{\partial}{\partial u}$$

where Z_3 and Z_4 can equivalently be written as

$$Z_3 = e^{x/2}\left(-\frac{\partial}{\partial x} + \frac{1}{2}u\frac{\partial}{\partial u}\right)$$

$$Z_4 = \frac{\partial}{\partial x}.$$

Case II: $b = 0$, $\sigma = 1/3$, r is arbitrary.

$$Z_1 = e^{4x/3}\dot{u}\left(u\ddot{u} - \dot{u}^2 + \frac{1}{4}u^4\right)\frac{\partial}{\partial u}$$

$$Z_2 = \dot{u}\frac{\partial}{\partial u}.$$

Case III: $b = 6\sigma - 2$, $r = 2\sigma - 1$, σ is arbitrary:

$$Z_1 = e^{4\sigma x}\dot{u}\left(u\ddot{u} - \dot{u}^2 + (3\sigma - 1)u\dot{u} + \frac{1}{4}u^4 + (3\sigma - 1)(1 - \sigma)u^2\right)\frac{\partial}{\partial u}$$

$$Z_2 = \dot{u}\frac{\partial}{\partial u}.$$

Case IV: $b = 4$, $\sigma = 1$, r is arbitrary.

$$Z_1 = e^{4x}\dot{u}\left(\left(u + \frac{4(1 - r)}{u}\right)\ddot{u} - \dot{u}^2 + 2\left(u + \frac{4(1 - r)}{u}\right)\dot{u} + \frac{1}{4}u^4\right.$$

$$\left. + \; 2(1 - r)u^2 + 4(1 - r)^2\right)\frac{\partial}{\partial u}$$

$$Z_2 = \dot{u}\frac{\partial}{\partial u}.$$

Case V: $b = 2\sigma$, σ and r are arbitrary

$$Z_1 = e^{2\sigma x}\dot{u}\left(\frac{\ddot{u}}{u} + (\sigma + 1)\frac{\dot{u}}{u} + \frac{1}{2}u^2 + \sigma(1 - r)\right)\frac{\partial}{\partial u}$$

$$Z_2 = \dot{u}\frac{\partial}{\partial u}.$$

Remark: *The vertical vector field of*

$$\frac{\partial}{\partial x}$$

is given by

$$-\dot{u}\frac{\partial}{\partial u}.$$

♣

9.3   Similarity Solutions

In this section we consider a reduction of the system

$$F_\nu \left(x_i, u_j, \frac{\partial u_j}{\partial x_i}, \ldots, \frac{\partial^r u_j}{\partial x_{i_1} \partial x_{i_2} \ldots \partial x_{i_r}} \right) = 0 \tag{24}$$

$(\nu = 1, \ldots, q, \; i = 1, \ldots, m, \; j = 1, \ldots, n)$ by its Lie point symmetry vector fields. This is made possible by constructing a so-called **symmetry ansatz**, i.e., a solution ansatz in terms of new dependent and new independent variables (also known as **similarity variables**) which should be such that the number of independent variables in (24) are reduced. Inserting this symmetry ansatz into (24) results in a **reduced system** of differential equations. Solutions of the reduced system (if any can be found) will then result in special solutions for system (24). These special solutions for system (24) are known as **similarity solutions** or L-invariant solutions, where L is the fundamental Lie algebra (or a Lie subalgebra) that is spanned by the symmetry vector fields.

In our first approach to find a reduced system and thus similarity solutions we consider L' to be a p-dimensional Lie subalgebra of the fundamental Lie algebra L for system (24) generated by the vector fields

$$Z_l = \sum_{i=1}^m \xi_{li} \frac{\partial}{\partial x_i} + \sum_{j=1}^n \eta_{lj} \frac{\partial}{\partial u_j}$$

where $l = 1, \ldots, p$. These vector fields form a basis of the Lie algebra L'.

For the invariance we consider the following

Definition 9.20 *A function $I(\mathbf{x}, \mathbf{u})$ is an **invariant of the Lie algebra L'** if*

$$L_Z I = 0 \tag{25}$$

for all $Z \in L'$ where I is not identically a constant.

Assume that the rank of the matrix

$$K = \begin{pmatrix} \xi_{11} & \cdots & \xi_{1m} & \eta_{1j} & \cdots & \eta_{1n} \\ \vdots & & & & & \vdots \\ \xi_{r1} & \cdots & \xi_{rm} & \eta_{rj} & \cdots & \eta_{rn} \end{pmatrix}$$

is equal to p_0 and that $p_0 < n$. There exist (locally) exactly $(m + n - p_0)$ functional independent invariants

$$I_1(\mathbf{x}, \mathbf{u}), \ldots, I_{m+n-p_0}(\mathbf{x}, \mathbf{u})$$

of the Lie algebra L' and any invariant of the Lie algebra L' is a function of them. We assume that the invariants can be selected in such a manner that the determinant of the matrix

$$\left(\frac{\partial I_l}{\partial u_j} \right)$$

(Jacobian) $l, j = 1, \ldots, n$, is different from zero. This is necessary for the existence of a similarity solution under the Lie algebra L'. We introduce new independent variables $\varsigma = (\varsigma_1, \ldots, \varsigma_{m-p_0})$ and dependent variables $\mathbf{v} = (v_1(\varsigma), \ldots, v_n(\varsigma))$ by

$$\varsigma_i = I_{n+i}(\mathbf{x}, \mathbf{u}) \tag{26}$$

$$v_j(\varsigma) = I_j(\mathbf{x}, \mathbf{u}) \tag{27}$$

where $i = 1, \ldots, m - p_0$ and $j = 1, \ldots, n$. Now we can express u_j in terms of $\mathbf{x}$, ς and $\mathbf{v}$, such that

$$u_j = \psi_j(\mathbf{x}, \varsigma, \mathbf{v}).$$

The derivatives $\partial u_j / \partial x_i$, $\partial^2 u_j / \partial x_{i_1} \partial x_{i_2}, \ldots$ are computed as derivatives of the composite function ψ_j. For example

$$\frac{\partial u_j}{\partial x_i} = \frac{\partial \psi_j}{\partial x_i} + \sum_{k=1}^{m-p_0} \sum_{l=1}^{n} \left(\frac{\partial \psi_j}{\partial \varsigma_k} + \frac{\partial \psi_j}{\partial v_l} \frac{\partial v_l}{\partial \varsigma_k} \right) \frac{\partial I_{n+k}}{\partial x_i}.$$

If we now insert the obtained expressions for u_j and its derivatives into the initial system, it reduces into a system for the dependent variable $\mathbf{v}(\varsigma)$

$$F'_\nu \left(\varsigma_i, v_j, \frac{\partial v_j}{\partial \varsigma_i}, \ldots, \frac{\partial^r v_j}{\partial \varsigma_{i_1} \ldots \partial \varsigma_{i_r}} \right) = 0,$$

where the variables $\mathbf{x}$ do not appear. We call this system the **reduced system**. The number of independent variables in the reduced system is smaller than in the initial system and in this sense system (24) is simplified. In particular, if a partial differential equation with two independent variables is considered which is invariant under a one-parameter transformation group, then this partial differential equation can be reduced to an ordinary differential equation.

If the system (24) has a solution $\mathbf{v}(\varsigma)$, then by making use of (25) and (26) one can find $\mathbf{u} = \psi(\mathbf{x})$, i.e., a similarity solution of the initial system. We must, however, note that the similarity solution thus obtained is only locally defined via the transformation group which leaves the differential equation invariant. We say that the solution is L'-invariant.

Example: Let us find the reduced system of the boundary layer equation given by

$$\frac{\partial u_1}{\partial x_3} + u_1 \frac{\partial u_1}{\partial x_1} + u_2 \frac{\partial u_1}{\partial x_2} = \frac{\partial^2 u_1}{\partial x_2^2} \tag{28}$$

$$\frac{\partial u_1}{\partial x_1} + \frac{\partial u_2}{\partial x_2} = 0. \tag{29}$$

This system admits a Lie algebra that is spanned by the vector fields

$$Z_1 = 2x_3 \frac{\partial}{\partial x_3} + x_2 \frac{\partial}{\partial x_2} - 2u_1 \frac{\partial}{\partial u_1} - u_2 \frac{\partial}{\partial u_2}$$

$$Z_2 = x_3 \frac{\partial}{\partial x_1} + \frac{\partial}{\partial u_1}.$$

Here $N = \mathbf{R}^2$ and $M = \mathbf{R}^3$. The matrix K is given by

$$K = \begin{pmatrix} 2x_3 & 0 & x_2 & -2u_1 & -u_2 \\ 0 & x_3 & 0 & 1 & 0 \end{pmatrix}.$$

We have rank$(K) = 2$ so that rank$(K) < 3$. We thus have $m + n - p_0 = 3$ functional independent invariants, where

$$L_{Z_1} I_1 = 0, \qquad L_{Z_2} I_1 = 0$$

$$L_{Z_1} I_2 = 0, \qquad L_{Z_2} I_2 = 0$$

$$L_{Z_1} I_3 = 0, \qquad L_{Z_2} I_3 = 0$$

so that

$$2x_3 \frac{\partial I_k}{\partial x_3} + x_2 \frac{\partial I_k}{\partial x_2} - 2u_1 \frac{\partial I_k}{\partial u_1} - u_2 \frac{\partial I_k}{\partial u_2} = 0 \tag{30a}$$

$$x_3 \frac{\partial I_k}{\partial x_1} + \frac{\partial I_k}{\partial u_1} = 0 \tag{30b}$$

with $k = 1, 2, 3$. System (30) has the following solutions:

$$\begin{aligned} I_1(x_1, x_2, x_3, u_1, u_2) &= x_3 u_1 - x_1 \\ I_2(x_1, x_2, x_3, u_1, u_2) &= x_2 u_2 \\ I_3(x_1, x_2, x_3, u_1, u_2) &= \frac{x_2^2}{x_3}. \end{aligned}$$

We now introduce the new variables from the given invariant functions

$$\varsigma_1 = \frac{x_2^2}{x_3}, \qquad v_1(\varsigma_1) = x_3 u_1 - x_1, \qquad v_2(\varsigma_1) = x_2 u_2$$

so that

$$u_1(x_1, x_2) = \frac{v_1(\varsigma_1) + x_1}{x_3}, \qquad u_2(x_1, x_2) = \frac{v_2(\varsigma_1)}{x_2} \tag{31}$$

and

$$\frac{\partial u_1}{\partial x_1} = \frac{1}{x_3}, \qquad \frac{\partial u_2}{\partial x_1} = 0 \tag{32a}$$

$$\frac{\partial u_1}{\partial x_2} = 2 \frac{x_2}{x_3^2} \frac{dv_1}{d\varsigma_1}, \qquad \frac{\partial u_2}{\partial x_2} = -\frac{v_2}{x_2^2} + \frac{2}{x_3} \frac{dv_2}{d\varsigma_1} \tag{32b}$$

$$\frac{\partial u_1}{\partial x_3} = -\frac{1}{x_3^2} \left(\varsigma_1 \frac{dv_1}{d\varsigma_1} + v_1 + x_1 \right), \qquad \frac{\partial^2 u_1}{\partial x_2^2} = \frac{2}{x_3^2} \left(2\varsigma_1 \frac{d^2 v_1}{d\varsigma_1^2} + \frac{dv_1}{d\varsigma_1} \right) \tag{32c}$$

where $i = 1, 2$. By inserting (31) and (32) into the boundary layer equation, the system of ordinary differential equations is obtained

$$4\varsigma_1\frac{d^2v_1}{d\varsigma_1^2} = (2v_2 - \varsigma_1 - 2)\frac{dv_1}{d\varsigma_1} \tag{33a}$$

$$2\frac{dv_2}{d\varsigma_1} = \frac{v_2}{\varsigma_1} - 1. \tag{33b}$$

System (33) can be integrated by quadratures to find a similarity solution $u_1(\mathbf{x}, v_1)$, $u_2(\mathbf{x}, v_2)$ for the boundary layer equation. ♣

We now describe the method of finding similarity solutions for a system of partial differential equations by considering the transformation group that leaves the system of equations invariant.

Let us consider the following partial differential equation,

$$F\left(x_1, x_2, u, \frac{\partial u}{\partial x_i}, \cdots, \frac{\partial^r u}{\partial x_{i_1}\partial x_{i_2}\ldots\partial x_{i_r}}\right) = 0 \tag{34}$$

where $i = 1, 2$ and $r = i_1 + i_2 + \cdots + i_r$. We assume that this equation admits the Lie point symmetry vector field

$$Z = \xi_1(\mathbf{x}, u)\frac{\partial}{\partial x_1} + \xi_2(\mathbf{x}, u)\frac{\partial}{\partial x_2} + \eta(\mathbf{x}, u)\frac{\partial}{\partial u}$$

where $\mathbf{x} = (x_1, x_2)$. This vector field can be a combination of different Lie point symmetry vector fields that span the fundamental Lie algebra L.

To find the associated one-parameter transformation group we have to solve the initial value problem (see chapter 4)

$$\frac{dx_1'}{d\varepsilon} = \xi_1'(\mathbf{x}', u'), \qquad \frac{dx_2'}{d\varepsilon} = \xi_2'(\mathbf{x}', u'), \qquad \frac{du'}{d\varepsilon} = \eta'(\mathbf{x}', u')$$

where ε is the group parameter and $\mathbf{x}' = \mathbf{x}$ and $u'(\mathbf{x}) = u(\mathbf{x})$ for $\varepsilon = 0$. Let the solution (one-parameter transformation group) be given by

$$x_1' = \varphi_1(\mathbf{x}, u, \varepsilon), \qquad x_2' = \varphi_2(\mathbf{x}, u, \varepsilon), \qquad u' = \phi(\mathbf{x}, u, \varepsilon).$$

We consider

$$x_i = \varsigma(x_1', x_2'), \qquad x_j = a$$

where $i \neq j = 1, 2$ and $a \in \mathbf{R}$, i.e. x_1 or x_2 can be chosen to be the similarity variable ς. u' will then result in

$$u' = \phi(\mathbf{x}, v(\varsigma)) \tag{35}$$

which is also known as the **similarity ansatz**, where

$$u(\mathbf{x})|_{x_i=\varsigma,\ x_j=a} \equiv v(\varsigma)$$

and ε was eliminated from x_1' and x_2'. By insertion of $u'(\mathbf{x}')$ into the equation

$$F\left(x_1', x_2', u', \frac{\partial u'}{\partial x_i'}, \ldots, \frac{\partial^r u'}{\partial x_{i_1}' \partial x_{i_2}' \ldots \partial x_{i_r}'}\right) = 0 \tag{36}$$

we find a reduced equation

$$F'\left(\varsigma, v(\varsigma), \frac{dv}{d\varsigma}, \ldots, \frac{d^r v}{d\varsigma^r}\right) = 0.$$

In solving this ordinary differential equation we find a similarity solution from (34).

Example: The nonlinear partial differential equation

$$\frac{\partial^2 u}{\partial x_1 \partial x_2} + \frac{\partial u}{\partial x_1} + u^2 = 0 \tag{37}$$

describes the relaxation to a Maxwell distribution. The symmetry vector fields are given by

$$Z_1 = \frac{\partial}{\partial x_1}, \qquad Z_2 = \frac{\partial}{\partial x_2}, \qquad Z_3 = -x_1\frac{\partial}{\partial x_1} + u\frac{\partial}{\partial u}, \qquad Z_4 = e^{x_2}\frac{\partial}{\partial x_2} - e^{x_2}u\frac{\partial}{\partial u}.$$

In order to find a similarity ansatz we consider the symmetry vector field

$$Z = c_1\frac{\partial}{\partial x_1} + c_2\frac{\partial}{\partial x_2} + c_3\left(-x_1\frac{\partial}{\partial x_1} + u\frac{\partial}{\partial u}\right)$$

where $c_1, c_2, c_3 \in \mathbf{R}$. The corresponding initial value problem is given by

$$\frac{dx_2'}{d\varepsilon} = c_2, \qquad \frac{dx_1'}{d\varepsilon} = c_1 - c_3 x_1', \qquad \frac{du'}{d\varepsilon} = c_3 u'.$$

The solution for this system provides the transformation group

$$x_1'(\mathbf{x}, u, \varepsilon) = \frac{c_1}{c_3} - \frac{c_1 - c_3 x_1}{c_3}e^{-c_3\varepsilon}$$
$$x_2'(\mathbf{x}, u, \varepsilon) = c_2\varepsilon + x_2$$
$$u'(\mathbf{x}, u, \varepsilon) = ue^{c_3\varepsilon}$$

where $c_3 \neq 0$. Now let $x_2 = \varsigma/c$ and $x_1 = 1$ with the constant $c \neq 0$. It follows that the similarity variable is

$$\varsigma = cx_2' + \frac{c_2 c}{c_3}\ln\frac{c_3 x_1' - c_1}{c_3 - c_1}$$

and the similarity ansatz

$$u'(x_1', x_2') = v(\varsigma)\frac{c_1 - c_3}{c_1 - c_3 x_1'}. \tag{38}$$

Inserting (38) into

$$\frac{\partial^2 u'}{\partial x_1' \partial x_2'} + \frac{\partial u'}{\partial x_1'} + u'^2 = 0$$

leads to the ordinary differential equation

$$c\frac{d^2 v}{d\varsigma^2} + (1-c)\frac{dv}{d\varsigma} - (1-v)v = 0.$$

This ordinary equation can be integrated by considering the change of dependent variables $dv/d\varsigma = V(v(\varsigma))$ whereby the similarity solution for the nonlinear partial differential equation (37) is found from (38). ♣

Example: The nonlinear partial differential equation

$$\frac{\partial^2 u}{\partial x_1 \partial x_2} + u^2 = 0 \tag{39}$$

is related to the partial differential equation in the previous example by the transformation

$$x_1 \rightarrow x_1$$
$$x_2 \rightarrow -e^{-x_2}$$
$$u \rightarrow e^{x_2} u.$$

The Lie point symmetry vector fields are given by

$$Z_1 = \frac{\partial}{\partial x_1}, \qquad Z_2 = \frac{\partial}{\partial x_2}, \qquad Z_3 = x_1\frac{\partial}{\partial x_1} - x_2\frac{\partial}{\partial x_2}, \qquad Z_4 = u\frac{\partial}{\partial u}.$$

We construct a similarity solution from this nonlinear partial differential equation by considering the symmetry vector field

$$Z = c_1\frac{\partial}{\partial x_1} + c_2\frac{\partial}{\partial x_2}$$

where $c_1, c_2 \in \mathbf{R}$. The associated initial value problem is

$$\frac{dx_1'}{d\varepsilon} = c_1, \qquad \frac{dx_2'}{d\varepsilon} = c_2, \qquad \frac{du'}{d\varepsilon} = 0$$

where $\mathbf{x}' = \mathbf{x}$ for $\varepsilon = 0$. The transformation group is thus given by

$$x_1' = c_1\varepsilon + x_1, \qquad x_2' = c_2\varepsilon + x_2, \qquad u' = u.$$

d Now let $x_1 = \varsigma$ and $x_2 = 0$. It follows that

$$\varsigma = x_1' - \frac{c_1}{c_2}x_2', \qquad u'(\mathbf{x}'(\mathbf{x})) = u(\mathbf{x})|_{x_1=\varsigma, \, x_2=0} \equiv v(\varsigma)$$

where $c_2 \neq 0$. The equation

$$\frac{\partial^2 u'}{\partial x_1' \partial x_2'} + u'^2 = 0$$

then becomes

$$\frac{c_1}{c_2}\frac{d^2 v}{d\varsigma^2} - v^2 = 0. \tag{40}$$

The general solution of (40) is given by

$$v(\varsigma) = c^2\left(\frac{-k}{1+k^2} + \frac{1}{\text{sn}^2(c(6c_1/c_2)^{-1/2}(\varsigma - c_3), k)}\right)$$

where c and c_3 are arbitrary constants, sn is a Jacobi elliptic function and k^2 is the root of the equation $1 - k^2 - k^4 = 0$. A similarity solution for $\partial^2 u/\partial x_1 \partial x_2 + u^2 = 0$ is then given by

$$u(\mathbf{x}) = c^2\left(\frac{-k}{1+k^2} + \frac{1}{\text{sn}^2(c(6c_1/c_2)^{-1/2}(x_1 - x_2 c_1/c_2 - c_3), k)}\right). \qquad \clubsuit$$

Remark: *The generalisation of the method, described above for the partial differential equation (34), is straightforward. It can be applied to systems of partial differential equations of the form (24).*

Note that the autonomous system of differential equations

$$\frac{d\mathbf{x}}{d\varepsilon} = \boldsymbol{\xi}(\mathbf{x}, \mathbf{u})$$

$$\frac{d\mathbf{u}}{d\varepsilon} = \boldsymbol{\eta}(\mathbf{x}, \mathbf{u})$$

can be written as

$$\frac{dx_1}{\xi_1(\mathbf{x}, \mathbf{u})} = \frac{dx_2}{\xi_2(\mathbf{x}, \mathbf{u})} = \cdots = \frac{dx_m}{\xi_m(\mathbf{x}, \mathbf{u})} = \frac{du_1}{\eta_1(\mathbf{x}, \mathbf{u})} = \cdots = \frac{du_n}{\eta_n(\mathbf{x}, \mathbf{u})} = \frac{d\varepsilon}{1} \tag{41}$$

where $\varphi_1(\mathbf{x}, \mathbf{u}), \ldots \varphi_{m-1}(\mathbf{x}, \mathbf{u}), \phi_1(\mathbf{x}, \mathbf{u}), \ldots, \phi_n(\mathbf{x}, \mathbf{u})$, are $m + n - 1$ independent invariants of (41), with the Jacobian

$$\frac{\partial(\phi_1, \ldots, \phi_n)}{\partial(u_1, \ldots, u_n)} \neq 0.$$

Now, from the invariant conditions of functions

$$u_i = s_i(\mathbf{x}), \qquad i = 1, \ldots, n, \tag{42}$$

it follows that (41) is invariant under a vector field) (or in vertical form) if and only if s satisfies the quasi-linear first order system of partial differential equations

$$\eta_i - \sum_{j=1}^{m} \xi_j(\mathbf{x}, \mathbf{u})\frac{\partial u_i}{\partial x_j} = 0, \qquad i = 1, 2, \ldots, n \tag{43}$$

known as the **invariant surface condition**. Thus

$$L_{Z_v} u_i = 0 \quad \Longleftrightarrow \quad \eta_i - \sum_{j=1}^{m} \xi_j(\mathbf{x}, \mathbf{u}) \frac{\partial u_i}{\partial x_j} = 0, \qquad (i = 1, \dots, n).$$

We say that the graph of a solution (42) to the system (14) defines an n-dimensional submanifold S_s of the space of dependent and independent variables. The solution s will be invariant under the one-paramenter subgroup generated by Z if and only if S_s is an invariant submanifold of this group. The invariant surface condition can be written as

$$j s^* (Z \lrcorner \theta_i) = 0, \qquad i = 1, \dots, n.$$

As a last example we consider a partial differential equation in four independent variables $(x_1, x_2, x_3. x_4)$.

Example: The three-dimensional linear diffusion equation is given by

$$\frac{\partial u}{\partial x_4} = \frac{\partial^2 u}{\partial x_1^2} + \frac{\partial^2 u}{\partial x_2^2} + \frac{\partial^2 u}{\partial x_3^2}. \tag{44}$$

The Lie algebra is spanned by thirteen vector fields which are listed in chapter 12. We make use of the following Abelian Lie subalgebra. The basis is spanned by the following vector fields

$$\{ \ Z_1, \ Z_4, \ Z_2 + c Z_3 \ \}$$

where $c \in \mathbf{R}$ and

$$Z_1 = \frac{\partial}{\partial x_3}, \qquad Z_2 = \frac{\partial}{\partial x_4}, \qquad Z_3 = u \frac{\partial}{\partial u}, \qquad Z_4 = x_1 \frac{\partial}{\partial x_2} - x_2 \frac{\partial}{\partial x_1}.$$

The transformation groups can easily be calculated. For Z_4 we find

$$\begin{pmatrix} x_1' \\ x_2' \end{pmatrix} = \begin{pmatrix} \cos \varepsilon_1 & -\sin \varepsilon_1 \\ \sin \varepsilon_1 & \cos \varepsilon_1 \end{pmatrix} \begin{pmatrix} x_1 \\ x_2 \end{pmatrix}, \quad x_3' = x_3, \quad x_4' = x_4, \quad u' = u.$$

For Z_1 we find

$$x_1'' = x_1', \quad x_2'' = x_2', \quad x_3'' = x_3' + \varepsilon_2, \quad x_4'' = x_4', \quad u'' = u'.$$

For $Z_2 + c Z_3$ we find

$$x_1''' = x_1'', \quad x_2''' = x_2'', \quad x_3''' = x_3'', \quad x_4''' = x_4'' + \varepsilon_3, \quad u''' = u'' \exp(c \varepsilon_3).$$

The composition of these transformation groups gives the three-parameter transformation group

$$\begin{pmatrix} x_1' \\ x_2' \end{pmatrix} = \begin{pmatrix} \cos\varepsilon_1 & -\sin\varepsilon_1 \\ \sin\varepsilon_1 & \cos\varepsilon_1 \end{pmatrix} \begin{pmatrix} x_1 \\ x_2 \end{pmatrix} \tag{45a}$$

$$x_3' = x_3 + \varepsilon_2 \tag{45b}$$

$$x_4' = x_4 + \varepsilon_3 \tag{45c}$$

$$u' = u \exp(c\varepsilon_3) \tag{45d}$$

where we identify $x_1''' \equiv x_1'$ etc. We choose $x_1 = \varsigma$ (similarity variable) and

$$x_2 = x_3 = x_4 = 0.$$

Then the above equations can be solved with respect to ε_1, ε_2, ε_3 and ς so that

$$\varepsilon_1 = \arctan\left(\frac{x_2'}{x_1'}\right), \qquad \varepsilon_2 = x_3', \qquad \varepsilon_3 = x_4'.$$

The similarity variable ς takes the form

$$\varsigma(x_1', x_2') = (x_1'^2 + x_2'^2)^{1/2}.$$

From (45d) we obtain the similarity ansatz

$$u'(\mathbf{x}') = v(\varsigma) \exp(cx_4').$$

Inserting this ansatz into the diffusion equation

$$\frac{\partial u'}{\partial x_4'} = \frac{\partial^2 u'}{\partial x_1'^2} + \frac{\partial^2 u'}{\partial x_2'^2} + \frac{\partial^2 u'}{\partial x_3'^2}$$

we find the linear ordinary differential equation

$$\frac{d^2v}{d\varsigma^2} + \frac{1}{\varsigma}\frac{dv}{d\varsigma} = cv$$

which is of Bessel type and can be solved in terms of Bessel functions.   ♣

9.4 Transforming Solutions to Solutions

Since the symmetry of a system of differential equations is a local group of transformations G, acting on some open set of the independent and dependent variables of that system in such a way that G transforms solutions of that system to solutions of the same system, we make the conclution that if

$$u_i = s_i(\mathbf{x}), \quad i = 1, \ldots, n$$

is a solution of the system, then

$$u_i' = s_i'(\mathbf{x}), \quad i = 1, \ldots, n$$

will also be a solution of the same system, where $g \in G$ acts such that

$$s_i' = g \cdot s_i, \qquad x_j' = g \cdot x_j, \quad i = 1, \ldots, n, \quad j = 1, \ldots, m.$$

Note that if Γ_s denotes the graph of the solution of a system of differential equations of the form (??), then the transform of Γ_s by $g \in G$ is not necessarily the graph of another single-valued function $u_i = s_i(\mathbf{x})$. However, since the Lie symmetry group G acts smoothly and the identily element of G leaves Γ_s unchanged, by suitably shrinking the domain of definition of s_i, we ensure that for elements g near the identity, the transform

$$g \cdot \Gamma_s = \Gamma_{s'}$$

is the graph of some single valued smooth function

$$u_i' = s_i'(\mathbf{x}), \quad i = 1, \ldots, n.$$

We now give some examples, where we transform solutions of equations by their Lie symmetry group.

Example: Consider

$$\frac{\partial^2 u}{\partial x_1 \partial x_2} + \frac{\partial u}{\partial x_1} + u^2 = 0$$

and the linear combination of its two Lie symmetry vector fields Z_1 and Z_4, i.e.,

$$cZ_1 + Z_4 \equiv c\frac{\partial}{\partial x_1} + e^{x_2}\frac{\partial}{\partial x_2} - e^{x_2}u\frac{\partial}{\partial u}.$$

The assosiated transformation is

$$x_1' = c\varepsilon + x_1, \qquad x_2' = \ln\left(e^{-x_2} - \varepsilon\right), \qquad u' = \left(1 - e^{x_2}\varepsilon\right)u(x_1, x_2),$$

where ε is the group parameter, i.e., a real constant. We can now state the following: Let $u = s(x_1, x_2)$ be a solution of the partial differential equation, then

$$u'(x_1', x_2') = \varepsilon\left(1 - \frac{1}{e^{x_2'} + \varepsilon}\right) s\left(x_1' - c\varepsilon, \ln\left(e^{x_2'} + \varepsilon\right)^{-1}\right)$$

is also a solution of the partial differential equation.   ♣

Example: We consider the two dimensional wave equation with $\lambda_1 = \lambda_2 = 0$, i.e.,

$$\left(\frac{\partial^2}{\partial x_1^2} - \frac{\partial^2}{\partial x_2^2}\right) u + \frac{1}{(x_1^2 - x_2^2)^2}\tilde{g}(u) = 0$$

where $\tilde{g}$ is an arbitrary smooth function. This equation admits the Lie symmetry vector field

$$\tilde{I}_2 = 2x_1 x_2 \frac{\partial}{\partial x_1} + (x_1 + x_2)\frac{\partial}{\partial x_2},$$

i.e., the partial differential equation is invariant under the transformation

$$x_1' = \frac{x_1}{\varepsilon^2(x_2^2 - x_1^2) - 2\varepsilon x_2 + 1}, \qquad x_2' = \frac{x_2 - \varepsilon(x_2^2 - x_1^2)}{\varepsilon^2(x_2^2 - x_1^2) - 2\varepsilon x_2 + 1}, \qquad u' = u(x_1, x_2).$$

Thus, if $u = s(x_1, x_2)$ is a solution of the partial differential equation, then

$$u'(x_1', x_2') = s\left(\frac{1}{2}\frac{x_1' + x_2'}{\varepsilon(x_1' + x_2') + 1} - \frac{1}{2}\frac{x_2' - x_1'}{\varepsilon(x_2' - x_1') + 1}, \ \frac{1}{2}\frac{x_1' + x_2'}{\varepsilon(x_1' + x_2') + 1} + \frac{1}{2}\frac{x_2' - x_1'}{\varepsilon(x_2' - x_1') + 1}\right)$$

is also a solution of the partial differential equation. As in the last example, ε is a real constant (group parameter).   ♣

Example: We consider the three-dimensional diffusion equation

$$\frac{\partial u}{\partial x_4} = \frac{\partial^2 u}{\partial x_1^2} + \frac{\partial^2 u}{\partial x_2^2} + \frac{\partial^2 u}{\partial x_3^2}$$

which is invariant under the transformation

$$\begin{pmatrix} x_1' \\ x_2' \end{pmatrix} = \begin{pmatrix} \cos\varepsilon_1 & -\sin\varepsilon_1 \\ \sin\varepsilon_1 & \cos\varepsilon_1 \end{pmatrix}\begin{pmatrix} x_1 \\ x_2 \end{pmatrix},$$

$$x_3' = x_3 + \varepsilon_2, \qquad x_4' = x_4 + \varepsilon_3$$

$$u' = u(x_1, x_2, x_3, x_4)\exp(c\varepsilon_3).$$

Thus, if $u = s(x_1, x_2, x_3, x_4)$ is a solution of the partial differential equation, then

$$u'(x_1', x_2', x_3', x_4') = e^{c\varepsilon_3}\, s\left(x_1'\cos\varepsilon_1 + x_2'\sin\varepsilon_1, \ -x_1'\sin\varepsilon_1 + x_2'\cos\varepsilon_1, \ x_3 - \varepsilon_2, \ x_4' - \varepsilon_3\right)$$

is also a solution of the above linear diffusion equation. Note that the transformed solution contains the real constants $c, \varepsilon_1, \varepsilon_2, \varepsilon_3 \in \mathbf{R}$.   ♣

9.5 Direct Method

The classical method for finding symmetry reduction of partial differential equations is the Lie grouyp method of infinitesimal transformations described above. An alternative method is the so-called **direct method** which use an ansatz for the solution and introduce a so-called reduced variable, which plays the role of the similiarity variable. We introduce the method with an example.

Example: Consider the **Fitzhugh-Nagumo equation**

$$\frac{\partial u}{\partial x_2} = \frac{\partial^2 u}{\partial x_1^2} + u(1 - u)(u - a),$$ (46)

where a is a constant. Without loss of generality we can set $-1 \leq a < 1$. Insert the ansatz

$$u(x_1, x_2) = f(x_1, x_2)w(z(x_1, x_2)) + g(x_1, x_2),$$ (47)

into (46) and require that $w(z)$ satisfies an ordinary differential equation. Here z is the so-called **reduced variable**. Substituting (47) in (46) we obtain

$$\left(f\frac{\partial^2 z}{\partial x_1^2}\right)\frac{d^2 w}{dz^2} + \left(2\frac{\partial f}{\partial x_1}\frac{\partial z}{\partial x_1} + f\frac{\partial^2 z}{\partial x_1^2} - f\frac{\partial z}{\partial x_2}\right)\frac{dw}{dz} - f^3 w^3 + ((a + 1 - 3g)f^2)w^2 +$$

$$\left(2(a + 1)gf - 3fg^2 - af + \frac{\partial^2 f}{\partial x_1^2} - \frac{\partial f}{\partial x_2}\right)w + \left(\frac{\partial^2 g}{\partial x_1^2} - \frac{\partial g}{\partial x_2} - g(g - a)(g - 1)\right) = 0.$$ (48)

Now we must require (48) to be an ordinary differential equation for $w(z)$. The usual procedure using the direct method is to impose that the different relationships among the coefficients of (48) to be a second order ordinary differential equation is the reduction to the travelling-wave ansatz. However one can equally consider it acceptable to reduce (48) to a first order ordinary differential equation. Setting in (48)

$$g = \frac{\partial z}{\partial x_1}$$ (49)

and demanding

$$3\frac{\partial^2 z}{\partial x_1^2} - \frac{\partial z}{\partial x_2} = \pm 2^{1/2}(a + 1 - 3g)\frac{\partial z}{\partial x_1}$$ (50a)

$$2(a + 1)g\frac{\partial z}{\partial x_1} - 3\frac{\partial z}{\partial x_1}g^2 - a\frac{\partial z}{\partial x_1} + \frac{\partial^3 z}{\partial x_1^3} - \frac{\partial^2 z}{\partial x_1 \partial x_2} = 0$$ (50b)

$$\frac{\partial^2 g}{\partial x_1^2} - \frac{\partial g}{\partial x_2} + g(g-1)(a-g) = 0 \tag{50c}$$

(48) becomes

$$\frac{d^2 w}{dz^2} - w^3 + \frac{a+1-3g}{\partial z/\partial x_1}\left(\pm 2^{1/2}\frac{dw}{dz} + w^2\right) = 0. \tag{51}$$

Equation (51) is satisfied if w satisfies the first order ordinary differential ordinary differential equation

$$\pm 2^{1/2}\frac{dw}{dz} + w^2 = 0. \tag{52}$$

Differentiating (52) once with respect to z and inserting (52) yields

$$\frac{d^2 w}{dz^2} - w^3 = 0.$$

Equation (52) can be integrated at once yielding

$$w(z) = \frac{+2^{1/2}}{z + z_0}. \tag{53}$$

By combining (47), (49) and (53) we can write the solution as

$$u(x_1, x_2) = \frac{\pm 2^{1/2}}{z + z_0}\frac{\partial z}{\partial x_1} + g. \tag{54}$$

There is a close connection of the Painlevé truncated expansion and the direct method.

9.6   Maps and Invariants

A one-dimensional map $f(x)$ is called an invariant of a two-dimensional map $g(x, y)$ if

$$g(x, f(x)) = f(f(x)).$$

The logistic map is an invariant of a class of two-dimensional maps. We construct a class of two-dimensional maps which admit the logistic maps as their invariant. Moreover we calculate their Ljapunov exponents. We show that the two-dimensional map can show hyperchaotic behaviour.

Example: The logistic equation

$$x_{t+1} = 2x_t^2 - 1, \quad t = 0, 1, 2, \ldots \qquad x_0 \in [-1, 1] \tag{55}$$

is the most studied equation with chaotic behaviour. All quantities of interest in chaotic dynamics can be calculated exactly. Examples are the fixed points and their stability, the periodic orbits and their stability, the moments, the invariant density, topological entropy, the metric entropy, Ljapunov exponent, autocorrelation function. The exact solution of (55) takes the form

$$x_t = \cos(2^t \arccos(x_0)) \tag{56}$$

since

$$\cos(2\alpha) = 2\cos^2(\alpha) - 1.$$

The Ljapunov exponent for almost all initial conditions is given by $\ln(2)$. The logistic equation is an invariant of a class of second order difference equations

$$x_{t+2} = g(x_t, x_{t+1}), \qquad t = 0, 1, 2, \ldots \tag{57}$$

This means that if (55) is satisfied for a pair (x_t, x_{t+1}), then (57) implies that (x_{t+1}, x_{t+2}) also satisfies (55). In other words, let

$$x_{t+1} = f(x_t), \qquad t = 0, 1, 2, \ldots \tag{58}$$

be a first order difference equation. Then (58) is called an **invariant** of (57) if

$$g(x, f(x)) = f(f(x)). \tag{59}$$

The second order difference equation (57) can be written as a first order system of difference equations $(x_{1,t} \equiv x_t)$

$$x_{1,t+1} = x_{2,t}, \qquad x_{2,t+1} = g(x_{1,t}, x_{2,t}). \tag{60}$$

If x_0 and x_1 are the initial conditions of (57) $(x_0, x_1 \in [-1, 1])$ and assuming that (55) is an invariant of (57) as well as that x_0 and x_1 satisfy the logistic equation, then a one-dimensional Ljapunov exponent of (60) is given by $\ln(2)$. Since system (60) is two-dimensional, we have a second one-dimensional Ljapunov exponent and a two-dimensional Ljapunov exponent. Let λ_1^I and λ_2^I be the two one-dimensional Ljapunov exponents. Let λ^{II} be the two-dimensional Ljapunov exponent. Then we have

$$\lambda^{II} = \lambda_1^I + \lambda_2^I. \tag{61}$$

Let us find the two-dimensional Ljapunov exponent. Consider the system of first order difference equations

$$x_{1,t+1} = f_1(x_{1,t}, x_{2,t}), \qquad x_{2,t+1} = f_2(x_{1,t}, x_{2,t}). \tag{62}$$

The variational equation is given by $(\mathbf{x}_t = (x_{1,t}, x_{2,t}))$

$$y_{1,t+1} = \frac{\partial f_1}{\partial x_1}(\mathbf{x}_t)y_{1,t} + \frac{\partial f_1}{\partial x_2}(\mathbf{x}_t)y_{2,t}, \quad y_{2,t+1} = \frac{\partial f_2}{\partial x_1}(\mathbf{x}_t)y_{1,t} + \frac{\partial f_2}{\partial x_2}(\mathbf{x}_t)y_{2,t}. \tag{63}$$

Let $\mathbf{y}_t$ and $\mathbf{v}_t$ be two quantities satisfying the variational equation (63). Let $\mathbf{e}_1$ and $\mathbf{e}_2$ be two unit vectors in $\mathbf{R}^2$ with

$$\mathbf{e}_1 \cdot \mathbf{e}_2 = 0$$

where $\cdot$ denotes the scalar product. Let $\wedge$ be the exterior product (Grassmann product). Then we find

$$\mathbf{y}_t \wedge \mathbf{v}_t = (y_{1,t}v_{2,t} - y_{2,t}v_{1,t})\mathbf{e}_1 \wedge \mathbf{e}_2. \tag{64}$$

Now we define

$$w_t := y_{1,t}v_{2,t} - y_{2,t}v_{1,t}. \tag{65}$$

Thus the time evolution of w_t is given by

$$w_{t+1} = \left(\frac{\partial f_1}{\partial x_1}(\mathbf{x}_t)\frac{\partial f_2}{\partial x_2}(\mathbf{x}_t) - \frac{\partial f_1}{\partial x_2}(\mathbf{x}_t)\frac{\partial f_2}{\partial x_1}(\mathbf{x}_t) \right) w_t. \tag{66}$$

The two-dimensional Ljapunov exponent is given by

$$\lambda^{II} = \lim_{T \to \infty} \frac{1}{T} \ln |w_T|. \tag{67}$$

Obviously, λ_1^I, λ_2^I and λ_{II} depend on the initial conditions of (62). If $f_1(x_1, x_2) = x_2$ and $f_2(x_1, x_2) = g(x_1, x_2)$ as in (60) we obtain from (66) that

$$w_{t+1} = -\frac{\partial g}{\partial x_1}(\mathbf{x}_t)w_t. \tag{68}$$

Without loss of generality we can set $w_0 = 1$.

We derive now a class of second order difference equation with the logistic map as an invariant. Our ansatz for $g(x_1, x_2)$ with $f(x) = 2x^2 - 1$ is given by

$$g(x_1, x_2) = a_{10}x_1 + a_{01}x_2 + a_{20}x_1^2 + a_{11}x_1x_2 + a_{02}x_2^2 + d. \tag{69}$$

Satisfying the condition (59) yields

$$g(x_1, x_2) = x_2 - 2x_1^2 + 2x_2^2 + d(1 + x_2 - 2x_1^2). \tag{70}$$

Since

$$\frac{\partial g}{\partial x_1} = -4x_1(d + 1). \tag{71}$$

we find that (68) takes the form

$$w_{t+1} = -4x_{1,t}(d + 1)w_t. \tag{72}$$

Let us now calculate the two-dimensional Ljapunov exponent λ^{II}. The initial values $x_{1,0}$, $x_{2,0}$ of the two-dimensional map $x_{1,t+1} = x_{2,t}$, $x_{2,t+1} = g(x_{1,t}, x_{2,t})$ satisfy the logistic map in our following calculations. Using (56) and (72), we obtain

$$\lambda^{II}(\theta_0) = \lim_{T \to \infty} \frac{1}{T} \ln \left(\prod_{t=1}^{T} 4|d + 1| \, |\cos(2^t \theta_0)| \right), \quad d \neq -1, \quad \theta_0 := \arccos(x_0) \tag{73}$$

or

$$\lambda^{II}(\theta_0) = 2 \ln 2 + \ln|d + 1| + \gamma(\theta_0) \tag{74}$$

where

$$\gamma(\theta_0) = \lim_{T \to \infty} \frac{1}{T} \sum_{t=1}^{T} \ln|\cos(2^t \theta_0)|. \tag{75}$$

Now, since

$$\cos(2^t \theta_0) = \cos(2^t \theta_0 \mod 2\pi), \tag{76}$$

we only need to study the Bernoulli shift map

$$\theta_{t+1} = 2\theta_t \mod 2\pi. \tag{77}$$

This map has the solution

$$\theta_t = 2^t \theta_0 \mod 2\pi. \tag{78}$$

The map (77) is ergodic with the invariant density

$$\rho(\theta) = \frac{1}{2\pi} \chi_{[0,2\pi)}(\theta) \tag{79}$$

where χ is the characteristic function. Thus we may apply **Birkhoff's ergodic theorem.** This then gives

$$\gamma(\theta_0) = \int_0^{2\pi} \rho(\theta) \ln|\cos\theta| d\theta = \frac{1}{2\pi} \int_0^{2\pi} \ln|\cos\theta| d\theta = \frac{2}{\pi} \int_0^{\pi/2} \ln(\cos\theta) d\theta. \tag{80}$$

It follows that

$$\gamma(\theta_0) = -\ln 2, \quad \text{for} \quad a.e \quad \theta_0 \in [0, 2\pi). \tag{81}$$

Thus

$$\lambda^{II} = \ln 2 + \ln|d + 1|, \quad d \neq -1 \tag{82}$$

Now, since one of the one-dimensional Ljapunov exponent is $\ln 2$, and

$$\lambda^{II} = \lambda_1^I + \lambda_2^I, \qquad \lambda_1^I \geq \lambda_2^I \tag{83}$$

we find the two one-dimensional Ljapunov exponent as

$$\lambda_1^I = \max\{\ln 2, \ \ln|d+1|\} \tag{84}$$

$$\lambda_2^I = \min\{\ln 2, \ \ln|d+1|\}. \tag{85}$$

Obviously λ^{II} can be made arbitrarily large positive or negative by appropriate choice of d. This implies that the spectrum of the one-dimensional Ljapunov exponents may be

$$(+,-), \quad (+,0), \qquad (+,+).$$

Thus hyperchaos can occur. Now, let $\{x_n(x_0)\}$ denote the orbit originating from x_0 for the logistic map. Then

$$\{x_n(x_0)\} \text{ is chaotic } \Leftrightarrow \arccos(x_0) \in \mathbf{R}\backslash\mathbf{Q}. \tag{86}$$

This follows from the fact that the orbit of the Bernoulli shift map is chaotic if and only if

$$\theta_0 \in \mathbf{R}\backslash\mathbf{Q}.$$

9.7 Computer Algebra Applications

Lie symmetries play a central role when we want to find out whether an ordinary or partial differential equation admits a first integral (constant of motion) or conservation law, respectively. Here we consider ordinary differential equations. To find the determining equations for the Lie symmetry vector field is a quite cumbersome task, since a number of prolongations have to be calculated. Here computer algebra is very helpful. As an example we consider the Lorenz model [6,7]

$$\frac{dx}{dt} = \sigma(y - x), \qquad \frac{dy}{dt} = -xz + rx - y, \qquad \frac{dz}{dt} = xy - bz.$$

The Lorenz system can be brought into the form of the third order differential equation

$$x\frac{d^3x}{dt^3} - \left(\frac{dx}{dt} - (\sigma + b + 1)x\right)\frac{d^2x}{dt^2} - (\sigma + 1)\left(\frac{dx}{dt}\right)^2 + (x^3 + b(\sigma + 1)x)\frac{dx}{dt} +$$

$$\sigma(x^4 + b(1 - r)x^2) = 0.$$

To study the Lie symmetries we adopt the jet bundle formalism. We consider symmetry vector fields of the form

$$V = (V_0(t, x, \dot{x}) + V_1(t, x, \dot{x})\ddot{x})\frac{\partial}{\partial x}.$$

To impose the symmetry condition we have to calculate the prolongation of V using the total differential operator

$$D := \frac{\partial}{\partial t} + \dot{x}\frac{\partial}{\partial x} + \ddot{x}\frac{\partial}{\partial \dot{x}} + \dots$$

To find the condition on V_0 and V_1 we consider the third order differential equation as a manifold

$$\Delta \equiv x\,\dddot{x} - (\dot{x} - (\sigma + b + 1)x)\ddot{x} - (\sigma + 1)\dot{x}^2 + (x^3 + b(\sigma + 1)x)\dot{x}$$

$$+ \sigma(x^4 + b(1 - r)x^2) = 0$$

together with its differential consequences $D(\Delta) = 0$, $D(D(\Delta)) = 0$. In the program we show that

$$V = \exp(2\sigma t)\dot{x}\left(x\ddot{x} - \dot{x}\dot{x} + \frac{x^4}{4}\right)\frac{\partial}{\partial x}$$

is a Lie symmetry vector field of the third order differential equation. In the program we set $x = u$. Replacing this vector field by the vector field $V = (V_0 + V_1)\partial/\partial x$ we find the condition on the functions V_0 and V_1 so that V is a Lie symmetry vector field.

```
operator D;

% Q depends on ts, u, ud, udd, uddd, udddd
depend Q, ts, u, ud, udd, uddd, udddd;

% total derivative
for all Q let
D(Q) = df(Q,ts) + ud*df(Q,u) + udd*df(Q,ud) + uddd*df(Q,udd)
       + udddd*df(Q,uddd) + uddddd*df(Q,udddd);

% Lorenz equation as third order equation
equ := u*uddd - (ud - (sig + b + 1)*u)*udd - (sig + 1)*ud*ud
+ (u*u*u + b*(sig + 1)*u)*ud + sig*(u*u*u*u + b*(1 - r)*u*u);

% prolongation of Lorenz equation
equp := D(equ);   equpp := D(D(equ));

% Lie symmetry, example
T1 := exp(4*ts/3)*ud*(u*udd - ud*ud + u**4/4);
% prolongation of Lie symmetry
T2 := D(T1);   T3 := D(T2);   T4 := D(T3);


res := T1*df(equ,u) + T2*df(equ,ud) + T3*df(equ,udd) +
T4*df(equ,uddd);

% finding uddd
list1 := solve(equ=0,uddd);
list2 := part(list1,1);
uddd := part(list2,2);
% finding udddd
list3 := solve(equp=0,udddd);
list4 := part(list3,1);
udddd := part(list4,2);
% finding uddddd
list5 := solve(equpp=0,uddddd);
list6 := part(list5,1);
uddddd := part(list6,2);

% uddd, udddd, uddddd is now replaced in res
res;

b := 0; sig := 1/3;

res;
```

Chapter 10

Lie-Bäcklund Vector Fields

10.1 Definitions and Examples

A symmetry of a differential equation was described as a transformation which maps any solution of a differential equation to another solution of the same equation. From this point of view we can extend the concept of symmetries of differential equations to Lie-Bäcklund symmetries (Bluman and Kumei [7], Olver [88], Olver [87], Anderson and Ibragimov [4], Kumei [72], Steeb and Euler [125], Steeb and Euler [125], Ibragimov and Shabat [59]). Invariant solution can also be constructed using Lie-Bäcklund vector fields (Euler *et al* [42]).

We consider the transformation

$$\mathbf{x}' = \mathbf{x} + \varepsilon \boldsymbol{\xi} \left(x_i, u_j, \frac{\partial u_j}{\partial x_i}, \frac{\partial^2 u_j}{\partial x_{i_1} \partial x_{i_2}}, \ldots, \frac{\partial^q u_j}{\partial x_{i_1} \cdots \partial x_{i_q}} \right) + O(\varepsilon^2)$$

$$\mathbf{u}' = \mathbf{u} + \varepsilon \boldsymbol{\eta} \left(x_i, u_j, \frac{\partial u_j}{\partial x_i}, \frac{\partial^2 u_j}{\partial x_{i_1} \partial x_{i_2}}, \ldots, \frac{\partial^q u_j}{\partial x_{i_1} \cdots \partial x_{i_q}} \right) + O(\varepsilon^2)$$

that leaves the system of partial differential equations

$$F_\nu \left(x_i, u_j, \frac{\partial u_j}{\partial x_i}, \frac{\partial^2 u_j}{\partial x_{i_1} \partial x_{i_2}}, \ldots, \frac{\partial^r u_j}{\partial x_{i_1} \cdots \partial x_{i_r}} \right) = 0$$

$(\nu = 1, \ldots, n)$ invariant where $\boldsymbol{\xi}[\mathbf{x}, \mathbf{u}]$ and $\boldsymbol{\eta}[\mathbf{x}, \mathbf{u}]$ are the infinitesimals. Here

$$\mathbf{x} = (x_1, \ldots, x_m), \quad \mathbf{u} = (u_1, \ldots, u_n), \quad \boldsymbol{\xi} = (\xi_1, \ldots, \xi_m), \quad \boldsymbol{\eta} = (\eta_1, \ldots, \eta_n)$$

where $i_1 \leq \ldots \leq i_r$ and $q > r$. The square brackets will serve to remind us that $\boldsymbol{\xi}$ and $\boldsymbol{\eta}$ depend on $\mathbf{x}$, $\mathbf{u}$ and derivatives of $\mathbf{u}$ with respect to x_i. In terms of the infinitesimals we have the following

Definition 10.1 *The infinitesimal generator*

$$Z_B := \sum_{j=1}^{m} \xi_j[\mathbf{x}, \mathbf{u}] \frac{\partial}{\partial x_j} + \sum_{i=1}^{n} \eta_i[\mathbf{x}, \mathbf{u}] \frac{\partial}{\partial u_i}$$

is a **Lie-Bäcklund symmetry vector field** *of the above system of partial differential equations if and only if*

$$L_{\bar{Z}_B} F_\nu \left(x_i, u_j, u_{j,i}, \ldots, u_{j,i_1 \cdots i_r} \right) \hat{=} 0$$

where $\bar{Z}_B$ is the prolongation of the given generator Z_B. Here $\hat{=}$ stands for the restriction to solutions of the equation.

The vertical Lie-Bäcklund symmetry vector field is given by

$$Z_{VB} := \sum_{j=1}^{n} U_j[\mathbf{x}, \mathbf{u}] \frac{\partial}{\partial u_j}$$

where

$$U_j[\mathbf{x}, \mathbf{u}] := \eta_j[\mathbf{x}, \mathbf{u}] - \sum_{i=1}^{m} \xi_i[\mathbf{x}, \mathbf{u}] \frac{\partial u_j}{\partial x_i}.$$

Thus one can, in general, consider a Lie-Bäcklund vector field in the form

$$Z_B = \sum_{j=1}^{n} g_j[\mathbf{x}, \mathbf{u}] \frac{\partial}{\partial u_j}$$

where g_j are arbitrary functions of their arguments which have to be determined for a system of partial differential equations.

The algorithm to determine the admitted Lie-Bäcklund symmetry vector fields of a given system of partial differential equations is essentially the same as in the case of the Lie point symmetry vector fields. A minor difference is that the equations resulting from the invariance condition $L_{\bar{Z}_B} F_\nu \hat{=} 0$ involve derivatives to order $q + r$. By making use of the differential consequences of the submanifold $F_\nu = 0$ the determining equations (a linear system of partial differential equations in $\boldsymbol{\xi}$ and $\boldsymbol{\eta}$) can be found.

Example: In this example we derive a class of analytic functions f for which the evolution equation

$$\frac{\partial^2 u}{\partial x_1 \partial x_2} = f(u) \tag{1}$$

admits Lie-Bäcklund vector fields. We apply the jet bundle formalism. Within this approach we consider the submanifold

$$F \equiv u_{12} - f(u) = 0$$

and all its differential consequences with respect to x_1, i.e.

$$F_1 \equiv u_{112} - u_1 f' = 0, \qquad F_{11} \equiv u_{1112} - u_{11} f' - u_1^2 f'' = 0$$

etc.. Let

$$Z_B = g(u, u_1, u_{11}, u_{111}) \frac{\partial}{\partial u}$$

be a Lie-Bäcklund vector field. The nonlinearity in the partial differential equation only appears in the function f which depends only on u and the term where the derivative appears is linear. Consequently we can assume that the vector field Z_B is linear in u_{111}, so that

$$Z_B = (g_1(u, u_1, u_{11}) + u_{111}) \frac{\partial}{\partial u}.$$

Furthermore, we can assume, without loss of generality, that the function g_1 does not depend on u. If we include the dependence of u, our calculations show that g_1 does not depend on u. Consequently,

$$Z_B = (g_2(u_1, u_{11}) + u_{111}) \frac{\partial}{\partial u}.$$

Owing to the structure of the given partial differential equation (1) we are only forced to include the term of the form $(\cdots)\partial/\partial u_{12}$ in the prolonged vector field $\bar{Z}_B$. From the invariance condition it follows that

$$u_{111}u_1 \frac{\partial^2 g_2}{\partial u_{11}^2} f' + u_{111} \frac{\partial^2 g_2}{\partial u_1 \partial u_{11}} f + u_{11} \frac{\partial^2 g_2}{\partial u_1^2} f + u_{11}u_1 \frac{\partial^2 g_2}{\partial u_1 \partial u_{11}} f' + (u_1^2 f'' + u_{11} f') \frac{\partial g_2}{\partial u_{11}}$$

$$+3u_{11}u_1 f'' + u_1^3 f''' + u_1 \frac{\partial g_2}{\partial u_1} f' - g_2 f' = 0 \qquad (2)$$

where $f' = df/du$. Separating out the terms with the factors $u_{111}u_1$ and u_{111} we obtain

$$u_{111}u_1 \frac{\partial^2 g_2}{\partial u_{11}^2} f' = 0 \qquad (3)$$

and

$$u_{111} \frac{\partial^2 g_2}{\partial u_1 \partial u_{11}} f = 0. \qquad (4)$$

If we assume that the function g_2 does not depend on u_{11}, then (3) and (4) are satisfied. From (2) it follows that

$$u_{11} \left(3u_1 f'' + f \frac{\partial^2 g_2}{\partial u_1^2} \right) = 0$$

and

$$u_1^3 f''' + u_1 f' \frac{\partial g_2}{\partial u_1} - g_2 f' = 0.$$

Since we assume that the function f is a nonlinear analytic function of u, it follows that the function g_2 must be of the form

$$g_2(u_1) = \frac{a}{3} u_1^3$$

where a is an arbitrary real parameter ($a \neq 0$). We obtain

$$u_{11}u_1(3f'' + 2af) = 0 \tag{5}$$

and

$$u_1^3(f''' + \frac{2a}{3}f') = 0. \tag{6}$$

For solving both (5) and (6) simultaneously, we have to solve

$$f'' + \frac{2a}{3}f = 0. \tag{7}$$

We can now state the following: If the function f satisfies the ordinary differential equation (7) then (1) admits Lie-Bäcklund vector fields. For the solution of (7) we have to distinguish between the cases $a > 0$ and $a < 0$. First let $a > 0$. We put $a = 3/2$. Then we obtain

$$f(u) = C_1 \sin u + C_2 \cos u,$$

where C_1 and C_2 are two real constants. Secondly let $a < 0$, we put $a = -3/2$. Then we obtain

$$f(u) = C_1 \cosh u + C_2 \sinh u.$$

To summarise, the evolution equations

$$\frac{\partial^2 u}{\partial x_1 \partial x_2} = C_1 \sin u + C_2 \cos u$$

and

$$\frac{\partial^2 u}{\partial x_1 \partial x_2} = C_1 \sinh u + C_2 \cosh u$$

admit Lie-Bäcklund vector fields. The simplest one is given by the vector field

$$Z_B = (g_2(u_1, u_{11}) + u_{111})\frac{\partial}{\partial u}$$

together with $g_2(u_1) = au_1^3/3$, i.e.

$$Z_B = (\frac{a}{3}u_1^3 + u_{111})\frac{\partial}{\partial u}.$$

A hierarchy of Lie-Bäcklund vector fields can now be obtained with the help of a recursion operator. Within the technique described above we find that the evolution equations of the form $\partial^2 u/\partial x_1 \partial x_2 = f(u)$, where f is a polynomial with degree higher than one in u, do not admit Lie-Bäcklund vector fields.   ♣

Example: We consider

$$\frac{\partial u}{\partial x_2} = \frac{\partial^2 u}{\partial x_1^2} + f(u) \tag{8}$$

where f is an analytic function. In the jet bundle formalism we have

$$F \equiv u_2 - u_{11} - f(u) = 0$$

with all the differential consequences with respect to x_1. Owing to the structure of (8) we consider the simplified Lie-Bäcklund vector field Z_B without loss of generality, namely

$$Z_B = (g_1(u, u_1, u_{11}) + u_{111}) \frac{\partial}{\partial u},$$

where g_1 is an analytic function. We need only the terms of the form $(\cdots) \partial/\partial u_2$ and $(\cdots) \partial/\partial u_{11}$ in the prolonged vector field $\bar{Z}_B$. From the invariance condition $L_{\bar{Z}_B} F \hat{=} 0$ it follows that

$$\frac{\partial g_1}{\partial u} f + u_1 \frac{\partial g_1}{\partial u_1} f' + (u_1^2 f'' + u_{11} f') \frac{\partial g_1}{\partial u_{11}} + 3 u_1 u_{11} f'' + u_1^3 f''' - u_1^2 \frac{\partial^2 g_1}{\partial u^2} - 2 u_1 u_{11} \frac{\partial^2 g_1}{\partial u \partial u_1}$$

$$- 2 u_1 u_{111} \frac{\partial^2 g_1}{\partial u \partial u_{11}} - u_{11}^2 \frac{\partial^2 g_1}{\partial u_1^2} - 2 u_{11} u_{111} \frac{\partial^2 g_1}{\partial u_1 \partial u_{11}} - u_{111}^2 \frac{\partial^2 g_1}{\partial u_{11}^2} - g_1 f' = 0. \qquad (9)$$

Separating out the term with the factor u_{111}^2 we obtain

$$u_{111}^2 \frac{\partial^2 g_1}{\partial u_{11}^2} = 0.$$

Consequently the function g_1 takes the form

$$g_1(u, u_1, u_{11}) = g_2(u, u_1) u_{11} + g_3(u, u_1). \qquad (10)$$

From (9) it also follows that

$$u_1 u_{111} \frac{\partial^2 g_1}{\partial u \partial u_{11}} = 0$$

and

$$u_{11} u_{111} \frac{\partial^2 g_1}{\partial u_1 \partial u_{11}} = 0.$$

With the help of (10) we find that the function g_2 does not depend on u_1 and u. Consequently,

$$g_1(u, u_1, u_{11}) = C_1 u_{11} + g_3(u, u_1). \qquad (11)$$

Inserting (11) into (9) it follows that

$$C_1 u_1^2 f'' + 3 u_1 u_{11} f'' + u_1 f'' - g_3 f' + f \frac{\partial g_3}{\partial u} + u_1 f' \frac{\partial g_3}{\partial u_1}$$

$$- u_1^2 \frac{\partial^2 g_3}{\partial u^2} - 2 u_1 u_{11} \frac{\partial^2 g_3}{\partial u \partial u_1} - u_{11}^2 \frac{\partial^2 g_3}{\partial u_1^2} = 0. \qquad (12)$$

From (12) we have

$$u_{11}^2 \frac{\partial^2 g_3}{\partial u_1^2} = 0 \qquad (13)$$

and therefore the function g_3 takes the form

$$g_3(u, u_1) = g_4(u)u_1 + g_5(u).$$

From (12) we obtain

$$u_1 u_{11}(3f'' - 2g_4') = 0$$

and therefore $g_4 = \frac{3}{2}f' + C_2$. It follows that

$$C_1 u_1^2 f'' + u_1 f''' - g_5 f' + \frac{3}{2} u_1 f f'' + f g_5' - \frac{3}{2} u_1^3 f''' - u_1^2 g_5'' = 0. \tag{14}$$

From (13) we see that the following statement holds: The diffusion equation (8) admits Lie-Bäcklund vector fields if and only if $f''(u) = 0$. Hence (8) becomes linear. The vector field Z_B takes the form

$$Z_B = (u_{111} + C_1 u_{11} + C_2 u_1)\frac{\partial}{\partial u}.$$

We obtain the same result when we extend the vector field to

$$Z_B = g(x_1, x_2, u, u_1, u_{11}, u_{111}, \ldots)\frac{\partial}{\partial u}.$$

Equation (8) belongs to the following class of partial differential equations which admit a hierarchy of Lie-Bäcklund vector fields, namely

$$\frac{\partial u}{\partial t} = \frac{\partial^2 u}{\partial x^2} + f_1(u)\left(\frac{\partial u}{\partial x}\right)^2 + f_2(u)\frac{\partial u}{\partial x} + f_3(u),$$

and the functions f_1, f_2 and f_3 satisfy the system of differential equations

$$f_2' f_3 = 0, \qquad f_2' f_1 = f_2'', \qquad f_3'' + (f_1 f_3)' = 0.$$

Thus, if $f_3(u) = 0$, $f_1(u) = 0$ and $f_2(u) = u$, (14) is satisfied and we obtain the well-known Burgers' equation. If we put $f_1(u) = 0$ and $f_2(u) = 0$, then it follows that $f_3''(u) = 0$. Consequently $f_3(u) = au + b$ $(a, b \in \mathbf{R})$.   ♣

We now give some more Lie-Bäcklund symmetry vector fields of well known partial differential equations.

Example: The **Burgers' equation**

$$\frac{\partial^2 u}{\partial x_1^2} - u\frac{\partial u}{\partial x_1} - \frac{\partial u}{\partial x_2} = 0$$

is written in the form

$$F \equiv u_{11} - uu_1 - u_2 = 0$$

in the jet bundle formalism. With the Lie-Bäcklund vector field ansatz

$$Z_B = g(x_1, x_2, u, u_1, u_{11}, u_{111})\frac{\partial}{\partial u}$$

Burgers' equation admits the following Lie-Bäcklund symmetry vector fields

$$Z_{B1} = \left(4u_{111} - 6uu_{11} - 6u_1^2 + 3u^2 u_1\right) \frac{\partial}{\partial u}$$

$$Z_{B2} = \left(4x_2 u_{111} + (2x - 6x_2 u)u_{11} - 6x_2 u_1^2 + (3x_2 u^2 - 2x_1 u)u_1 - u^2\right) \frac{\partial}{\partial u}$$

$$Z_{B3} = \left(4x_2^2 u_{111} + (4x_2 x_1 - 6x_2^2 u)u_{11} - 6x_2^2 u_1^2 + (3x_2^2 u^2 - 4x_2 x_1 u + x_1^2)u_1 \right.$$
$$\left. -2x_2 u^2 + 2x_1 u + 6\right) \frac{\partial}{\partial u}$$

$$Z_{B4} = \left(4x_2^3 u_{111} + (6x_2^2 x_1 - 6x_2^3 u)u_{11} - 6x_2^3 u_1^2 + (3x_2^3 u^2 - 6x_2^2 x_1 u + 3x_2 x_1^2 + 12x_2^2)u_1 \right.$$
$$\left. -3x_2^2 u^2 + 6x_2 x_1 u - 3x_1^2 - 6x_2\right) \frac{\partial}{\partial u}.$$

There is a hierarchy of Lie-Bäcklund vector fields. ♣

Example: The Korteweg-de Vries equation

$$\frac{\partial^3 u}{\partial x_1^3} + u\frac{\partial u}{\partial x_1} + \frac{\partial u}{\partial x_2} = 0$$

is written in the form

$$F \equiv u_{111} + uu_1 + u_2 = 0$$

in the jet bundle formalism. With the Lie-Bäcklund vector field ansatz

$$Z_B = g(x_1, x_2, u, u_1, u_{11}, u_{111}, u_{1111}, u_{11111})\frac{\partial}{\partial u}$$

The Korteweg-de Vries equation admits the following generators (the first four are Lie point symmetry vector fields in vertical form)

$$Z_{V1} = (x_2 u_1 - 1)\frac{\partial}{\partial u}, \quad Z_{V2} = \frac{1}{3}(x_1 u_1 + 3x_2 u_2 + 2u)\frac{\partial}{\partial u}, \quad Z_{V3} = u_1\frac{\partial}{\partial u}, \quad Z_{V4} = -u_2\frac{\partial}{\partial u}$$

$$Z_B = \left(\frac{3}{5}u_{11111} + uu_{111} + 2u_1 u_{11} + \frac{1}{2}u^2 u_1\right)\frac{\partial}{\partial u}.$$

For the modified Korteweg-de Vries equation

$$F \equiv u_{111} + u^2 u_1 + u_2 = 0$$

the following are the first two Lie-Bäcklund vector fields in a hierarchy of Lie-Bäcklund vector fields (with the same ansatz given above)

$$Z_{B1} = \left(u_{111} + u^2 u_1\right)\frac{\partial}{\partial u}$$
$$Z_{B2} = \left(u_{11111} + \frac{5}{3}u^2 u_{111} + \frac{20}{3}uu_1 u_{11} + \frac{5}{3}u_1^3 + \frac{5}{6}u^4 u_1\right)\frac{\partial}{\partial u}.$$

♣

Example: The **Harry-Dym equation** ($\lambda \in \mathbf{R}$)

$$\frac{\partial u}{\partial x_2} - \lambda u^3 \frac{\partial^3 u}{\partial x_1^3} = 0$$

is written in the form

$$F \equiv u_2 - \lambda u^3 u_{111} = 0$$

applying the jet bundle formalism. A Lie-Bäcklund vector field is given by

$$Z_B = \left(u^5 u_{11111} + 5u^4 u_1 u_{1111} + 5u^4 u_{11} u_{111} + \frac{5}{3} u^3 u_1^2 u_{111} \right) \frac{\partial}{\partial u}. \qquad \clubsuit$$

Example: The system

$$\frac{\partial u_1}{\partial x_2} - \frac{\partial^2 u_1}{\partial x_1^2} - \frac{1}{2} u_2^2 = 0, \qquad \frac{\partial u_2}{\partial x_2} - \frac{\partial^2 u_2}{\partial x_1^2} = 0$$

is written in the form

$$F_1 \equiv u_{1,2} - u_{1,11} - \frac{1}{2} u_2^2, \qquad F_2 \equiv u_{2,2} - 2u_{2,11}$$

in the jet bundle formalism. There exists only the one Lie-Bäcklund symmetry vector field

$$Z_B = (u_{1,111} + 3u_2 u_{2,1}) \frac{\partial}{\partial u_1} + 4u_{2,111} \frac{\partial}{\partial u_2}. \qquad \clubsuit$$

Let us now discuss the question of obtaining the **Lie-Bäcklund transformation group** by considering the Lie-Bäcklund symmetry vector field

$$Z_B = \sum_{j=1}^{n} g_j[\mathbf{x}, \mathbf{u}] \frac{\partial}{\partial u_j}.$$

This implies that one has to consider the evolution equation

$$\frac{\partial u_j}{\partial \varepsilon} = g_j[\mathbf{x}, \mathbf{u}]$$

where $j = 1, \ldots, n$. The solution (provided it exists) of the initial value problem $u(x, 0) = f(x)$ will determine the group action

$$[\exp(\varepsilon Z_B) f](x) \equiv u(x, \varepsilon).$$

Here we are forced to assume that the solution to this Cauchy problem is uniquely determined provided the initial data $f(x)$ is chosen in some appropriate space of functions, at least for ε sufficiently small. The resulting flow $\exp(\varepsilon Z)$ will then be on the given function space. The verification of this hypothesis leads to some very difficult problems on existence and uniqueness of solutions to systems of evolution equations. The problem will become apparent in the following example.

Example: Consider the Lie-Bäcklund vector field

$$Z_B = u_{11} \frac{\partial}{\partial u}.$$

The corresponding one-parameter group will be obtained by solving the Cauchy problem

$$\frac{\partial u}{\partial \varepsilon} = \frac{\partial^2 u}{\partial x^2}, \qquad u(x,0) = f(x).$$

The solution can be given as Lie series

$$u(x,\varepsilon) = \exp(\varepsilon Z) f(x).$$

Thus exponenting the Lie-Bäcklund vector field is equivalent to solving the heat equation. Several difficulties are immediately apparent. Firstly, for $\varepsilon < 0$ we are dealing with the "backwards heat equation" which is a classic ill-posed problem and may not even have solutions. Thus we should only expect to have a "semi-group" of transformations generated by Z. Secondly, unless we impose some growth conditions the solution will not in general be unique. The formal series solution to the evolution equation is

$$u(x,\varepsilon) = f(x) + \varepsilon \frac{\partial^2 f}{\partial x^2} + \frac{\varepsilon^2}{2!} \frac{\partial^4 f}{\partial x^4} + \cdots.$$

However, even if f is analytic, this Lie series for u may not converge. In fact, it will converge only if f is an entire analytic function satisfying the growth condition

$$|f(x)| \le C \exp(Kx^2)$$

for positive constants C, K, which are the same growth conditions needed to ensure uniqueness of solutions. ♣

10.2   Invariant Solutions

Lie-Bäcklund vector fields can also be used to construct **similarity solutions**. We show this with the following example. The jet bundle formalism is applied.

Example: The nonlinear diffusion equation

$$\frac{\partial u}{\partial x_2} = \frac{\partial}{\partial x_1}\left(\frac{1}{u^2}\frac{\partial u}{\partial x_1}\right) \tag{15}$$

is considered. Within the jet bundle formalism we have the submanifold

$$F(x_1, x_2, u, u_1, u_{11}) \equiv u_2 - \frac{u_{11}}{u^2} + 2\frac{u_1^2}{u^3} = 0$$

and its differential consequences. The contact form is

$$\theta = du - u_1 dx_1 - u_2 dx_2.$$

Equation (15) admits the following Lie point symmetry vector fields (in vertical form)

$$X_v = -u_1\frac{\partial}{\partial u}, \qquad T_v = -u_2\frac{\partial}{\partial u}$$

$$S_v = (-xu_1 - 2tu_2)\frac{\partial}{\partial u}$$

$$V_v = (xu_1 + u)\frac{\partial}{\partial u}.$$

By considering the ansatz

$$Z_B = (g_1(u, u_1, u_{11}) + g_2(u)u_{111})\frac{\partial}{\partial u}$$

we find the following Lie-Bäcklund vector field

$$Z_B = \left(\frac{u_{111}}{u^3} - \frac{9u_1 u_{11}}{u^4} + \frac{12u_1^3}{u^5}\right)\frac{\partial}{\partial u}.$$

As a subalgebra we consider a linear combination of the vector fields T_v and Z_B, i.e., $aT_v + Z_B$ ($a \in \mathbf{R}$). The invariant solution for this Lie subalgebra will now be discussed. The condition

$$(aT_v + Z_B)\lrcorner\theta = 0$$

where $\lrcorner$ denotes the contraction, leads to the submanifold

$$-au_2 + \frac{u_{111}}{u^3} - \frac{9u_1 u_{11}}{u^4} + \frac{12u_1^3}{u^5} = 0.$$

Consequently, it follows that

$$js^* \left[-a \left(\frac{u_{11}}{u^2} - \frac{2u_1^2}{u^3} \right) + \frac{u_{111}}{u^3} - \frac{9u_1 u_{11}}{u^4} + \frac{12u_1^3}{u^5} \right] \equiv \frac{\partial^3}{\partial x_1^3} \left(\frac{1}{2u^2} \right) - a \frac{\partial^2}{\partial x_1^2} \left(\frac{1}{u} \right) = 0 \quad (16)$$

where s is the cross section

$$s(x_1, x_2) = (x_1, x_2, u(x_1, x_2))$$

with

$$js^*\theta = 0, \qquad js^*\theta_1 = 0, \ldots$$

and js the extension of s up to infinite order. For deriving (16) we have taken into account the identity

$$\frac{1}{u^3} \frac{\partial^3 u}{\partial x_1^3} - \frac{9}{u^4} \frac{\partial u}{\partial x_1} \frac{\partial^2 u}{\partial x_1^2} + \frac{12}{u^5} \left(\frac{\partial u}{\partial x_1} \right)^3 \equiv -\frac{\partial^3}{\partial x_1^3} \left(\frac{1}{2u^2} \right)$$

and (15). Since derivatives of u with respect to x_2 do not appear in (16) we are able to consider (16) as an ordinary differential equation of third order

$$\frac{d^3}{dx_1^3} \left(\frac{1}{2u^2} \right) - a \frac{d^2}{dx_1^2} \frac{1}{u} = 0 \quad (17)$$

where x_2 plays the role of a parameter and occurs in the constants of integration. The integration of (17) yields

$$\frac{du}{dx_1} + au^2 = (C_1(x_2)x_1 + C_2(x_2))u^3. \quad (18)$$

In order to determine the constants of integration $C_1(x_2)$ and $C_2(x_2)$ we must first solve the ordinary differential equation (18), where a new constant of integration appears which also depends on x_2. Then we insert the solution into the partial differential equation (15) and determine the quantities $C_1(x_2)$, $C_2(x_2)$ and $C_3(x_2)$. Equation (18) is a special case of **Abel's equation** of the first kind and is written in the form

$$\frac{dz}{dy} = z^3 + P(x_1) \quad (19)$$

where

$$x_1(y) = \frac{1}{C_1(x_2)} \left[y(6C_1(x_2) - 2a^2) \right]^{3C_1(x_2)/(6C_1(x_2)-2a^2)}$$

$$u(x_1(y), x_2) = \frac{z(y)}{C_1(x_2)x_1 + C_2(x_2)} + \frac{3}{a} [C_1(x_2)x_1(y) + C_2(x_2)]^{(a^2-6C_1(x_2))/(3C_1(x_2))}$$

$$P(x_1(y)) = \frac{a}{3} \left[C_1(x_2) - \frac{2a^2}{9} \right] [C_1(x_2)x_1(y) + C_2(x_2)]^{(a^2-3C_1(x_2))/C_1(x_2)} .$$

We consider the case

$$C_1(x_2) = \frac{2a^2}{9}$$

so that

$$P(x_1) = 0.$$

By integrating equation (19) we find

$$z(y) = [-2y - 2C_3(x_2)]^{-\frac{1}{2}}.$$

By now expressing $u(x_1, x_2)$ in terms of x_1, $C_2(x_2)$ and $C_3(x_2)$ we insert this expression into the diffusion equation (15) and find a condition on $C_2(x_2)$ and $C_3(x_2)$. We consider

$$C_2(x_2) = 0$$

so that the determining equation reduces to

$$9\frac{dC_3}{dx_2} - 2a^2 C_3(x_2) = 0,$$

with solution

$$C_3(x_2) = k_1 \exp\left(\frac{2}{9}a^2 x_2\right).$$

With the given restrictions we find the invariant (similarity) solution of the diffusion equation (15) to be

$$u(x_1, x_2) = \frac{3\left(81 - 4a^4 k_1 x_1 \exp\left(\frac{2}{9}a^2 x_2\right)\right)^{\frac{1}{2}} + 27}{2ax_1 \left(81 - 4a^4 k_1 x_1 \exp\left(\frac{2}{9}a^2 x_2\right)\right)^{\frac{1}{2}}}.$$

♣

10.3 Computer Algebra Applications

We show that the sine-Gordon equation

$$\frac{\partial^2 u}{\partial x_1 \partial x_2} = \sin(u)$$

admits the Lie-Bäcklund symmetry vector fields

$$Z = (u_{x_1 x_1 x_1} + (2u_x)^3)\frac{\partial}{\partial u}.$$

From the sine-Gordon equation we obtain the manifold

$$u_{x_1 x_2} = \sin(u).$$

Taking the derivatives of the sine-Gordon equation with respect to x_1 leads to the manifolds

$$u_{x_1 x_1 x_2} = u_{x_1}\cos(u)$$

$$u_{x_1 x_1 x_1 x_2} = u_{x_1 x_1}\cos(u) - (u_{x_1})^2\cos(u)$$

$$u_{x_1 x_1 x_1 x_1 x_2} = u_{x_1 x_1 x_1}\cos(u) - 3u_{x_1}u_{x_1 x_1}\sin(u) - (u_{x_1})^3\cos(u).$$

We calculate the prolongated vector field $\bar{Z}$

$$\bar{Z} = Z + (4u_{x_1 x_1 x_1 x_2} + 6(u_{x_1})^2 u_{x_1 x_2})\frac{\partial}{\partial u_{x_2}} + (4u_{x_1 x_1 x_1 x_1 x_2} + 12u_{x_1}u_{x_1 x_1}u_{x_1 x_2} + 6(u_{x_1})^2\frac{\partial}{\partial u_{x_1 x_2}}$$

Taking the Lie derivative yields

$$L_{\bar{Z}}(u_{x_1 x_2} - \sin u) =$$

$$4u_{x_1 x_1 x_1 x_1 x_2} + 12u_{x_1}u_{x_1 x_1} + u_{x_1 x_2} + 6(u_{x_1})^2 u_{x_1 x_1 x_2} - (4u_{x_1 x_1 x_1} + 2(u_{x_1})^3)\cos u.$$

Inserting the manifolds into the right-hand side of this equation leads to

$$L_{\bar{Z}}(u_{x_1 x_2} - \sin(u)) = 0.$$

The REDUCE program is

```
% lb.red


operator D1, D2;  % total differential operators

depend Q, x1, x2, u, u1, u2, u12, u112, u1112, u11112;

for all Q let
D1(Q) = df(Q,x1) + u1*df(Q,u) + u11*df(Q,u1) + u111*df(Q,u11)
+ u1111*df(Q,u111) + u12*df(Q,u2) + u112*df(Q,u12)
+ u1112*df(Q,u112) + u11112*df(Q,u1112);

for all Q let
D2(Q) = df(Q,x2) + u2*df(Q,u) + u22*df(Q,u2) + u222*df(Q,u22)
+ u12*df(Q,u1) + u112*df(Q,u11) + u1112*df(Q,u111)
+ u11112*df(Q,u1111);

equ := u12 - sin(u);

V := (4*u111 + 2*u1*u1*u1);

V12 := D1(D2(V));
write V12;

result := V*df(equ,u) + V12*df(equ,u12);

u11112 := u111*cos(u) - 3*u1*u11*sin(u) - u1*u1*u1*cos(u);
u1112 := u11*cos(u) - u1*u1*sin(u);
u112 := u1*cos(u);
u12 := sin(u);

result;
```

Chapter 11

Differential Equation for a Given Lie Algebra

Thus far we have constructed Lie point symmetry vector fields and Lie-Bäcklund symmetry vector fields of a given partial differential equation. In this chapter we study the inverse problem, i.e., we construct partial differential equations that admit a given symmetry vector field (Rosenhaus [100]).

Let $Z_1, \ldots, Z_q$ be a set of Lie point symmetry vector fields or Lie Bäcklund symmetry vector fields. The question is: What is the differential equation that admits these Lie symmetry vector fields ? We restrict ourselves to partial differential equations of second order. An example for a Lie-Bäcklund vector field is also investigated.

11.1 Lie Point Symmetry Vector Fields

Let us consider the case with one dependent variable u and two independent variables x_1 and x_2. We assume that the Lie algebra is spanned by the following Lie point symmetry vector fields

$$\left\{ \frac{\partial}{\partial x_1}, \frac{\partial}{\partial x_2}, \frac{\partial}{\partial u}, u\frac{\partial}{\partial x_1}, u\frac{\partial}{\partial x_2}, u\frac{\partial}{\partial u}, x_1\frac{\partial}{\partial x_1} + x_2\frac{\partial}{\partial x_2} \right\}$$

and that the partial differential equation is of the form

$$F\left(x_1, x_2, u, \frac{\partial u}{\partial x_1}, \frac{\partial u}{\partial x_2}, \frac{\partial^2 u}{\partial x_1^2}, \frac{\partial^2 u}{\partial x_1 \partial x_2}, \frac{\partial^2 u}{\partial x_2^2}\right) = 0.$$

In the jet bundle $J^2(\mathbf{R}^2, \mathbf{R}^1)$ we write

$$F(x_1, x_2, u, u_1, u_2, u_{11}, u_{12}, u_{22}) = 0$$

where we assume that F is an analytic function of its arguments.

The invariance condition is

$$L_{\bar{Z}_V} F \hat{=} 0.$$

The prolonged vector field $\bar{Z}_V$ of the vertical vector field

$$Z_V = U(x_1, x_2, u, u_1, u_2)\frac{\partial}{\partial u}$$

is given by

$$\bar{Z}_V = U\frac{\partial}{\partial u} + D_1(U)\frac{\partial}{\partial u_1} + D_2(U)\frac{\partial}{\partial u_2} + D_1(D_1(U))\frac{\partial}{\partial u_{11}}$$
$$+ D_1(D_2(U))\frac{\partial}{\partial u_{12}} + D_2(D_2(U))\frac{\partial}{\partial u_{22}}$$

where D_i is the total derivative operator and

$$U = b(x_1, x_2, u) - a_1(x_1, x_2, u)u_1 - a_2(x_1, x_2, u)u_2.$$

Recall that this vertical vector field corresponds to the symmetry vector field

$$Z = a_1(x_1, x_2, u)\frac{\partial}{\partial x_1} + a_2(x_1, x_2, u)\frac{\partial}{\partial x_2} + b(x_1, x_2, u)\frac{\partial}{\partial u}.$$

The prolongation of the symmetry vector field $\partial/\partial x$ (in vertical form) is given by

$$\bar{Z}_V = -u_1\frac{\partial}{\partial u} - u_{11}\frac{\partial}{\partial u_1} - u_{12}\frac{\partial}{\partial u_2} - u_{111}\frac{\partial}{\partial u_{11}} - u_{112}\frac{\partial}{\partial u_{12}} - u_{122}\frac{\partial}{\partial u_{22}}.$$

From the invariance condition it follows that

$$u_1\frac{\partial F}{\partial u} + u_{11}\frac{\partial F}{\partial u_1} + u_{12}\frac{\partial F}{\partial u_2} + u_{111}\frac{\partial F}{\partial u_{11}} + u_{112}\frac{\partial F}{\partial u_{12}} + u_{122}\frac{\partial F}{\partial u_{22}} \hat{=} 0. \qquad (1)$$

The first prolongation of the partial differential equation $F = 0$ is obtained by applying the operator D_1 so that

$$\frac{\partial F}{\partial x_1} + u_1\frac{\partial F}{\partial u} + u_{11}\frac{\partial F}{\partial u_1} + u_{12}\frac{\partial F}{\partial u_2} + u_{111}\frac{\partial F}{\partial u_{11}} + u_{112}\frac{\partial F}{\partial u_{12}} + u_{122}\frac{\partial F}{\partial u_{22}} = 0. \qquad (2)$$

Inserting (2) into (1) yields

$$\frac{\partial F}{\partial x_1} = 0.$$

Analogously it can be shown that

$$\frac{\partial F}{\partial x_2} = 0, \qquad \frac{\partial F}{\partial u} = 0$$

which follows from the symmetry vector fields $\partial/\partial x_2$ and $\partial/\partial u$ respectively. We conclude that F does not depend explicitly on x_1, x_2 and u. Thus we can assume that F is of the form

$$F(u_1, u_2, u_{11}, u_{12}, u_{22}) = 0.$$

For the remaining Lie point symmetry vector fields $u\partial/\partial x_1$, $u\partial/\partial x_2$, $u\partial/\partial u$, and $x_1\partial/\partial x_1 + x_2\partial/\partial x_2$ we find:

$$u_1^2\frac{\partial F}{\partial u_1} + u_1 u_2\frac{\partial F}{\partial u_2} + 3u_1 u_{11}\frac{\partial F}{\partial u_{11}} + (u_{11}u_2 + 2u_1 u_{12})\frac{\partial F}{\partial u_{12}} + (2u_{12}u_2 + u_1 u_{22})\frac{\partial F}{\partial u_{22}} = 0 \quad (3)$$

$$u_1 u_2\frac{\partial F}{\partial u_1} + u_2^2\frac{\partial F}{\partial u_2} + (2u_1 u_{12} + u_{11}u_2)\frac{\partial F}{\partial u_{11}} + (2u_2 u_{12} + u_1 u_{22})\frac{\partial F}{\partial u_{12}} + 3u_2 u_{22}\frac{\partial F}{\partial u_{22}} = 0 \quad (4)$$

$$u_1\frac{\partial F}{\partial u_1} + u_2\frac{\partial F}{\partial u_2} + u_{11}\frac{\partial F}{\partial u_{11}} + u_{12}\frac{\partial F}{\partial u_{12}} + u_{22}\frac{\partial F}{\partial u_{22}} = 0, \quad (5)$$

$$u_1\frac{\partial F}{\partial u_1} + u_2\frac{\partial F}{\partial u_2} + 2u_{11}\frac{\partial F}{\partial u_{11}} + 2u_{12}\frac{\partial F}{\partial u_{12}} + 2u_{22}\frac{\partial F}{\partial u_{22}} = 0. \quad (6)$$

From (5) and (6) we obtain

$$u_2\frac{\partial F}{\partial u_2} = -u_1\frac{\partial F}{\partial u_1}. \quad (7)$$

By making use of (7) we find the following conditions on F:

$$3u_1 u_{11}\frac{\partial F}{\partial u_{11}} + (u_{11}u_2 + 2u_1 u_{12})\frac{\partial F}{\partial u_{12}} + (2u_{12}u_2 + u_1 u_{22})\frac{\partial F}{\partial u_{22}} = 0 \quad (8)$$

$$(2u_1 u_{12} + u_{11}u_2)\frac{\partial F}{\partial u_{11}} + (2u_2 u_{12} + u_1 u_{22})\frac{\partial F}{\partial u_{12}} + 3u_2 u_{22}\frac{\partial F}{\partial u_{22}} = 0 \quad (9)$$

$$u_{11}\frac{\partial F}{\partial u_{11}} + u_{12}\frac{\partial F}{\partial u_{12}} + u_{22}\frac{\partial F}{\partial u_{22}} = 0. \quad (10)$$

Equations (8), (9) and (10) can be expressed in matrix form

$$\begin{pmatrix} 3u_1 u_{11} & (u_{11}u_2 + 2u_1 u_{12}) & (2u_{12}u_2 + u_1 u_{22}) \\ (2u_1 u_{12} + u_{11}u_2) & (2u_2 u_{12} + u_1 u_{22}) & 3u_2 u_{22} \\ u_{11} & u_{12} & u_{22} \end{pmatrix} \begin{pmatrix} \partial F/\partial u_{11} \\ \partial F/\partial u_{12} \\ \partial F/\partial u_{22} \end{pmatrix} = \begin{pmatrix} 0 \\ 0 \\ 0 \end{pmatrix}. \quad (11)$$

Let

$$A := \begin{pmatrix} 3u_1u_{11} & (u_{11}u_2 + 2u_1u_{12}) & (2u_{12}u_2 + u_1u_{22}) \\ (2u_1u_{12} + u_{11}u_2) & (2u_2u_{12} + u_1u_{22}) & 3u_2u_{22} \\ u_{11} & u_{12} & u_{22} \end{pmatrix}$$

with

$$\det A = 2(u_{22}u_{11} - u_{12}^2)(u_2^2u_{11} + u_1^2u_{22} - 2u_2u_1u_{12}).$$

If $\det A \neq 0$, we obtain the trivial solution. The nontrivial solution for F is given by

$$F \equiv \det A = 0 \tag{12}$$

since (12) is the consistency condition of system (11). Thus, a partial differential equation that admits the given Lie point symmetry vector fields is given by

$$\frac{\partial^2 u}{\partial x_1^2}\frac{\partial^2 u}{\partial x_2^2} - \left(\frac{\partial^2 u}{\partial x_1 \partial x_2}\right)^2 = 0. \tag{13}$$

This equation is known as the **Monge-Ampére equation** for the surface $u = u(x_1, x_2)$ with the Gaussian curvature $K = 0$. The partial differential equation

$$\left(\frac{\partial u}{\partial x_1}\right)^2 \frac{\partial^2 u}{\partial x_2^2} - 2\frac{\partial u}{\partial x_1}\frac{\partial u}{\partial x_2}\frac{\partial^2 u}{\partial x_1 \partial x_2} + \left(\frac{\partial u}{\partial x_2}\right)^2 \frac{\partial^2 u}{\partial x_1^2} = 0$$

admits the same Lie point symmetry vector fields. It can be transformed, via the Legendre transformation, to

$$\frac{\partial^2 x_1(x_2, u)}{\partial x_2^2} = 0 \quad \text{or} \quad \frac{\partial^2 x_2(x_1, u)}{\partial x_1^2} = 0.$$

By studying the Lie point symmetry vector fields

$$Z_1 = \frac{\partial}{\partial x_1}, \quad Z_4 = x_1\frac{\partial}{\partial x_1}, \quad Z_7 = x_2\frac{\partial}{\partial x_1}, \quad Z_{10} = u\frac{\partial}{\partial x_2}$$

$$Z_2 = \frac{\partial}{\partial x_2}, \quad Z_5 = x_2\frac{\partial}{\partial x_2}, \quad Z_8 = x_1\frac{\partial}{\partial x_2}, \quad Z_{11} = x_1\frac{\partial}{\partial u}$$

$$Z_3 = \frac{\partial}{\partial u}, \quad Z_6 = u\frac{\partial}{\partial u}, \quad Z_9 = u\frac{\partial}{\partial x_1}, \quad Z_{12} = x_2\frac{\partial}{\partial u} \tag{14}$$

with the analytic function $F(x_1, x_2, u, u_1, u_{11}, u_{12}, u_{22}) = 0$, we find the Monge-Ampére equation (13). The fundamental Lie algebra of the Monge-Ampére equation is a fifteen dimensional Lie algebra spanned by the twelve Lie symmetry vector fields (14) with the following three symmetry vector fields

$$Z_{13} = x_1 P, \qquad Z_{14} = x_2 P, \qquad Z_{15} = u P$$

where

$$P := x_1\frac{\partial}{\partial x_1} + x_2\frac{\partial}{\partial x_2} + u\frac{\partial}{\partial u}.$$

We now consider m independent variables $x_1, x_2, \ldots, x_m$ and one dependent variable u. The Lie algebra is spanned by the vector fields

$$\left\{ \frac{\partial}{\partial x_i}, \frac{\partial}{\partial u}, u\frac{\partial}{\partial u}, x_i\frac{\partial}{\partial x_j}, x_i\frac{\partial}{\partial u}, u\frac{\partial}{\partial x_i} \; : \; i,j = 1,\ldots,m \right\}.$$

The applied algorithm above with

$$F(u_{i_1 i_2}) = 0$$

$i_1 \leq i_2 = 1, \ldots, m$ leads to the following system of equations

$$2u_{11}\frac{\partial F}{\partial u_{11}} + u_{12}\frac{\partial F}{\partial u_{12}} + u_{13}\frac{\partial F}{\partial u_{13}} + \cdots + u_{1n}\frac{\partial F}{\partial u_{1n}} \hat{=} 0$$

$$2u_{21}\frac{\partial F}{\partial u_{11}} + u_{22}\frac{\partial F}{\partial u_{12}} + u_{23}\frac{\partial F}{\partial u_{13}} + \cdots + u_{2n}\frac{\partial F}{\partial u_{1n}} \hat{=} 0$$

$$\cdots\cdots\cdots\cdots\cdots\cdots\cdots\cdots\cdots\cdots\cdots\cdots\cdots\cdots\cdots\cdots\cdots$$

$$2u_{n1}\frac{\partial F}{\partial u_{11}} + u_{n2}\frac{\partial F}{\partial u_{12}} + u_{n3}\frac{\partial F}{\partial u_{13}} + \cdots + u_{nn}\frac{\partial F}{\partial u_{1n}} \hat{=} 0.$$

The only nontrivial F is given by

$$F := \det \begin{pmatrix} u_{11} & u_{12} & \cdots & u_{1n} \\ u_{21} & u_{22} & \cdots & u_{2n} \\ \cdots\cdots\cdots\cdots\cdots\cdots \\ u_{n1} & u_{n2} & \cdots & u_{nn} \end{pmatrix} = 0.$$

Example: We consider two independent variables x_1, x_2 and two dependent variables u, v. For the Lie algebra which is assumed to be spanned by the vector fields

$$\left\{ \frac{\partial}{\partial x_i}, \frac{\partial}{\partial u}, \frac{\partial}{\partial v}, x_i\frac{\partial}{\partial x_i}, u\frac{\partial}{\partial v}, v\frac{\partial}{\partial u}, u\frac{\partial}{\partial u}, v\frac{\partial}{\partial v}, x_i\frac{\partial}{\partial x_j}, x_i\frac{\partial}{\partial u}, x_i\frac{\partial}{\partial v}, u\frac{\partial}{\partial x_i}, v\frac{\partial}{\partial x_i} \right\}$$

$(i,j = 1,2)$ the only nontrivial solution of the conditional equations for

$$F(u_{11}, u_{12}, u_{22}, v_{11}, v_{12}, v_{22}) = 0$$

is given by

$$\left(\frac{\partial^2 u}{\partial x_2^2}\frac{\partial^2 v}{\partial x_1^2} - \frac{\partial^2 u}{\partial x_1^2}\frac{\partial^2 v}{\partial x_2^2} \right)^2$$

$$+4\left(\frac{\partial^2 u}{\partial x_2^2}\frac{\partial^2 v}{\partial x_1 \partial x_2} - \frac{\partial^2 u}{\partial x_1 \partial x_2}\frac{\partial^2 v}{\partial x_2^2} \right)\left(\frac{\partial^2 u}{\partial x_1^2}\frac{\partial^2 v}{\partial x_1 \partial x_2} - \frac{\partial^2 u}{\partial x_1 \partial x_2}\frac{\partial^2 v}{\partial x_1^2} \right) = 0. \quad \clubsuit$$

11.2   Lie-Bäcklund Vector Fields

In this section we study the question of finding a partial differential equation that admits not only Lie point symmetry vector fields but also Lie-Bäcklund vector fields.

We consider, within the jet bundle $J^2(\mathbf{R}^2, \mathbf{R}^1)$, the Lie point symmetry vector field

$$Z = \xi_1(x_1, x_2)\frac{\partial}{\partial x_1} + \xi_2(x_1, x_2)\frac{\partial}{\partial x_2} \tag{15}$$

where

$$\xi_1 = \alpha x_1 + \beta_1, \qquad \xi_2 = -\alpha x_2 + \beta_2$$

with α and β_1, β_2 constants, as well as the Lie-Bäcklund symmetry vector field

$$Z_B = \left(2u_{111} + u_1^3\right)\frac{\partial}{\partial u}. \tag{16}$$

First we consider the Lie point symmetry vector fields (15) with the analytic function

$$F(u, u_{12}) = 0.$$

From the invariance condition

$$L_Z F \hat{=} 0$$

we find

$$(-(\alpha x_1 + \beta_1)u_1 - (-\alpha x_2 + \beta_2)u_2)\frac{\partial F}{\partial u} - ((\alpha x_1 + \beta_1)u_{112} + (-\alpha x_2 + \beta_2)u_{122})\frac{\partial F}{\partial u_{12}} \hat{=} 0. \tag{17}$$

By considering the differential consequence of $F = 0$ we obtain

$$\frac{\partial F}{\partial u}u_1 + \frac{\partial F}{\partial u_{12}}u_{112} = 0 \tag{18}$$

$$\frac{\partial F}{\partial u}u_2 + \frac{\partial F}{\partial u_{12}}u_{122} = 0 \tag{19}$$

$$\frac{\partial^2 F}{\partial u^2}u_1^2 + \frac{\partial F}{\partial u}u_{11} + \frac{\partial^2 F}{\partial u_{12}^2}u_{112}^2 + \frac{\partial F}{\partial u_{12}}u_{1112} = 0 \tag{20}$$

$$\frac{\partial^3 F}{\partial u^3}u_1^3 + 3\frac{\partial^2 F}{\partial u^2}u_1 u_{11} + \frac{\partial F}{\partial u}u_{111} + \frac{\partial^3 F}{\partial u_{12}^3}u_{112}^3 + 3\frac{\partial^2 F}{\partial u_{12}^2}u_{112}u_{1112} + \frac{\partial F}{\partial u_{12}}u_{11112} = 0. \tag{21}$$

From (18), (19) and (17) we find that

$$(\alpha x_1 + \beta_1)\left(\frac{\partial F}{\partial u}u_1 + \frac{\partial F}{\partial u_{12}}u_{112}\right) + (-\alpha x_2 + \beta_2)\left(\frac{\partial F}{\partial u}u_2 + \frac{\partial F}{\partial u_{12}}u_{122}\right)$$

$$+(-(\alpha x_1 + \beta_1)u_1 - (-\alpha x_2 + \beta_2)u_2)\frac{\partial F}{\partial u} - ((\alpha x_1 + \beta_1)u_{112} + (-\alpha x_2 + \beta_2)u_{122})\frac{\partial F}{\partial u_{12}} = 0$$

so that no conditions are imposed on F. This means that the partial differential equation

$$F\left(u(x_1, x_2), \frac{\partial^2 u(x_1, x_2)}{\partial x_1 \partial x_2}\right) = 0$$

admits the symmetry vector field

$$Z = (\alpha x_1 + \beta_1)\frac{\partial}{\partial x_1} + (-\alpha x_2 + \beta_2)\frac{\partial}{\partial x_2}.$$

Let us now consider the Lie-Bäcklund vector field (16) with the special case

$$F \equiv u_{12} - f(u) = 0 \tag{22}$$

where f is an arbitrary function of u. The invariance condition

$$L_{\tilde{Z}_B}F \hat{=} 0$$

results in the following condition on F

$$(2u_{111} + u_1^3)\frac{\partial F}{\partial u} + (2u_{11112} + 6u_1 u_{11} u_{12} + 3u_1^2 u_{112})\frac{\partial F}{\partial u_{12}} \hat{=} 0. \tag{23}$$

By inserting (22) into (23) and (18)~(21) we obtain a condition on $f(u)$, namely

$$2u_1^3\left(\frac{df}{du} + \frac{d^3 f}{du^3}\right) + 6u_1 u_{11}\left(f + \frac{d^2 f}{du^2}\right) = 0. \tag{24}$$

From (24) it follows that

$$\frac{d^2 f}{du^2} + f = 0.$$

Thus f is given by

$$f(u) = c_1 \sin u + c_2 \cos u,$$

where c_1, c_2 are arbitrary constants. The partial differential equation

$$\frac{\partial^2 u}{\partial x_1 \partial x_2} = c_1 \sin u + c_2 \cos u$$

known as the sine-Gordon equation thus admits the Lie-Bäcklund symmetry vector field (16). Z_B is the first vector field in a hierarchy of Lie-Bäcklund vector fields for the sine-Gordon equation. The higher Lie-Bäcklund vector fields of the sine-Gordon equation can be found with the help of the recursion operator

$$R := \left(\frac{\partial}{\partial x_1}\right)^2 + \left(\frac{\partial u}{\partial x_1}\right)^2 - \frac{\partial u}{\partial x_1}D_{x_1}^{-1} \cdot \frac{\partial^2 u}{\partial x_1^2}.$$

11.3   Computer Algebra Applications

We find the determinant of the matrix A. The REDUCE program is

```
% detmat.red

depend u, x1, x2;

matrix A(3,3);

A(1,1) := 3*df(u,x1)*df(u,x1,2);
A(1,2) := df(u,x1,2)*df(u,x2) + 2*df(u,x1)*df(u,x1,x2);
A(1,3) := 2*df(u,x1,x2)*df(u,x2) + df(u,x1)*df(u,x2,2);
A(2,1) := 2*df(u,x1)*df(u,x1,x2) + df(u,x1,2)*df(u,x2);
A(2,2) := 2*df(u,x2)*df(u,x1,x2) + df(u,x1)*df(u,x2,2);
A(2,3) := 3*df(u,x2)*df(u,x2,2);
A(3,1) := df(u,x1,2);
A(3,2) := df(u,x1,x2);
A(3,3) := df(u,x2,2);

equ := det(A);

% check result
fac := 2*(df(u,x2,2)*df(u,x1,2) - df(u,x1,x2)*df(u,x1,x2))*
 (df(u,x2)*df(u,x2)*df(u,x1,2) + df(u,x1)*df(u,x1)*df(u,x2,2)
 - 2*df(u,x2)*df(u,x1)*df(u,x1,x2));

result := equ - fac;
```

Chapter 12

A List of Lie Symmetry Vector Fields

Lie symmetry groups for a large class of linear and nonlinear partial differential equations have been calculated and studied in the literature. Many mathematicians and physicists have contributed to these studies and the calculations are distributed in a large number of different journals. Because of the large amount of information contained in the infinitesimal group generators we give a list of some of the most important partial differential equations in physics and their Lie symmetry vector fields. It is, of course, impossible to make such a list complete.

We adopt the following notation: For the symmetry vector field we write

$$Z = \sum_{j=1}^{m} \xi_j(\mathbf{x}, \mathbf{u}) \frac{\partial}{\partial x_j} + \sum_{j=1}^{n} \eta_j(\mathbf{x}, \mathbf{u}) \frac{\partial}{\partial u_j}$$

where ξ_j and η_j are the coefficient functions.

The **linear wave equation** in three spatial dimensions is given by

$$\frac{\partial^2 u}{\partial x_1^2} + \frac{\partial^2 u}{\partial x_2^2} + \frac{\partial^2 u}{\partial x_3^2} - \frac{\partial^2 u}{\partial x_4^2} = 0.$$

The Lie algebra of the symmetry vector fields is spanned by the fifteen vector fields

$$Z_{x_i} = \frac{\partial}{\partial x_i} \qquad i = 1, \dots, 4$$

$$Z_{r_{ij}} = x_i \frac{\partial}{\partial x_j} - x_j \frac{\partial}{\partial x_i} \qquad i \neq j = 1, 2, 3$$

$$Z_{d_i} = x_i \frac{\partial}{\partial x_4} + x_4 \frac{\partial}{\partial x_i} \qquad i = 1, 2, 3$$

$$Z_l = x_1 \frac{\partial}{\partial x_1} + x_2 \frac{\partial}{\partial x_2} + x_3 \frac{\partial}{\partial x_3} + x_4 \frac{\partial}{\partial x_4}$$

$$Z_{l_i} = -2x_i \left(x_1 \frac{\partial}{\partial x_1} + x_2 \frac{\partial}{\partial x_2} + x_3 \frac{\partial}{\partial x_3} + x_4 \frac{\partial}{\partial x_4} \right)$$

$$+ (x_1^2 + x_2^2 + x_3^2 - x_4^2) \frac{\partial}{\partial x_i} + 2x_i u \frac{\partial}{\partial u}, \qquad i = 1, 2, 3$$

$$Z_{l_4} = 2x_4 \left(x_1 \frac{\partial}{\partial x_1} + x_2 \frac{\partial}{\partial x_2} + x_3 \frac{\partial}{\partial x_3} \right)$$

$$+ (x_1^2 + x_2^2 + x_3^2 + x_4^2) \frac{\partial}{\partial x_4} - 2x_4 u \frac{\partial}{\partial u}$$

which generate the conformal algebra c_4 for $\mathbf{R}^4$ with the Lorentz metric

$$g = dx_1 \otimes dx_1 + dx_2 \otimes dx_2 + dx_3 \otimes dx_3 - dx_4 \otimes dx_4$$

and which include the ten vector fields of the Poincaré group and the five vector fields of the conformal transformations. The additional vector fields

$$u \frac{\partial}{\partial u}, \qquad f(x_1, x_2, x_3, x_4) \frac{\partial}{\partial u}$$

where f is an arbitrary solution of the wave equation, reflect the linearity of the equation.

The three-dimensional **linear diffusion equation** (or **linear heat equation**) is given by

$$\frac{\partial u}{\partial x_4} = \frac{\partial^2 u}{\partial x_1^2} + \frac{\partial^2 u}{\partial x_2^2} + \frac{\partial^2 u}{\partial x_3^2}.$$

The diffusion constant D, which is assumed to be constant is included in the time $t \to t/D$. The Lie algebra is spanned by the thirteen vector fields

$$Z_1 = \frac{\partial}{\partial x_1}, \qquad Z_2 = \frac{\partial}{\partial x_2}, \qquad Z_3 = \frac{\partial}{\partial x_3}, \qquad Z_4 = \frac{\partial}{\partial x_4}$$

$$Z_5 = u \frac{\partial}{\partial u}, \qquad Z_6 = x_1 \frac{\partial}{\partial x_1} + x_2 \frac{\partial}{\partial x_2} + x_3 \frac{\partial}{\partial x_3} + 2x_4 \frac{\partial}{\partial x_4}$$

$$Z_7 = x_4 \frac{\partial}{\partial x_1} - \frac{x_1 u}{2} \frac{\partial}{\partial u}, \qquad Z_8 = x_4 \frac{\partial}{\partial x_2} - \frac{x_2 u}{2} \frac{\partial}{\partial u}, \qquad Z_9 = x_4 \frac{\partial}{\partial x_3} - \frac{x_3 u}{2} \frac{\partial}{\partial u}$$

$$Z_{10} = x_1 \frac{\partial}{\partial x_2} - x_2 \frac{\partial}{\partial x_1}, \qquad Z_{11} = x_2 \frac{\partial}{\partial x_3} - x_3 \frac{\partial}{\partial x_2}, \qquad Z_{12} = x_3 \frac{\partial}{\partial x_1} - x_1 \frac{\partial}{\partial x_3}$$

$$Z_{13} = x_1 x_4 \frac{\partial}{\partial x_1} + x_2 x_4 \frac{\partial}{\partial x_2} + x_3 x_4 \frac{\partial}{\partial x_3} + x_4^2 \frac{\partial}{\partial x_4} - \left(\frac{x_1^2}{4} + \frac{x_2^2}{4} + \frac{x_3^2}{4} + \frac{3x_4}{2} \right) u \frac{\partial}{\partial u}.$$

The **Potential Burgers' equation** represents the simplest wave equation combining both dissipative and nonlinear effects and appears in a wide variety of physical applications. The equation is given by

$$\frac{\partial u}{\partial x_2} = \frac{\partial^2 u}{\partial x_1^2} + \left(\frac{\partial u}{\partial x_1}\right)^2.$$

The Lie algebra is spanned by the following six vector fields

$$Z_1 = \frac{\partial}{\partial x_1}, \qquad Z_4 = x_1 \frac{\partial}{\partial x_1} + 2x_2 \frac{\partial}{\partial x_2}$$

$$Z_2 = \frac{\partial}{\partial x_2}, \qquad Z_5 = 2x_2 \frac{\partial}{\partial x_1} - x_1 \frac{\partial}{\partial u}$$

$$Z_3 = \frac{\partial}{\partial u}, \qquad Z_6 = 4x_2 x_1 \frac{\partial}{\partial x_1} + 4x_2^2 \frac{\partial}{\partial x_2} - (x_1^2 + 2x_2)\frac{\partial}{\partial u}$$

and the infinite dimensional subalgebra

$$Z_f = f(x_1, x_2) e^{-u} \frac{\partial}{\partial u}$$

where $f(x_1, x_2)$ is any solution to the linear diffusion equation.

The **Korteweg-de Vries equation** arises in the theory of long waves in shallow water and other physical systems in which both nonlinear and dispersive effects are relevant. It is given by

$$\frac{\partial u}{\partial x_2} + u\frac{\partial u}{\partial x_1} + \frac{\partial^3 u}{\partial x_1^3} = 0.$$

The Lie algebra is spanned by the four vector fields

$$Z_1 = \frac{\partial}{\partial x_1}, \quad Z_2 = \frac{\partial}{\partial x_2}, \quad Z_3 = x_2 \frac{\partial}{\partial x_1} + \frac{\partial}{\partial u}, \quad Z_4 = x_1 \frac{\partial}{\partial x_1} + 3x_2 \frac{\partial}{\partial x_2} - 2u\frac{\partial}{\partial u}.$$

The two-dimensional generalisation of the Korteweg-de Vries equation or **Kadomtsev-Petviashvili equation** is given by

$$\frac{\partial}{\partial x_1}\left(\frac{\partial u}{\partial x_3} + \frac{3}{2}u\frac{\partial u}{\partial x_1} + \frac{1}{4}\frac{\partial^3 u}{\partial x_1^3}\right) + \frac{3}{4}\sigma\frac{\partial^2 u}{\partial x_2^2} = 0$$

where $\sigma = \pm 1$. The infinite parameter group is generated by

$$
\begin{aligned}
Z_1 &= f(x_3)\frac{\partial}{\partial x_3} + \left[\frac{1}{3}xf'(x_3) - \frac{2}{9}\sigma x_2^2 f''(x_3)\right]\frac{\partial}{\partial x_1} + \frac{2}{3}x_2 f'(x_3)\frac{\partial}{\partial x_2} \\
&\quad + \left[-\frac{2}{3}uf'(x_3) + \frac{2}{9}xf''(x_3) - \frac{4}{27}\sigma x_2^2 f''(x_3)\right]\frac{\partial}{\partial u} \\
Z_2 &= g(x_3)\frac{\partial}{\partial u} + \frac{2}{3}g'(x_3)\frac{\partial}{\partial u} \\
Z_3 &= -\frac{2}{3}\sigma x_2 h'(x_3)\frac{\partial}{\partial x_1} + h(x_3)\frac{\partial}{\partial x_2} - \frac{4}{9}\sigma x_2 h''(x_3)\frac{\partial}{\partial u}
\end{aligned}
$$

where $f(x_3)$, $g(x_3)$ and $h(x_3)$ are three arbitrary smooth function of the variable x_3 and $f'(x_3) \equiv df/dx_3$, $f''(x_3) = d^2 f/dx_3^2$ etc. For the linear ansatz

$$
f(x_3) = c_1 + c_2 x_3, \qquad g(x_3) = c_3 + c_4 x_3, \qquad h(x_3) = c_5 + c_6 x_3
$$

the Lie algebra is spanned by the following six infinitesimal generators

$$
Z_1 = \frac{\partial}{\partial x_3}, \qquad Z_4 = 2\sigma x_2 \frac{\partial}{\partial x_1} + 3x_3\frac{\partial}{\partial x_2}
$$

$$
Z_2 = \frac{\partial}{\partial x_1}, \qquad Z_5 = 3x_3\frac{\partial}{\partial x_1} - 2\frac{\partial}{\partial u}
$$

$$
Z_3 = \frac{\partial}{\partial x_2}, \qquad Z_6 = 3x_3\frac{\partial}{\partial x_3} + x_1\frac{\partial}{\partial x_1} + 2x_2\frac{\partial}{\partial x_2} - 2u\frac{\partial}{\partial u}.
$$

This equation also admits Lie-Bäcklund vector fields.

The **modified Kadomtsev-Petviashvili equation** is given by

$$
\frac{\partial^2 u}{\partial x_1 \partial x_3} = \frac{\partial^2 u}{\partial x_1^2} + 3\frac{\partial^2 u}{\partial x_2^2} - 6\left(\frac{\partial u}{\partial x_1}\right)^2 \frac{\partial^2 u}{\partial x_1^2} - 6\frac{\partial u}{\partial x_2}\frac{\partial^2 u}{\partial x_1^2}.
$$

The infinite parameter group is generated by

$$
\begin{aligned}
Z_1 &= f(x_3)\frac{\partial}{\partial x_3} + \left[\frac{1}{3}xf'(x_3) + \frac{1}{18}x_2^2 f''(x_3)\right]\frac{\partial}{\partial x_1} \\
&\quad + \frac{2}{3}x_2 f'(x_3)\frac{\partial}{\partial x_2} + \left[\frac{1}{18}x_1 x_2 f''(x_3) + \frac{1}{324}x_2^3 f'''(x_3)\right]\frac{\partial}{\partial u} \\
Z_2 &= \frac{1}{6}x_2 g'(x_3)\frac{\partial}{\partial x_1} + g(x_3)\frac{\partial}{\partial x_2} + \left[\frac{1}{12}x_1 g'(x_3) + x_2^2 g''(x_3)\right]\frac{\partial}{\partial u} \\
Z_3 &= h(x_3)\frac{\partial}{\partial x_1} + \frac{1}{6}x_2 h'(x_3)\frac{\partial}{\partial u} \\
Z_4 &= k(x_3)\frac{\partial}{\partial u}.
\end{aligned}
$$

The 3 + 1 dimensional Kadomtsev-Petviashvili equation, also known as **Jimbo-Miwa equation** is given by

$$2\frac{\partial^2 u}{\partial x_4 \partial x_2} - 3\frac{\partial^2 u}{\partial x_1 \partial x_3} + 3\frac{\partial u}{\partial x_2}\frac{\partial^2 u}{\partial x_1^2} + 3\frac{\partial u}{\partial x_1}\frac{\partial^2 u}{\partial x_1 \partial x_2} + \frac{\partial^4 u}{\partial x_1^3 \partial x_2} = 0.$$

The Lie algebra is spanned by the following ten generators

$$Z_1 = \frac{\partial}{\partial x_4} \qquad Z_3 = \frac{\partial}{\partial x_2}$$

$$Z_2 = \frac{\partial}{\partial x_1} \qquad Z_4 = \frac{\partial}{\partial x_3}$$

$$Z_5 = x_2\frac{\partial}{\partial x_2} + x_3\frac{\partial}{\partial x_3}$$

$$Z_6 = x_1\frac{\partial}{\partial x_1} + 2x_3\frac{\partial}{\partial x_3} + 3x_4\frac{\partial}{\partial x_4} - u\frac{\partial}{\partial u}$$

$$Z_7 = f_1(x_4)\frac{\partial}{\partial x_1} + \frac{2}{3}f_1'(x_4)x_1\frac{\partial}{\partial u}$$

$$Z_8 = f_2(x_3)\frac{\partial}{\partial x_1} - f_2'(x_3)x_2\frac{\partial}{\partial u}$$

$$Z_9 = f_3(x_3)\frac{\partial}{\partial x_2} + \frac{3}{4}f_3'(x_3)x_4\frac{\partial}{\partial x_1} - \left(\frac{1}{2}f_3'(x_3)x_1 + \frac{3}{4}f_3''(x_3)x_4x_2\right)\frac{\partial}{\partial u}$$

$$Z_{10} = f_4(x_3, x_4)\frac{\partial}{\partial u}.$$

f' denotes the derivative of f, where $f_1'(x_4) \neq 0$, $f'(x_3) \neq 0$ and f_1, f_2, f_3, f_4 are arbitrary smooth functions.

Remark: *The loop group structure that occurs for the Kadomtsev-Petviashvili equation and other integrable equations in more than 1 + 1 dimensions is absent here. The reason is that the loop group structure requires the presence of terms of the type*

$$h(x_4)\frac{\partial}{\partial x_4} + \cdots$$

or

$$k(x_3)\frac{\partial}{\partial x_3} + \cdots$$

that are absent in this case.

In the investigation of the formation of Maxwellian tails the following nonlinear partial differential equation

$$\frac{\partial^2 u}{\partial x_1 \partial x_2} + \frac{\partial u}{\partial x_1} + u^2 = 0$$

arises. The Lie algebra is spanned by the vector fields

$$Z_1 = \frac{\partial}{\partial x_1}, \quad Z_2 = \frac{\partial}{\partial x_2}, \quad Z_3 = -x_1\frac{\partial}{\partial x_1} + u\frac{\partial}{\partial u}, \quad Z_4 = e^{x_2}\frac{\partial}{\partial x_2} - e^{x_2}u\frac{\partial}{\partial u}.$$

The equation can be simplified using the transformation

$$\tilde{X}_1(x_1, x_2) = x_1$$
$$\tilde{X}_2(x_1, x_2,) = -e^{-x_2}$$
$$\tilde{U}(\tilde{X}_1(x_1, x_2,), \tilde{X}_2(x_1, x_2)) = e^{x_2}u(x_1, x_2).$$

We thus obtain the equation

$$\frac{\partial^2 \tilde{U}}{\partial \tilde{X}_1 \partial \tilde{X}_2} + \tilde{U}^2 = 0$$

where $-1 \leq \tilde{X}_2 < 0$ since $0 \leq X_1 < \infty$. The equation

$$\frac{\partial^? \tilde{U}}{\partial \tilde{X}_1 \partial \tilde{X}_2} + \tilde{U}^n = 0$$

($n = 2, 3, \ldots$) admits the following symmetry vector fields

$$\tilde{Z}_1 = \frac{\partial}{\partial \tilde{X}_1}, \qquad \tilde{Z}_2 = \frac{\partial}{\partial \tilde{X}_2}$$

$$\tilde{Z}_3 = -\tilde{T}\frac{\partial}{\partial \tilde{X}_2} + \frac{\tilde{U}}{n-1}\frac{\partial}{\partial \tilde{U}}, \qquad \tilde{Z}_4 = -\tilde{X}_1\frac{\partial}{\partial \tilde{X}_1} + \frac{\tilde{U}}{n-1}\frac{\partial}{\partial \tilde{U}}.$$

The **Carleman-Boltzmann equation** can be written in the form

$$\frac{\partial^2 u}{\partial x_1^2} - \frac{\partial^2 u}{\partial x_2^2} - 2\frac{\partial u}{\partial x_1}\frac{\partial u}{\partial x_2} = 0.$$

The Lie algebra is spanned by the four vector fields

$$Z_1 = \frac{\partial}{\partial x_2}, \qquad Z_3 = \frac{\partial}{\partial u}, \qquad Z_2 = \frac{\partial}{\partial x_1}, \qquad Z_4 = x_1\frac{\partial}{\partial x_1} + x_2\frac{\partial}{\partial x_2}.$$

The **stochastic Fokker-Planck equation** in one-space dimension is given by

$$\frac{\partial u}{\partial x_2} = A(x_1)u + B(x_1)\frac{\partial u}{\partial x_1} + C(x_1)\frac{\partial^2 u}{\partial x_1^2}.$$

Let f, g be smooth functions of x_1. It follows that

$$A(x_1) = -\frac{df}{dx_1} + \left(\frac{dg}{dx_1}\right)^2 + \frac{d^2g}{dx_1^2}$$

$$B(x_1) = -f(x_1) + 2g\frac{dg}{dx_1}$$

$$C(x_1) = \frac{1}{2}g(x_1) \neq 0.$$

The following cases are considered:

Case 1: $A = A(x_1)$, $B = B(x_1)$, $C = C(x_1)$. The Lie symmetry generators are

$$Z_1 = \frac{\partial}{\partial x_2}, \qquad Z_2 = u\frac{\partial}{\partial u}, \qquad Z_3 = \alpha(x_1, x_2)\frac{\partial}{\partial u}$$

where α is a solution of the Fokker-Planck equation.

Case 2: $f(x_1) = 1$ and $g(x_1) = x_1$. The only Lie symmetry generators are $\{Z_1, Z_2, Z_3\}$ which are given above.

Case 3: $f(x_1) = x_1$ and $g(x_1) = 1$. In addition to Z_1, Z_2 and Z_3 we have

$$Z_4 = e^{x_2}\frac{\partial}{\partial x_2}, \qquad Z_5 = e^{-x_2}\frac{\partial}{\partial x_1} + 2x_1 e^{-x_2}u\frac{\partial}{\partial u}$$

$$Z_6 = x_1 e^{2x_2}\frac{\partial}{\partial x_1} + e^{2x_2}\frac{\partial}{\partial x_2} - e^{2x_2}u\frac{\partial}{\partial u}$$

$$Z_7 = x_1 e^{-2x_2}\frac{\partial}{\partial x_1} - e^{-2x_2}\frac{\partial}{\partial x_2} + 2x_1^2 e^{-2x_1}u\frac{\partial}{\partial u}.$$

Case 4: $f(x_1) = x_1$ and $g(x_1) = x_1$. In addition to Z_1, Z_2 and Z_3 we have

$$Z_4 = x_1\frac{\partial}{\partial x_1}, \qquad Z_5 = x_1 t\frac{\partial}{\partial x_1} - \left(\ln x_1 + \frac{x_2}{2}\right)u\frac{\partial}{\partial u}$$

$$Z_6 = \frac{x_1}{2}\ln x_1\frac{\partial}{\partial x_1} + x_2\frac{\partial}{\partial x_2} - \frac{1}{4}\left(\ln x_1 + \frac{x_2}{2}\right)u\frac{\partial}{\partial u}$$

$$Z_7 = x_1 x_2 \ln x_1\frac{\partial}{\partial x_1} + x_2^2\frac{\partial}{\partial x_2} - \frac{1}{2}\left(x_2 \ln x_1 + \ln^2 x_1 + x_2 + \frac{x_2^2}{4}\right)u\frac{\partial}{\partial u}.$$

Case 5: $f(x_1) = x_1$ and $g(x_1) = x_2 + 1$. The only Lie symmetry generators are $\{Z_1, Z_2, Z_3\}$.

The generalized **nonlinear Schrödinger equation** is given by

$$i\frac{\partial\psi}{\partial x_4} + \frac{\partial^2\psi}{\partial x_1^2} + \frac{\partial^2\psi}{\partial x_2^2} + \frac{\partial^2\psi}{\partial x_3^2} = a_0\psi + a_1|\psi|^2\psi + a_2|\psi|^4\psi$$

where ψ is a complex valued function. We can write

$$\psi(x_1, x_2, x_3, x_4) = u_1(x_1, x_2, x_3, x_4) + iu_2(x_1, x_2, x_3, x_4)$$

where u_1 and u_2 are real-valued functions and $a_j \in \mathbf{R}$ $j = 0, 1, 2$.

Case 1: $a_1 \neq 0$, $a_2 \neq 0$. The Lie algebra is spanned by the following eleven generators:

$$Z_1 = \frac{\partial}{\partial x_4} + a_0\left(u_2\frac{\partial}{\partial u_1} - u_1\frac{\partial}{\partial u_2}\right)$$

$$Z_2 = \frac{\partial}{\partial x_1}, \qquad Z_5 = x_3\frac{\partial}{\partial x_2} - x_2\frac{\partial}{\partial x_3}$$

$$Z_3 = \frac{\partial}{\partial x_2}, \qquad Z_6 = x_1\frac{\partial}{\partial x_3} - x_3\frac{\partial}{\partial x_1}$$

$$Z_4 = \frac{\partial}{\partial x_3}, \qquad Z_7 = x_2\frac{\partial}{\partial x_1} - x_1\frac{\partial}{\partial x_2}$$

$$Z_8 = x_4\frac{\partial}{\partial x_1} - \frac{1}{2}x_1\left(u_2\frac{\partial}{\partial u_1} - u_1\frac{\partial}{\partial u_2}\right), \quad Z_9 = x_4\frac{\partial}{\partial x_2} - \frac{1}{2}x_2\left(u_2\frac{\partial}{\partial u_1} - u_1\frac{\partial}{\partial u_2}\right)$$

$$Z_{10} = x_4\frac{\partial}{\partial x_3} - \frac{1}{2}x_3\left(u_2\frac{\partial}{\partial u_1} - u_1\frac{\partial}{\partial u_2}\right), \quad Z_{11} = u_2\frac{\partial}{\partial u_1} - u_1\frac{\partial}{\partial u_2}.$$

Case 2: $a_1 = 0$, $a_2 \neq 0$ or $a_1 \neq 0$, $a_2 = 0$. The Lie algebra is spanned by the eleven generators from case 1 and the generator

$$Z_{12} = 2x_4\frac{\partial}{\partial x_4} + \left(x_1\frac{\partial}{\partial x_1} + x_2\frac{\partial}{\partial x_2} + x_3\frac{\partial}{\partial x_3}\right) - \delta\left(u_1\frac{\partial}{\partial u_1} + u_2\frac{\partial}{\partial u_2}\right)$$

$$+ 2a_0x_4\left(u_2\frac{\partial}{\partial u_1} - u_1\frac{\partial}{\partial u_2}\right)$$

where

$$\delta = \begin{cases} \frac{1}{2} & \text{for } a_2 \neq 0 \\ 1 & \text{for } a_1 \neq 0 \end{cases}.$$

Remark: *The group associated with case 1 is known as the extended Galilei group $\tilde{G}$, and plays a fundamental role in non-relativistic quantum mechanics. A large class of* equations are invariant under $\tilde{G}$, in particular any nonlinear Schrödinger equation of the form

$$i\frac{\partial\psi}{\partial x_4} + \frac{\partial^2\psi}{\partial x_1^2} + \frac{\partial^2\psi}{\partial x_2^2} + \frac{\partial^2\psi}{\partial x_3^2} = F(|\psi|)\psi$$

where F is a smooth function.

The **von Karman equation** is given by

$$\Delta^2 F = \left(\frac{\partial^2 w}{\partial x_1 \partial x_2}\right)^2 - \frac{\partial^2 w}{\partial x_1^2}\frac{\partial^2 w}{\partial x_2^2}$$

$$\Delta^2 w = -\frac{\partial^2 w}{\partial x_3^2} + \frac{\partial^2 F}{\partial x_2^2}\frac{\partial^2 w}{\partial x_1^2} + \frac{\partial^2 F}{\partial x_1^2}\frac{\partial^2 w}{\partial x_2^2} - 2\frac{\partial^2 F}{\partial x_1 \partial x_2}\frac{\partial^2 w}{\partial x_1 \partial x_2}$$

where

$$\Delta^2 := \frac{\partial^4}{\partial x_1^4} + 2\frac{\partial^4}{\partial x_1^2 \partial x_2^2} + \frac{\partial^4}{\partial x_2^4}.$$

There are eleven infinitesimal generators of a finite symmetry group

$$Z_1 = \frac{\partial}{\partial x_1}, \qquad Z_5 = x_2\frac{\partial}{\partial x_1} - x_1\frac{\partial}{\partial x_2}$$

$$Z_2 = \frac{\partial}{\partial x_2}, \qquad Z_6 = x_3\frac{\partial}{\partial w}$$

$$Z_3 = \frac{\partial}{\partial x_3}, \qquad Z_7 = x_2\frac{\partial}{\partial w}$$

$$Z_4 = \frac{\partial}{\partial w}, \qquad Z_8 = x_1\frac{\partial}{\partial w}$$

$$Z_9 = 2x_3\frac{\partial}{\partial x_3} + x_1\frac{\partial}{\partial x_1} + x_2\frac{\partial}{\partial x_2}$$

$$Z_{10} = x_3 x_2\frac{\partial}{\partial w}, \qquad Z_{11} = x_3 x_2\frac{\partial}{\partial w}.$$

In addition there are three generators depending on three unconstrained smooth functions of the variable x_3:

$$Z_{12} = f(x_3)\frac{\partial}{\partial F}, \qquad Z_{13} = x_1 g(x_3)\frac{\partial}{\partial F}, \qquad Z_{14} = x_2 h(x_3)\frac{\partial}{\partial F}.$$

The **three-wave resonant process** in $(2+1)$ dimensions in the case of explosive instability is described by the equations

$$\frac{\partial u_1}{\partial x_3} + c_1\frac{\partial u_1}{\partial x_1} + d_1\frac{\partial u_1}{\partial x_2} - iu_2^* u_3^* = 0$$

$$\frac{\partial u_2}{\partial x_3} + c_2\frac{\partial u_2}{\partial x_1} + d_2\frac{\partial u_2}{\partial x_2} - iu_1^*u_3^* = 0$$

$$\frac{\partial u_3}{\partial x_3} + c_3\frac{\partial u_3}{\partial x_1} + d_3\frac{\partial u_2}{\partial x_2} - iu_2^*u_1^* = 0$$

where $u_j(x_1, x_2, x_3)$ $(j = 1, 2, 3)$ are the complex amplitudes of the wave packets, c_j, d_j are the group velocities and the asterisk denotes complex conjugate. The infinite dimensional Lie algebra is given by

$$Z_1 = f_1(\eta_1)\alpha_{23}^{-1}\left(c_1\frac{\partial}{\partial x_1} + d_1\frac{\partial}{\partial x_2} + \frac{\partial}{\partial x_3}\right) - \frac{1}{2}f_1'(\eta_1)\left(u_2\frac{\partial}{\partial u_2} + u_2^*\frac{\partial}{\partial u_2^*} + u_3\frac{\partial}{\partial u_3} + u_3^*\frac{\partial}{\partial u_3^*}\right)$$

$$Z_2 = f_2(\eta_2)\alpha_{31}^{-1}\left(c_2\frac{\partial}{\partial x_1} + d_2\frac{\partial}{\partial x_2} + \frac{\partial}{\partial x_3}\right) - \frac{1}{2}f_2'(\eta_2)\left(u_1\frac{\partial}{\partial u_1} + u_1^*\frac{\partial}{\partial u_1^*} + u_3\frac{\partial}{\partial u_3} + u_3^*\frac{\partial}{\partial u_3^*}\right)$$

$$Z_3 = f_3(\eta_3)\alpha_{12}^{-1}\left(c_3\frac{\partial}{\partial x_1} + d_3\frac{\partial}{\partial x_2} + \frac{\partial}{\partial x_3}\right) - \frac{1}{2}f_3'(\eta_3)\left(u_1\frac{\partial}{\partial u_1} + u_1^*\frac{\partial}{\partial u_1^*} + u_2\frac{\partial}{\partial u_2} + u_2^*\frac{\partial}{\partial u_2^*}\right)$$

where f_j are arbitrary smooth functions depending on the variable

$$\eta_j = x_1 - \frac{c_k - c_l}{d_k - d_l}x_2 + \frac{c_kd_l - c_ld_k}{d_k - d_l}$$

and

$$\alpha_{kl} = \frac{d_1(c_3 - c_2) + d_2(c_1 - c_3) + d_3(c_2 - c_1)}{d_k - d_l}$$

with the cyclic indices $j, k, l = 1, 2, 3$.

The **stream function equation** describes the motion of an incompressible constant-property fluid in two dimensions and is given by

$$\nabla^2\frac{\partial u}{\partial x_3} + \frac{\partial u}{\partial x_2}\nabla^2\frac{\partial u}{\partial x_1} - \frac{\partial u}{\partial x_1}\nabla^2\frac{\partial u}{\partial x_2} = \nu\nabla^4u$$

where

$$\nabla^2 := \frac{\partial^2}{\partial x_1^2} + \frac{\partial^2}{\partial x_2^2}, \qquad \nabla^4 := \frac{\partial^4}{\partial x_1^4} + 2\frac{\partial^4}{\partial x_1^2\partial x_2^2} + \frac{\partial^4}{\partial x_2^4}.$$

For the case $\nu = 0$ the coefficient functions are

$$\begin{aligned} \xi_1 &= c_1x_1 + c_2x_2 + c_3x_3x_2 + f_1(x_3) + c_4 \\ \xi_2 &= -c_2x_1 + c_1x_2 - c_3x_3x_1 + f_2(x_3) + c_5 \\ \xi_3 &= c_1c_8x_3 + c_6 \end{aligned}$$

$$\eta = c_1(2 - c_8)u + \frac{1}{2}c_3(x_1^2 + x_2^2) - f_2'(x_3)x_1 + f_1'(x_3)x_2 + f_3(x_3) + c_7$$

where f_1, f_2 and f_3 are arbitrary smooth functions of x_3. The coefficient functions are the same for the full viscous equation, where $\nu =$ constant with $c_8 = 2$. The infinite parameter group is generated by

$$Z_1 \;=\; \frac{\partial}{\partial x_1}, \qquad Z_2 = \frac{\partial}{\partial x_2}, \qquad Z_3 = \frac{\partial}{\partial x_3}, \qquad Z_4 = \frac{\partial}{\partial u}$$

$$Z_5 \;=\; x_1 \frac{\partial}{\partial x_1} + x_2 \frac{\partial}{\partial x_2} + c_8 x_3 \frac{\partial}{\partial x_3} + (2 - c_8) u \frac{\partial}{\partial u}$$

$$Z_6 \;=\; x_2 \frac{\partial}{\partial x_1} - x_1 \frac{\partial}{\partial x_2}$$

$$Z_7 \;=\; x_3 x_2 \frac{\partial}{\partial x_1} - x_3 x_1 \frac{\partial}{\partial x_2} + \frac{1}{2}(x_1^2 + x_2^2) \frac{\partial}{\partial u}$$

$$Z_8 \;=\; f_1(x_3) \frac{\partial}{\partial x_1} + f_2(x_3) \frac{\partial}{\partial x_2} + (-f_2'(x_3) x_1 + f_1'(x_3) x_2 + f_3(x_3)) \frac{\partial}{\partial u}.$$

The **Khokhlov-Zabolotskaya equation** in two space dimensions is given by

$$\frac{\partial^2 u}{\partial x_1 \partial x_3} - \frac{\partial}{\partial x_1}\left(u \frac{\partial u}{\partial x_1} \right) = \frac{\partial^2 u}{\partial x_2^2}.$$

It describes the prolongation of a sound beam in a non-linear medium. The infinite dimensional Lie algebra is given by the vector fields

$$Z_1 \;=\; -\frac{1}{3}(2uf' + x_1 f'' + \frac{1}{2} x_2^2 f''') \frac{\partial}{\partial u} + \frac{1}{6}(2x_1 f' + x_2 f'') \frac{\partial}{\partial x_1} + f \frac{\partial}{\partial x_2} + \frac{3}{2} x_2 f' \frac{\partial}{\partial x_2}$$

$$Z_2 \;=\; -g' \frac{\partial}{\partial u} + g \frac{\partial}{\partial x_1}$$

$$Z_3 \;=\; -\frac{1}{2} x_2 h'' \frac{\partial}{\partial u} + \frac{1}{2} x_2 h' \frac{\partial}{\partial x_1} + h \frac{\partial}{\partial x_2}$$

$$Z_4 \;=\; u \frac{\partial}{\partial u} + x_1 \frac{\partial}{\partial x_1} + \frac{1}{2} x_2 \frac{\partial}{\partial x_2}$$

where f, g and h are arbitrary smooth functions of x_3.

The real **three-dimensional Landau-Ginzburg equation** is given by

$$\frac{\partial^2 u}{\partial x_1^2} + \frac{\partial^2 u}{\partial x_2^2} + \frac{\partial^2 u}{\partial x_3^2} + \frac{\partial u}{\partial x_4} = a_1 + a_2 u + a_3 u^3 + a_4 u^5.$$

It describes the kinetics of phase transitions. There are three general cases to be considered

Case 1: For the case

1a) $a_4 \neq 0$ and at least one of a_1, a_2 and a_3 is nonzero, or

1b) $a_4 = 0$ and $a_3 \neq 0$ and at least one of a_1 and a_2 is nonzero, the Lie algebra is spanned by the following seven vector fields

$$
\begin{aligned}
Z_1 &= \frac{\partial}{\partial x_4}, & Z_5 &= x_2 \frac{\partial}{\partial x_1} - x_1 \frac{\partial}{\partial x_2} \\
Z_2 &= \frac{\partial}{\partial x_1}, & Z_6 &= x_2 \frac{\partial}{\partial x_3} - x_3 \frac{\partial}{\partial x_2} \\
Z_3 &= \frac{\partial}{\partial x_2}, & Z_7 &= x_3 \frac{\partial}{\partial x_1} - x_1 \frac{\partial}{\partial x_3} \\
Z_4 &= \frac{\partial}{\partial x_3}.
\end{aligned}
$$

Case 2: For the case when $a_3 \neq 0$ and $a_1 = a_2 = a_4 = 0$ the Lie algebra is spanned by the following eight vector fields $Z_1, Z_2, Z_3, Z_4, Z_5, Z_6, Z_7$ and

$$
Z_8 = x_1 \frac{\partial}{\partial x_1} + x_2 \frac{\partial}{\partial x_2} + x_3 \frac{\partial}{\partial x_3} + 2x_4 \frac{\partial}{\partial x_4} - u \frac{\partial}{\partial u}.
$$

Case 3: For the case when $a_4 \neq 0$ and $a_1 = a_2 = a_3 = 0$ the Lie algebra is spanned by the following eight vector fields $Z_1, Z_2, Z_3, Z_4, Z_5, Z_6, Z_7$ and

$$
Z_8 = x_1 \frac{\partial}{\partial x_1} + x_2 \frac{\partial}{\partial x_2} + x_3 \frac{\partial}{\partial x_3} + 2x_4 \frac{\partial}{\partial x_4} - \frac{1}{2} u \frac{\partial}{\partial u}.
$$

The nonlinear Klein-Gordon equation

$$
\left(\frac{\partial^2}{\partial x_m^2} + \delta \sum_{i=1}^{m-1} \frac{\partial^2}{\partial x_i^2} \right) \psi = -2(a_2 \psi + 2a_4 \psi^3 + 3a_6 \psi^5)
$$

where ψ is a complex-valued function and $\delta = \pm 1$. The differential operator on the left-hand side is the Laplace-Beltrami operator in Minkowski space $M(m-1,1)$ $(\delta = -1)$ or in Euclidean space $E(n+1)$ $(\delta = +1)$. ψ is a complex valued function. a_2, a_4 and $a_6 \neq 0$ are constants. This is the equation of motion of the classical ψ^6-fields theory.

Case 1: $a_2 \neq 0$, $a_4 \neq 0$, $a_6 \neq 0$. For the case $\delta = -1$ the Lie algebra is spanned by

$$
\begin{aligned}
Z_\mu &= \frac{\partial}{\partial x_\mu}, & \mu &= 1, 2, \ldots, m \\
R_{ij} &= x_i \frac{\partial}{\partial x_j} - x_j \frac{\partial}{\partial x_i}, & i, j &= 1, 2, \ldots, m-1 \\
L_{0j} &= x_0 \frac{\partial}{\partial x_j} + x_j \frac{\partial}{\partial x_0} & j &= 1, 2, \ldots, m-1.
\end{aligned}
$$

For the case $\delta = +1$ the Lie algebra is spanned by

$$Z_\mu = \frac{\partial}{\partial x_\mu}, \qquad \mu = 1, 2, \ldots, n$$

$$R_{\mu\nu} = x_\mu \frac{\partial}{\partial x_\nu} - x_\nu \frac{\partial}{\partial x_\mu}, \qquad \mu, \nu = 1, 2, \ldots, m - 1.$$

Case 2: $a_2 = a_4 = 0$, $-6a_6 = a$. For the case $\delta = -1$ the Lie algebra is spanned by

$$Z_\mu = \frac{\partial}{\partial x_\mu}, \qquad \mu = 1, 2, \ldots, m$$

$$R_{ij} = x_i \frac{\partial}{\partial x_j} - x_j \frac{\partial}{\partial x_i} \qquad i, j = 1, 2, \ldots, n$$

$$L_{0j} = x_0 \frac{\partial}{\partial x_j} + x_j \frac{\partial}{\partial x_0}, \qquad j = 1, 2, \ldots, m - 1$$

$$P = \sum_{\mu=1}^{m-1} x_\mu \frac{\partial}{\partial x_\mu} - \frac{\psi}{2} \frac{\partial}{\partial \psi}.$$

For the case $\delta = +1$ the Lie algebra is spanned by

$$Z_\mu = \frac{\partial}{\partial x_\mu}$$

$$R_{\mu\nu} = x_\mu \frac{\partial}{\partial x_\nu} - x_\nu \frac{\partial}{\partial x_\mu}, \qquad \mu, \nu = 1, \ldots, m - 1$$

$$P = \sum_{\mu=1}^{m-1} x_\mu \frac{\partial}{\partial x_\mu} - \frac{\psi}{2} \frac{\partial}{\partial \psi}.$$

The **Navier-Stokes equations** for an incompressible viscous fluid, with velocity field (u, v, w) and pressure p, is given by

$$\frac{\partial u}{\partial x_4} + u \frac{\partial u}{\partial x_1} + v \frac{\partial u}{\partial x_2} + w \frac{\partial u}{\partial x_3} + \frac{\partial p}{\partial x_1} - \nu \left(\frac{\partial^2 u}{\partial x_1^2} + \frac{\partial^2 u}{\partial x_2^2} + \frac{\partial^2 u}{\partial x_3^2} \right) = 0$$

$$\frac{\partial v}{\partial x_4} + u \frac{\partial v}{\partial x_1} + v \frac{\partial v}{\partial x_2} + w \frac{\partial v}{\partial x_3} + \frac{\partial p}{\partial x_2} - \nu \left(\frac{\partial^2 v}{\partial x_1^2} + \frac{\partial^2 v}{\partial x_2^2} + \frac{\partial^2 v}{\partial x_3^2} \right) = 0$$

$$\frac{\partial w}{\partial x_4} + u \frac{\partial w}{\partial x_1} + v \frac{\partial w}{\partial x_2} + w \frac{\partial w}{\partial x_3} + \frac{\partial p}{\partial x_3} - \nu \left(\frac{\partial^2 w}{\partial x_1^2} + \frac{\partial^2 w}{\partial x_2^2} + \frac{\partial^2 w}{\partial x_3^2} \right) = 0$$

$$\frac{\partial u}{\partial x_1} + \frac{\partial v}{\partial x_2} + \frac{\partial w}{\partial x_3} = 0.$$

The modified pressure has absorbed the density factor, assumed to be constant, and includes also the potential of the external force field, if such is present. The kinematic viscosity ν is assumed to be constant. The following Lie symmetry vector fields are found

$$Z_1 = \frac{\partial}{\partial x_4}$$

$$Z_2 = f_1(x_4)\frac{\partial}{\partial p}$$

$$Z_3 = x_1\frac{\partial}{\partial x_1} + \frac{\partial}{\partial x_2} + x_3\frac{\partial}{\partial x_3} + 2x_4\frac{\partial}{\partial x_4} - u\frac{\partial}{\partial u} - v\frac{\partial}{\partial v} - w\frac{\partial}{\partial w} - 2p\frac{\partial}{\partial p}$$

$$Z_4 = x_1\frac{\partial}{\partial x_2} - x_2\frac{\partial}{\partial x_1} + u\frac{\partial}{\partial v} - v\frac{\partial}{\partial u}$$

$$Z_5 = x_2\frac{\partial}{\partial x_3} - x_3\frac{\partial}{\partial x_2} + v\frac{\partial}{\partial w} - w\frac{\partial}{\partial v}$$

$$Z_6 = x_3\frac{\partial}{\partial x_1} - x_1\frac{\partial}{\partial x_3} + w\frac{\partial}{\partial u} - u\frac{\partial}{\partial w}$$

$$Z_7 = g_1(x_4)\frac{\partial}{\partial x_1} + g_1'(x_4)\frac{\partial}{\partial u} - x_1 g_1''(x_4)\frac{\partial}{\partial p}$$

$$Z_8 = g_2(x_4)\frac{\partial}{\partial x_2} + g_2'(x_4)\frac{\partial}{\partial v} - x_2 g_2''(x_4)\frac{\partial}{\partial p}$$

$$Z_9 = g_3(x_4)\frac{\partial}{\partial x_3} + g_3'(x_4)\frac{\partial}{\partial w} - x_3 g_3''(x_4)\frac{\partial}{\partial p}.$$

Under the presence of rotational symmetry the Navier-Stokes equations can be transformed into the following system of two equations

$$\frac{\partial}{\partial x_4}(D\psi) - \nu D^2\psi - r\frac{\partial(\psi, D\psi/r^2)}{\partial(r, x_3)} - \frac{1}{r^2}\frac{\partial}{\partial x_3}(W^2) = 0$$

$$\frac{\partial W}{\partial x_4} - \nu DW - \frac{1}{r}\frac{\partial(\psi, W)}{\partial(r, W)} = 0$$

where $\psi(r, x_3, x_4)$ is the stream function defined, by virtue of the continuity equation $\text{div}\mathbf{V} = 0$, by the relation

$$V^r := \frac{1}{r}\frac{\partial\psi}{\partial x_3} \;; \qquad V^{x_3} := -\frac{1}{r}\frac{\partial\psi}{\partial r}$$

and the function W is defined by

$$W := rV^\varphi$$

with

$$D := \frac{\partial^2}{\partial r^2} - \frac{1}{r}\frac{\partial}{\partial r} + \frac{\partial^2}{\partial x_3^2}$$

$$\frac{\partial(f, g)}{\partial(r, x_3)} := \frac{\partial f}{\partial r}\frac{\partial g}{\partial x_3} - \frac{\partial f}{\partial x_3}\frac{\partial g}{\partial r}.$$

The following Lie symmetry vector fields are found

$$Z_1 = \frac{\partial}{\partial x_4}$$

$$Z_2 = x_4\frac{\partial}{\partial x_4} + \frac{1}{2}r\frac{\partial}{\partial r} + \frac{1}{2}x_3\frac{\partial}{\partial x_3} + \frac{1}{2}\psi\frac{\partial}{\partial \psi}$$

$$Z_3 = h_1(x_4)\frac{\partial}{\partial x_3} - \frac{1}{2}r^2h_1'(x_4)\frac{\partial}{\partial \psi}$$

$$Z_4 = h_2(x_4)\frac{\partial}{\partial \psi}$$

where h_1 and h_2 are arbitrary smooth functions of x_4.

For a relativistic particle of mass m in flat space-time the **Hamilton-Jacobi equation** is given by

$$\left(\frac{\partial S}{\partial \tau}\right)^2 - \left(\frac{\partial S}{\partial q_1}\right)^2 - \left(\frac{\partial S}{\partial q_2}\right)^2 - \left(\frac{\partial S}{\partial q_3}\right)^2 - (mc)^2 = 0.$$

Here the independent coordinates are q_α ($\alpha = 1, 2, 3$) and $\tau = ct$, the action S is the dependent variable and c is the speed of light in vacuum. If we divide S by mc and call the resulting quantity the new normalised action (retaining the same symbol S), the differential equation is

$$\left(\frac{\partial S}{\partial \tau}\right)^2 - \left(\frac{\partial S}{\partial q_1}\right)^2 - \left(\frac{\partial S}{\partial q_2}\right)^2 - \left(\frac{\partial S}{\partial q_3}\right)^2 - 1 = 0.$$

The Lie algebra is spanned by the following twenty-one generators

$$Z_j = \frac{\partial}{\partial q_j}, \qquad Z_4 = \frac{\partial}{\partial \tau}, \qquad Z_5 = \frac{\partial}{\partial S}$$

$$Z_6 = \left(S\frac{\partial}{\partial S} + \tau\frac{\partial}{\partial \tau} + \sum_{j=1}^{3}q_j\frac{\partial}{\partial q_j}\right)$$

$$Z_{r_{jk}} = \left(q_j\frac{\partial}{\partial q_k} - q_k\frac{\partial}{\partial q_j}\right)$$

$$Z_{l_j} = \left(q_j\frac{\partial}{\partial \tau} + \tau\frac{\partial}{\partial q_j}\right)$$

$$Z_{q_j} = \frac{1}{2}\left(r^2 - \tau^2 + S^2\right)\frac{\partial}{\partial q_j} - q_j\left(S\frac{\partial}{\partial S} + \tau\frac{\partial}{\partial \tau} + \sum_{k=1}^{3}q_k\frac{\partial}{\partial q_k}\right)$$

$$Z_\tau = -\frac{1}{2}\left(r^2 - \tau^2 + S^2\right)\frac{\partial}{\partial \tau} - \tau\left(S\frac{\partial}{\partial S} + \tau\frac{\partial}{\partial \tau} + \sum_{k=1}^{3}q_k\frac{\partial}{\partial q_k}\right)$$

$$Z_S = \frac{1}{2}\left(r^2 - \tau^2 + S^2\right)\frac{\partial}{\partial S} - S\left(S\frac{\partial}{\partial S} + \tau\frac{\partial}{\partial \tau} + \sum_{k=1}^{3} q_k\frac{\partial}{\partial q_k}\right)$$

$$Z_{Sj} = \left(q_j\frac{\partial}{\partial S} - S\frac{\partial}{\partial q_j}\right), \qquad Z_{S\tau} = \left(\tau\frac{\partial}{\partial S} + S\frac{\partial}{\partial \tau}\right)$$

where $j, k = 1, 2, 3$.

The **classical Euclidean** $SU(2)$ **Yang-Mills equations** in the Lorentz gauge can be written in the following form:

$$\sum_{\mu=1}^{4}\left(\frac{\partial^2 A_{a\nu}}{\partial x_\mu^2} - \frac{\partial^2 A_{a\mu}}{\partial x_\mu \partial x_\nu} - g\sum_{b=1}^{3}\sum_{c=1}^{3}\epsilon_{abc}A_{c\nu}\frac{\partial A_{b\nu}}{\partial x_\mu}\right.$$

$$\left. + 2g\sum_{b=1}^{3}\sum_{c=1}^{3}\epsilon_{abc}A_{b\mu}\frac{\partial A_{c\nu}}{\partial x_\mu} + g\sum_{b=1}^{3}\sum_{c=1}^{3}\epsilon_{abc}A_{b\mu}\frac{\partial A_{c\mu}}{\partial x_\nu}\right) = 0$$

$$\sum_{\mu=1}^{4}\frac{\partial A_{a\mu}}{\partial x_\mu} = 0$$

where $\nu = 1, \ldots, 4$ and $a, b, c, = 1, 2, 3$ with ϵ_{abc} the totally antisymmetric tensor $\epsilon_{123} = +1$. The Lie algebra is spanned by

$$Z_\mu = \frac{\partial}{\partial x_\mu}$$

$$Z_{\mu\nu} = x_\mu\frac{\partial}{\partial x_\nu} - x_\nu\frac{\partial}{\partial x_\mu} + \sum_{a=1}^{3}\left(A_{a\mu}\frac{\partial}{\partial A_{a\nu}} - A_{a\nu}\frac{\partial}{\partial A_{a\mu}}\right)$$

$$Z_{ab} = \sum_{\mu=1}^{4}\left(A_{a\mu}\frac{\partial}{\partial A_{b\mu}} - A_{b\mu}\frac{\partial}{\partial A_{a\mu}}\right)$$

$$Z = \sum_{a=1}^{3}\sum_{\mu=1}^{4}\left(x_\mu\frac{\partial}{\partial x_\mu} - A_{a\mu}\frac{\partial}{\partial A_{a\mu}}\right),$$

where $\mu, \nu = 1, \ldots, 4$ and $a, b = 1, 2, 3$.

The **linear Dirac equation** with nonvanishing rest mass is given by

$$\left(i\hbar\left(\gamma_0\frac{\partial}{\partial x_4} + \gamma_1\frac{\partial}{\partial x_1} + \gamma_2\frac{\partial}{\partial x_2} + \gamma_3\frac{\partial}{\partial x_3}\right) - m_0 c\right)\psi(\mathbf{x}) = 0$$

where $\mathbf{x} = (x_1, x_2, x_3, x_4)$, $x_4 = ct$ and

$$\psi(\mathbf{x}) = \begin{pmatrix} \psi_1(\mathbf{x}) \\ \psi_2(\mathbf{x}) \\ \psi_3(\mathbf{x}) \\ \psi_4(\mathbf{x}) \end{pmatrix}.$$

The **gamma matrices** are defined by

$$\gamma_0 \equiv \beta := \begin{pmatrix} 1 & 0 & 0 & 0 \\ 0 & 1 & 0 & 0 \\ 0 & 0 & -1 & 0 \\ 0 & 0 & 0 & -1 \end{pmatrix} \qquad \gamma_1 := \begin{pmatrix} 0 & 0 & 0 & 1 \\ 0 & 0 & 1 & 0 \\ 0 & -1 & 0 & 0 \\ -1 & 0 & 0 & 0 \end{pmatrix}$$

$$\gamma_2 := \begin{pmatrix} 0 & 0 & 0 & -i \\ 0 & 0 & i & 0 \\ 0 & i & 0 & 0 \\ -i & 0 & 0 & 0 \end{pmatrix} \qquad \gamma_3 := \begin{pmatrix} 0 & 0 & 1 & 0 \\ 0 & 0 & 0 & -1 \\ -1 & 0 & 0 & 0 \\ 0 & 1 & 0 & 0 \end{pmatrix}.$$

Since $\psi_k(\mathbf{x})$ $(k = 1, 2, 3, 4)$ is a complex-valued function we put $\psi_k(\mathbf{x}) = u_k(\mathbf{x}) + iv_k(\mathbf{x})$, where u_k and v_k are real valued functions. Then we obtain the following coupled system of eight partial differential equations

$$\lambda\frac{\partial u_1}{\partial x_4} + \lambda\frac{\partial u_3}{\partial x_3} + \lambda\frac{\partial u_4}{\partial x_1} + \lambda\frac{\partial v_4}{\partial x_2} - v_1 = 0$$

$$-\lambda\frac{\partial v_1}{\partial x_4} - \lambda\frac{\partial v_3}{\partial x_3} - \lambda\frac{\partial v_4}{\partial x_1} + \lambda\frac{\partial u_4}{\partial x_2} - u_1 = 0$$

$$\lambda\frac{\partial u_2}{\partial x_4} + \lambda\frac{\partial u_3}{\partial x_1} - \lambda\frac{\partial v_3}{\partial x_2} - \lambda\frac{\partial u_4}{\partial x_3} - v_2 = 0$$

$$-\lambda\frac{\partial v_2}{\partial x_4} - \lambda\frac{\partial v_3}{\partial x_1} - \lambda\frac{\partial u_3}{\partial x_2} + \lambda\frac{\partial v_4}{\partial x_3} - u_2 = 0$$

$$-\lambda\frac{\partial u_1}{\partial x_3} - \lambda\frac{\partial u_2}{\partial x_1} - \lambda\frac{\partial v_2}{\partial x_2} - \lambda\frac{\partial u_3}{\partial x_4} - v_3 = 0$$

$$\lambda\frac{\partial v_1}{\partial x_3} + \lambda\frac{\partial v_2}{\partial x_1} - \lambda\frac{\partial u_2}{\partial x_2} + \lambda\frac{\partial v_3}{\partial x_4} - u_3 = 0$$

$$-\lambda\frac{\partial u_1}{\partial x_1} + \lambda\frac{\partial v_1}{\partial x_2} + \lambda\frac{\partial u_2}{\partial x_3} - \lambda\frac{\partial u_4}{\partial x_4} - v_4 = 0$$

$$\lambda\frac{\partial v_1}{\partial x_1} + \lambda\frac{\partial u_1}{\partial x_2} - \lambda\frac{\partial v_2}{\partial x_3} + \lambda\frac{\partial v_4}{\partial x_4} - u_4 = 0$$

where $\lambda = \hbar/(m_0 c)$. The Lie point symmetry vector fields of this system are given by the ten generators of the field extended Poincaré group

$$Z_i = \frac{\partial}{\partial x_i}, \quad i = 1, \ldots, 4$$

$$Z_{r_{12}} = x_2\frac{\partial}{\partial x_1} - x_1\frac{\partial}{\partial x_2} - \frac{v_1}{2}\frac{\partial}{\partial u_1} + \frac{v_2}{2}\frac{\partial}{\partial u_2} - \frac{v_3}{2}\frac{\partial}{\partial u_3} + \frac{v_4}{2}\frac{\partial}{\partial u_4}$$

$$+ \frac{u_1}{2}\frac{\partial}{\partial v_1} - \frac{u_2}{2}\frac{\partial}{\partial v_2} + \frac{u_3}{2}\frac{\partial}{\partial v_3} - \frac{u_4}{2}\frac{\partial}{\partial v_4}$$

$$Z_{r_{13}} = x_3\frac{\partial}{\partial x_1} - x_1\frac{\partial}{\partial x_3} - \frac{u_2}{2}\frac{\partial}{\partial u_1} + \frac{u_1}{2}\frac{\partial}{\partial u_2} - \frac{u_4}{2}\frac{\partial}{\partial u_3} + \frac{u_3}{2}\frac{\partial}{\partial u_4}$$

$$- \frac{v_2}{2}\frac{\partial}{\partial v_1} + \frac{v_1}{2}\frac{\partial}{\partial v_2} - \frac{v_4}{2}\frac{\partial}{\partial v_3} + \frac{v_3}{2}\frac{\partial}{\partial v_4}$$

$$Z_{r_{23}} = x_3\frac{\partial}{\partial x_2} - x_2\frac{\partial}{\partial x_3} - \frac{v_2}{2}\frac{\partial}{\partial u_1} - \frac{v_1}{2}\frac{\partial}{\partial u_2} - \frac{v_4}{2}\frac{\partial}{\partial u_3} - \frac{v_3}{2}\frac{\partial}{\partial u_4}$$

$$+ \frac{u_2}{2}\frac{\partial}{\partial v_1} + \frac{u_1}{2}\frac{\partial}{\partial v_2} + \frac{u_4}{2}\frac{\partial}{\partial v_3} + \frac{u_3}{2}\frac{\partial}{\partial v_4}$$

$$Z_{l_{14}} = x_4\frac{\partial}{\partial x_1} + x_1\frac{\partial}{\partial x_4} + \frac{u_4}{2}\frac{\partial}{\partial u_1} + \frac{u_3}{2}\frac{\partial}{\partial u_2} + \frac{u_2}{2}\frac{\partial}{\partial u_3} + \frac{u_1}{2}\frac{\partial}{\partial u_4}$$

$$+ \frac{v_4}{2}\frac{\partial}{\partial v_1} + \frac{v_3}{2}\frac{\partial}{\partial v_2} + \frac{v_2}{2}\frac{\partial}{\partial v_3} + \frac{v_1}{2}\frac{\partial}{\partial v_4}$$

$$Z_{l_{24}} = x_4\frac{\partial}{\partial x_2} + x_2\frac{\partial}{\partial x_4} + \frac{v_4}{2}\frac{\partial}{\partial u_1} - \frac{v_3}{2}\frac{\partial}{\partial u_2} + \frac{v_2}{2}\frac{\partial}{\partial u_3} - \frac{v_1}{2}\frac{\partial}{\partial u_4}$$

$$- \frac{u_4}{2}\frac{\partial}{\partial v_1} + \frac{u_3}{2}\frac{\partial}{\partial v_2} - \frac{u_2}{2}\frac{\partial}{\partial v_3} + \frac{u_1}{2}\frac{\partial}{\partial v_4}$$

$$Z_{l_{34}} = x_4\frac{\partial}{\partial x_3} + x_3\frac{\partial}{\partial x_4} + \frac{u_3}{2}\frac{\partial}{\partial u_1} - \frac{u_4}{2}\frac{\partial}{\partial u_2} + \frac{u_1}{2}\frac{\partial}{\partial u_3} - \frac{u_2}{2}\frac{\partial}{\partial u_4}$$

$$+ \frac{v_3}{2}\frac{\partial}{\partial v_1} - \frac{v_4}{2}\frac{\partial}{\partial v_2} + \frac{v_1}{2}\frac{\partial}{\partial v_3} - \frac{v_2}{2}\frac{\partial}{\partial v_4}$$

and the four additional generators

$$Z_{11} = u_4\frac{\partial}{\partial u_1} - u_3\frac{\partial}{\partial u_2} - u_2\frac{\partial}{\partial u_3} + u_1\frac{\partial}{\partial u_4} - v_4\frac{\partial}{\partial v_1} + v_3\frac{\partial}{\partial v_2} + v_2\frac{\partial}{\partial v_3} - v_1\frac{\partial}{\partial v_4}$$

$$Z_{12} = v_1\frac{\partial}{\partial u_1} + v_2\frac{\partial}{\partial u_2} + v_3\frac{\partial}{\partial u_3} + v_4\frac{\partial}{\partial u_4} - u_1\frac{\partial}{\partial v_1} - u_2\frac{\partial}{\partial v_2} - u_3\frac{\partial}{\partial v_3} - u_4\frac{\partial}{\partial v_4}$$

$$Z_{13} = v_4\frac{\partial}{\partial u_1} - v_3\frac{\partial}{\partial u_2} - v_2\frac{\partial}{\partial u_3} + v_1\frac{\partial}{\partial u_4} + u_4\frac{\partial}{\partial v_1} - u_3\frac{\partial}{\partial v_2} - u_2\frac{\partial}{\partial v_3} + u_1\frac{\partial}{\partial v_4}$$

$$Z_{14} = u_1\frac{\partial}{\partial u_1} + u_2\frac{\partial}{\partial u_2} + u_3\frac{\partial}{\partial u_3} + u_4\frac{\partial}{\partial u_4} + v_1\frac{\partial}{\partial v_1} + v_2\frac{\partial}{\partial v_2} + v_3\frac{\partial}{\partial v_3} + v_4\frac{\partial}{\partial v_4}.$$

The **Maxwell-Dirac equation**

$$\left(\frac{\partial^2}{\partial x_4^2} - \sum_{j=1}^{3}\frac{\partial^2}{\partial x_j^2}\right) A_\mu = \mu_0 ec\bar{\psi}\gamma_\mu\psi$$

$$\left(i\hbar\left(\gamma_0\frac{\partial}{\partial x_4}+\sum_{j=1}^{3}\gamma_j\frac{\partial}{\partial x_j}\right)-m_0c\right)\psi = e\left(\sum_{j=1}^{3}\gamma_jA^j+\gamma_0A_0\right)\psi$$

$$\frac{\partial A_0}{\partial x_4}+\frac{\partial A_1}{\partial x_1}+\frac{\partial A_2}{\partial x_2}+\frac{\partial A_3}{\partial x_3} = 0,$$

is the Dirac equation which is coupled with the vector potential A_μ where $A_0 = U/c$ and U denotes the scalar potential. Here $\mu = 0, 1, 2, 3$ and

$$\bar\psi = (\psi_1^*, \ \psi_2^*, \ -\psi_3^*, \ -\psi_4^*).$$

c is the speed of light, e the charge, μ_0 the permeability of free space, $\hbar = h/(2\pi)$, where h is Planck's constant and m_0 the particle rest mass. This can be written as the following coupled system of thirteen partial differential equations

$$\left(\frac{\partial^2}{\partial x_4^2}-\sum_{j=1}^{3}\frac{\partial^2}{\partial x_j^2}\right)A_0 = (\mu_0 ec)(u_1^2+v_1^2+u_2^2+v_2^2+u_3^2+v_3^2+u_4^2+v_4^2)$$

$$\left(\frac{\partial^2}{\partial x_4^2}-\sum_{j=1}^{3}\frac{\partial^2}{\partial x_j^2}\right)A_1 = (2\mu_0 ec)(u_1u_4+v_1v_4+u_2u_3+v_2v_3)$$

$$\left(\frac{\partial^2}{\partial x_4^2}-\sum_{j=1}^{3}\frac{\partial^2}{\partial x_j^2}\right)A_2 = (2\mu_0 ec)(u_1v_4-u_4v_1+u_3v_2-u_2v_3)$$

$$\left(\frac{\partial^2}{\partial x_4^2}-\sum_{j=1}^{3}\frac{\partial^2}{\partial x_j^2}\right)A_3 = (2\mu_0 ec)(u_1u_3+v_3v_1-u_2u_4-v_2v_4)$$

$$\lambda\frac{\partial u_1}{\partial x_4}+\lambda\frac{\partial u_3}{\partial x_3}+\lambda\frac{\partial u_4}{\partial x_1}+\lambda\frac{\partial v_4}{\partial x_2}-v_1 = \frac{e}{m_0c}(A_0v_1-A_3v_3-A_1v_4+A_2u_4)$$

$$-\lambda\frac{\partial v_1}{\partial x_4}-\lambda\frac{\partial v_3}{\partial x_3}-\lambda\frac{\partial v_4}{\partial x_1}+\lambda\frac{\partial u_4}{\partial x_2}-u_1 = \frac{e}{m_0c}(A_0u_1-A_3u_3-A_1u_4-A_2v_4)$$

$$\lambda\frac{\partial u_2}{\partial x_4}+\lambda\frac{\partial u_3}{\partial x_1}-\lambda\frac{\partial v_3}{\partial x_2}-\lambda\frac{\partial u_4}{\partial x_3}-v_2 = \frac{e}{m_0c}(A_0v_2-A_1v_3-A_2u_3+A_3v_4)$$

$$-\lambda\frac{\partial v_2}{\partial x_4}-\lambda\frac{\partial v_3}{\partial x_1}-\lambda\frac{\partial u_3}{\partial x_2}+\lambda\frac{\partial v_4}{\partial x_3}-u_2 = \frac{e}{m_0c}(A_0u_2-A_1u_3+A_2v_3+A_3u_4)$$

$$-\lambda\frac{\partial u_1}{\partial x_3}-\lambda\frac{\partial u_2}{\partial x_1}-\lambda\frac{\partial v_2}{\partial x_2}-\lambda\frac{\partial u_3}{\partial x_4}-v_3 = \frac{e}{m_0c}(A_3v_1+A_1v_2-A_2u_2-A_0v_3)$$

$$\lambda\frac{\partial v_1}{\partial x_3}+\lambda\frac{\partial v_2}{\partial x_1}-\lambda\frac{\partial u_2}{\partial x_2}+\lambda\frac{\partial v_3}{\partial x_4}-u_3 = \frac{e}{m_0c}(A_3u_1+A_1u_2+A_2v_2-A_0u_3)$$

$$-\lambda\frac{\partial u_1}{\partial x_1} + \lambda\frac{\partial v_1}{\partial x_2} + \lambda\frac{\partial u_2}{\partial x_3} - \lambda\frac{\partial u_4}{\partial x_4} - v_4 = \frac{e}{m_0 c}(A_1 v_1 + A_2 u_1 - A_3 v_2 - A_0 v_4)$$

$$\lambda\frac{\partial v_1}{\partial x_1} + \lambda\frac{\partial u_1}{\partial x_2} - \lambda\frac{\partial v_2}{\partial x_3} + \lambda\frac{\partial v_4}{\partial x_4} - u_4 = \frac{e}{m_0 c}(A_1 u_1 - A_2 v_1 - A_3 u_2 - A_0 u_4)$$

$$\frac{\partial A_0}{\partial x_4} + \frac{\partial A_1}{\partial x_1} + \frac{\partial A_2}{\partial x_2} + \frac{\partial A_3}{\partial x_3} = 0.$$

We found the following ten field extended generators for the Poincaré group that leave the coupled Maxwell-Dirac equation invariant:

$$Z_i = \frac{\partial}{\partial x_i}, \qquad i = 1, \ldots, 4$$

$$Z_{r_{12}} = x_2\frac{\partial}{\partial x_1} - x_1\frac{\partial}{\partial x_2} - \frac{v_1}{2}\frac{\partial}{\partial u_1} + \frac{v_2}{2}\frac{\partial}{\partial u_2} - \frac{v_3}{2}\frac{\partial}{\partial u_3} + \frac{v_4}{2}\frac{\partial}{\partial u_4}$$

$$+ \frac{u_1}{2}\frac{\partial}{\partial v_1} - \frac{u_2}{2}\frac{\partial}{\partial v_2} + \frac{u_3}{2}\frac{\partial}{\partial v_3} - \frac{u_4}{2}\frac{\partial}{\partial v_4} + A_2\frac{\partial}{\partial A_1} - A_1\frac{\partial}{\partial A_2}$$

$$Z_{r_{13}} = x_3\frac{\partial}{\partial x_1} - x_1\frac{\partial}{\partial x_3} - \frac{u_2}{2}\frac{\partial}{\partial u_1} + \frac{u_1}{2}\frac{\partial}{\partial u_2} - \frac{u_4}{2}\frac{\partial}{\partial u_3} + \frac{u_3}{2}\frac{\partial}{\partial u_4}$$

$$- \frac{v_2}{2}\frac{\partial}{\partial v_1} + \frac{v_1}{2}\frac{\partial}{\partial v_2} - \frac{v_4}{2}\frac{\partial}{\partial v_3} + \frac{v_3}{2}\frac{\partial}{\partial v_4} + A_3\frac{\partial}{\partial A_1} - A_1\frac{\partial}{\partial A_3}$$

$$Z_{r_{23}} = x_3\frac{\partial}{\partial x_2} - x_2\frac{\partial}{\partial x_3} - \frac{v_2}{2}\frac{\partial}{\partial u_1} - \frac{v_1}{2}\frac{\partial}{\partial u_2} - \frac{v_4}{2}\frac{\partial}{\partial u_3} - \frac{v_3}{2}\frac{\partial}{\partial u_4}$$

$$+ \frac{u_2}{2}\frac{\partial}{\partial v_1} + \frac{u_1}{2}\frac{\partial}{\partial v_2} + \frac{u_4}{2}\frac{\partial}{\partial v_3} + \frac{u_3}{2}\frac{\partial}{\partial v_4} + A_3\frac{\partial}{\partial A_2} - A_2\frac{\partial}{\partial A_3}$$

$$Z_{l_{14}} = x_4\frac{\partial}{\partial x_1} + x_1\frac{\partial}{\partial x_4} + \frac{u_4}{2}\frac{\partial}{\partial u_1} + \frac{u_3}{2}\frac{\partial}{\partial u_2} + \frac{u_2}{2}\frac{\partial}{\partial u_3} + \frac{u_1}{2}\frac{\partial}{\partial u_4}$$

$$+ \frac{v_4}{2}\frac{\partial}{\partial v_1} + \frac{v_3}{2}\frac{\partial}{\partial v_2} + \frac{v_2}{2}\frac{\partial}{\partial v_3} + \frac{v_1}{2}\frac{\partial}{\partial v_4} + A_0\frac{\partial}{\partial A_1} + A_1\frac{\partial}{\partial A_0}$$

$$Z_{l_{24}} = x_4\frac{\partial}{\partial x_2} + x_2\frac{\partial}{\partial x_4} + \frac{v_4}{2}\frac{\partial}{\partial u_1} - \frac{v_3}{2}\frac{\partial}{\partial u_2} + \frac{v_2}{2}\frac{\partial}{\partial u_3} - \frac{v_1}{2}\frac{\partial}{\partial u_4}$$

$$- \frac{u_4}{2}\frac{\partial}{\partial v_1} + \frac{u_3}{2}\frac{\partial}{\partial v_2} - \frac{u_2}{2}\frac{\partial}{\partial v_3} + \frac{u_1}{2}\frac{\partial}{\partial v_4} + A_0\frac{\partial}{\partial A_2} + A_2\frac{\partial}{\partial A_0}$$

$$Z_{l_{34}} = x_4\frac{\partial}{\partial x_3} + x_3\frac{\partial}{\partial x_4} + \frac{u_3}{2}\frac{\partial}{\partial u_1} - \frac{u_4}{2}\frac{\partial}{\partial u_2} + \frac{u_1}{2}\frac{\partial}{\partial u_3} - \frac{u_2}{2}\frac{\partial}{\partial u_4}$$

$$+ \frac{v_3}{2}\frac{\partial}{\partial v_1} - \frac{v_4}{2}\frac{\partial}{\partial v_2} + \frac{v_1}{2}\frac{\partial}{\partial v_3} - \frac{v_2}{2}\frac{\partial}{\partial v_4} + A_0\frac{\partial}{\partial A_3} + A_3\frac{\partial}{\partial A_0}.$$

Chapter 13

Recursion Operators

13.1 Gateaux Derivative

In this chapter we study symmetries of partial differential equations in connection with the Gateaux derivative (Fokas [47], Fokas and Fuchssteiner [48], Kowalski and Steeb [71]). In particular we are concerned with the study of recursion operators. Recursion operators have been used in the early work on soliton theory for generating symmetries (see Olver [88], Wadati [142], Ibragimov and Shabat [59], Bluman and Kumei [7]).

We consider the manifold N as a topological vector space W. The space of linear continuous mappings of W into some topological vector space W_1 will be denoted by $L(W, W_1)$. We consider $L(W, W_1)$ as a topological vector space with the topology of bounded convergence.

Let f be a (nonlinear) mapping $f : W \to W_1$. The Gateaux derivative is defined as follows:

Definition 13.1 f *is* **Gateaux differentiable** *in* $\mathbf{u} \in W$ *if there exists a mapping* $\vartheta \in L(W, W_1)$ *such that for all* $\mathbf{v} \in W$

$$\lim_{\epsilon \to 0} \frac{1}{\epsilon}(f(\mathbf{u} + \epsilon \mathbf{v}) - f(\mathbf{u}) - \epsilon \vartheta \mathbf{v}) = 0 \tag{1}$$

in the topology of W_1. *The linear mapping* $\vartheta \in L(W, W_1)$ *is called the* **Gateaux derivative** *of* f *in* $\mathbf{u}$ *and is written as* $\vartheta \mathbf{v} = f'(\mathbf{u})[\mathbf{v}]$, *where*

$$f'(\mathbf{u})[\mathbf{v}] := \frac{\partial}{\partial \epsilon} f(\mathbf{u} + \epsilon \mathbf{v})|_{\epsilon=0}.$$

Here $\mathbf{u} = (u_1, \ldots, u_n)$ and $\mathbf{v} = (v_1, \ldots, v_n)$. If f is Gateaux differentiable in all points $\mathbf{u} \in W$, we can consider the Gateaux derivative as a (in general nonlinear) mapping

$$f' : W \to L(W, W_1).$$

Suppose f' is again Gateaux differentiable in $\mathbf{u} \in W$. The second derivative of f in $\mathbf{u} \in W$ is a linear mapping

$$f''(\mathbf{u}) \in L(W, L(W, W_1)).$$

The mapping $f''(\mathbf{u})$ is considered to be bilinear, i.e.,

$$f''(\mathbf{u}) : W \times W \to W_1.$$

Under certain assumptions it can be shown that this mapping is symmetric:

$$f''(\mathbf{u})(\mathbf{v}, \mathbf{w}) = f''(\mathbf{u})(\mathbf{w}, \mathbf{v})$$

for all $\mathbf{w}, \mathbf{v} \in W$.

Definition 13.2 *A mapping $f : W \to W_1$ is called twice differentiable if its first and second Gateaux derivatives exist and if $f''(\mathbf{u})$ is a symmetric bilinear mapping for all $\mathbf{u} \in W$.*

Remark: *In the limit given in (1) a uniformity in $\mathbf{v}$ is not required. If this limit is uniform on all sequentially compact subsets of W, the mapping f is called* **Hadamard differentiable**. *If the limit is uniform on all bounded subsets of W, the mapping f is called* **Fréchet differentiable**.

Definition 13.3 *An evolution equation is a partial differential equation of the form*

$$\frac{\partial \mathbf{u}}{\partial x_{m+1}} = \mathbf{F}\left(\mathbf{x}, x_{m+1}, \mathbf{u}, \frac{\partial \mathbf{u}}{\partial \mathbf{x}}, \ldots, \frac{\partial^r \mathbf{u}}{\partial \mathbf{x}^r}\right) \tag{2}$$

where $\mathbf{F} = (F_1, \ldots, F_n)$, $\mathbf{x} = (x_1, \ldots, x_m)$ and $\mathbf{u} = (u_1, \ldots, u_n)$. In physics the independent variable x_{m+1} is associated with time.

Here $\mathbf{F}$ is assumed to be a smooth function with respect to $\mathbf{x}, x_{m+1}, \mathbf{u}, \partial \mathbf{u}/\partial \mathbf{x}, \ldots, \partial^r \mathbf{u}/\partial \mathbf{x}^r$.

Definition 13.4 *Consider the map $f(\mathbf{u})$*

$$f(\mathbf{u}) := \frac{\partial \mathbf{u}}{\partial x_{m+1}} - \mathbf{F}\left(\mathbf{x}, x_{m+1}, \mathbf{u}, \frac{\partial \mathbf{u}}{\partial \mathbf{x}}, \ldots, \frac{\partial^r \mathbf{u}}{\partial \mathbf{x}^r}\right).$$

The equation

$$\vartheta \mathbf{v} = 0$$

is called the **variational equation** *(or* **linearized equation***) of the evolution equation (2).*

From definition 13.4 it follows that

$$
\begin{aligned}
\vartheta\mathbf{v} &= \lim_{\epsilon\to 0}\left[\frac{f(\mathbf{u}+\epsilon\mathbf{v})-f(\mathbf{u})}{\epsilon}\right] \\
&= \lim_{\epsilon\to 0}\left[\frac{1}{\epsilon}\left(\frac{\partial(\mathbf{u}+\epsilon\mathbf{v})}{\partial x_{m+1}}-\mathbf{F}\left(\mathbf{x},x_{m+1},\mathbf{u}+\epsilon\mathbf{v},\frac{\partial(\mathbf{u}+\epsilon\mathbf{v})}{\partial\mathbf{x}},\dots,\frac{\partial^r(\mathbf{u}+\epsilon\mathbf{v})}{\partial\mathbf{x}^r}\right)\right.\right. \\
&\quad -\left.\left.\frac{\partial\mathbf{u}}{\partial x_{m+1}}+\mathbf{F}\left(\mathbf{x},x_{m+1},\mathbf{u},\frac{\partial\mathbf{u}}{\partial\mathbf{x}},\dots,\frac{\partial^r\mathbf{u}}{\partial\mathbf{x}^r}\right)\right)\right] \\
&= \frac{\partial\mathbf{v}}{\partial x_{m+1}}-\lim_{\epsilon\to 0}\left[\frac{1}{\epsilon}\left(\mathbf{F}\left(\mathbf{x},x_{m+1},\mathbf{u}+\epsilon\mathbf{v},\frac{\partial(\mathbf{u}+\epsilon\mathbf{v})}{\partial\mathbf{x}},\dots,\frac{\partial^r(\mathbf{u}+\epsilon\mathbf{v})}{\partial\mathbf{x}^r}\right)\right.\right. \\
&\quad -\left.\left.\mathbf{F}\left(\mathbf{x},x_{m+1},\mathbf{u},\frac{\partial\mathbf{u}}{\partial\mathbf{x}},\dots,\frac{\partial^r\mathbf{u}}{\partial\mathbf{x}^r}\right)\right)\right]
\end{aligned}
$$

so that the variational equation can be written as

$$
\frac{\partial\mathbf{v}}{\partial x_{m+1}}=\mathbf{F}'(\mathbf{u})[\mathbf{v}]. \tag{3}
$$

Example: Consider the **Lorenz model**

$$
\frac{du_1}{dx_1}-\sigma(u_2-u_1)=0
$$

$$
\frac{du_2}{dx_1}+u_1u_3-ru_2=0
$$

$$
\frac{du_3}{dx_1}-u_1u_2+bu_3=0
$$

where σ, b and r are positive constants. Here $m=0$ and $n=3$ so that there is one independent variable x_1 and three dependent variables u_1, u_2 and u_3. The left hand side defines a map f. Since

$$
f(\mathbf{u}+\epsilon\mathbf{v})-f(\mathbf{u})=\begin{pmatrix}\epsilon\dfrac{dv_1}{dx_1}-\epsilon\sigma(v_2-v_1)\\[2mm]\epsilon\dfrac{dv_2}{dx_1}+\epsilon u_1v_3+\epsilon u_3v_1+\epsilon^2 v_1v_3-\epsilon rv_2\\[2mm]\epsilon\dfrac{dv_3}{dx_1}-\epsilon u_1v_2-\epsilon u_2v_1-\epsilon^2 v_1v_2+bv_3\end{pmatrix}
$$

we obtain

$$
\vartheta\mathbf{v}=\begin{pmatrix}\dfrac{dv_1}{dx_1}-\sigma(v_2-v_1)\\[2mm]\dfrac{dv_2}{dx_1}+u_1v_3+u_3v_1-rv_2\\[2mm]\dfrac{dv_3}{dx_1}-u_1v_2-u_2v_1+bv_3\end{pmatrix}.
$$

The variational equation $\vartheta \mathbf{v} = 0$ takes the form

$$\frac{dv_1}{dx_1} = \sigma(v_2 - v_1)$$

$$\frac{dv_2}{dx_1} = -u_1 v_3 - u_3 v_1 + r v_2$$

$$\frac{dv_3}{dx_1} = u_1 v_2 + u_2 v_1 - b v_3.$$

In order to solve the variational equation, we have to solve the Lorenz model.   ♣

Example: Consider the Korteweg-de Vries equation

$$\frac{\partial u}{\partial x_2} - 6u \frac{\partial u}{\partial x_1} - \frac{\partial^3 u}{\partial x_1^3} = 0.$$

Here $m = 2$ and $n = 1$. We put $u_1 = u$ and $v_1 = v$. The left hand side defines the map

$$f(u) = \frac{\partial u}{\partial x_2} - 6u \frac{\partial u}{\partial x_1} - \frac{\partial^3 u}{\partial x_1^3}.$$

It follows that

$$f(u + \epsilon v) - f(u) = \epsilon \frac{\partial v}{\partial x_2} - 6\epsilon v \frac{\partial u}{\partial x_1} - 6\epsilon u \frac{\partial v}{\partial x_1} - 6\epsilon^2 v \frac{\partial v}{\partial x_1} - \epsilon \frac{\partial^3 v}{\partial x_1^3}$$

so that

$$\lim_{\epsilon \to 0} \frac{1}{\epsilon} (f(u + \epsilon v) - f(u)) = \frac{\partial v}{\partial x_2} - 6 \frac{\partial u}{\partial x_1} v - 6u \frac{\partial v}{\partial x_1} - \frac{\partial^3 v}{\partial x_1^3}.$$

Therefore the variational equation of the Korteweg-de Vries equation is

$$\frac{\partial v}{\partial x_2} - 6v \frac{\partial u}{\partial x_1} - 6u \frac{\partial v}{\partial x_1} - \frac{\partial^3 v}{\partial x_1^3} = 0.$$ ♣

Definition 13.5 *The Gateaux derivative of an operator valued function $R(\mathbf{u})$ is defined as*

$$R'(\mathbf{u})[\mathbf{v}]\mathbf{w} := \left. \frac{\partial [R(\mathbf{u} + \epsilon \mathbf{v})\mathbf{w}]}{\partial \epsilon} \right|_{\epsilon = 0}$$

where $R'(\mathbf{u})[\mathbf{v}]\mathbf{w}$ is the derivative of $R(\mathbf{u})$ evaluated at $\mathbf{v}$ and then applied to $\mathbf{w}$, where $\mathbf{w}, \mathbf{v}$ and $\mathbf{u}$ are smooth functions of $x_1, \ldots, x_m, x_{m+1}$ in the vector space W.

Example: Let $m = 2$ and

$$R(\mathbf{u}) = \frac{\partial}{\partial x_1} + \mathbf{u} + \frac{\partial \mathbf{u}}{\partial x_1} D_{x_1}^{-1}$$

where $\mathbf{u}$ is a smooth function of x_1 and x_2, and

$$D_{x_1}^{-1} f(x_1) := \int^{x_1} f(s) ds.$$

We calculate the Gateaux derivative of $R(\mathbf{u})$. It follows that

$$
\begin{aligned}
R(\mathbf{u} + \epsilon\mathbf{v})\mathbf{w} &= \frac{\partial \mathbf{w}}{\partial x_1} + (\mathbf{u} + \epsilon\mathbf{v})\mathbf{w} + \left(\frac{\partial}{\partial x_1}(\mathbf{u} + \epsilon\mathbf{v}) \right) D_{x_1}^{-1}\mathbf{w} \\
&= \frac{\partial \mathbf{w}}{\partial x_1} + \mathbf{u}\mathbf{w} + \epsilon\mathbf{v}\mathbf{w} + \frac{\partial \mathbf{u}}{\partial x_1} D_{x_1}^{-1}\mathbf{w} + \epsilon\frac{\partial \mathbf{v}}{\partial x_1} D_{x_1}^{-1}\mathbf{w}.
\end{aligned}
$$

Therefore

$$R'(\mathbf{u})[\mathbf{v}]\mathbf{w} = \mathbf{v}\mathbf{w} + \frac{\partial \mathbf{v}}{\partial x_1} D_{x_1}^{-1}\mathbf{w}. \qquad \clubsuit$$

Let $f, g, h : W \to W$ be three maps, where W is a topological vector space ($\mathbf{u} \in W$). Assume that the Gateaux derivatives of f, g and h exists up to infinite order. Let

$$f'(\mathbf{u})[\mathbf{v}] := \left. \frac{\partial f(\mathbf{u} + \epsilon\mathbf{v})}{\partial \epsilon} \right|_{\epsilon=0}$$

$$g'(\mathbf{u})[\mathbf{v}] := \left. \frac{\partial g(\mathbf{u} + \epsilon\mathbf{v})}{\partial \epsilon} \right|_{\epsilon=0}$$

$$h'(\mathbf{u})[\mathbf{v}] := \left. \frac{\partial h(\mathbf{u} + \epsilon\mathbf{v})}{\partial \epsilon} \right|_{\epsilon=0}.$$

An example of W is the Schwartzian space $S(\mathbf{R}^n)$.

Definition 13.6 *The* **Lie product** *(or* **commutator***) of f and g is defined by*

$$[f, g] := f'(\mathbf{u})[g] - g'(\mathbf{u})[f].$$

Let f, g and $h \in W$. Then we find

$$[f, g] = -[g, f]$$
$$[f, [g, h]] + [g, [h, f]] + [h, [f, g]] = 0 \quad \text{(Jacobi identity)}.$$

From $[f, g] = -[g, f]$ it follows that $[f, f] = 0$. Thus the maps form a Lie algebra.

Example: Consider $m = 2$ and $n = 1$. We put $u_1 = u$ and

$$f(u) := \frac{\partial u}{\partial x_2} - u\frac{\partial u}{\partial x_1} - \frac{\partial^3 u}{\partial x_1^3}$$

$$g(u) := \frac{\partial u}{\partial x_2} - \frac{\partial^2 u}{\partial x_1^2}.$$

We calculate the Lie product $[f, g]$. We find

$$
\begin{aligned}
f'(u)[g] &= \frac{\partial}{\partial \epsilon} f(u + \epsilon g(u))|_{\epsilon=0} \\
&= \frac{\partial}{\partial \epsilon}\left[\frac{\partial}{\partial x_2}(u + \epsilon g(u)) - (u + \epsilon g(u))\frac{\partial}{\partial x_1}(u + \epsilon g(u)) - \frac{\partial^3}{\partial x_1^3}(u + \epsilon g(u))\right]\Bigg|_{\epsilon=0}.
\end{aligned}
$$

Consequently,

$$f'(u)[g] = \frac{\partial g(u)}{\partial x_2} - g(u)\frac{\partial u}{\partial x_1} - u\frac{\partial g(u)}{\partial x_1} - \frac{\partial^3 g(u)}{\partial x_1^3}.$$

Also

$$
\begin{aligned}
g'(u)[f] &= \frac{\partial}{\partial \epsilon}g(u + \epsilon f(u))|_{\epsilon=0} \\
&= \frac{\partial}{\partial \epsilon}\left[\frac{\partial}{\partial x_2}(u + \epsilon f(u)) - \frac{\partial^2}{\partial x_1^2}(u + \epsilon f(u))\right]\Bigg|_{\epsilon=0} \\
&= \frac{\partial f(u)}{\partial x_2} - \frac{\partial^2 f(u)}{\partial x_1^2}.
\end{aligned}
$$

Hence

$$
\begin{aligned}
[f, g] &= f'(u)[g] - g'(u)[f] \\
&= \frac{\partial}{\partial x_2}(g(u) - f(u)) - g(u)\frac{\partial u}{\partial x_1} - u\frac{\partial g(u)}{\partial x_1} - \frac{\partial^3 g(u)}{\partial x_1^3} + \frac{\partial^2 f(u)}{\partial x_1^2}.
\end{aligned}
$$

Inserting the given f and g we obtain

$$
\begin{aligned}
[f, g] = {}& \frac{\partial}{\partial x_2}\left(\frac{\partial u}{\partial x_2} - \frac{\partial^2 u}{\partial x_1^2}\right) - \left(\frac{\partial u}{\partial x_2} - \frac{\partial^2 u}{\partial x_1^2}\right)\frac{\partial u}{\partial x_1} \\
& - u\frac{\partial}{\partial x_1}\left(\frac{\partial u}{\partial x_2} - \frac{\partial^2 u}{\partial x_1^2}\right) - \frac{\partial^3}{\partial x_1^3}\left(\frac{\partial u}{\partial x_2} - \frac{\partial^2 u}{\partial x_1^2}\right) \\
& - \frac{\partial}{\partial x_2}\left(\frac{\partial u}{\partial x_2} - u\frac{\partial u}{\partial x_1} - \frac{\partial^3 u}{\partial x_1^3}\right) + \frac{\partial^2}{\partial x_1^2}\left(\frac{\partial u}{\partial x_2} - u\frac{\partial u}{\partial x_1} - \frac{\partial^3 u}{\partial x_1^3}\right)
\end{aligned}
$$

Therefore

$$[f, g] = -2 \frac{\partial u}{\partial x_1} \frac{\partial^2 u}{\partial x_1^2}.$$ ♣

We now give the connection with the Lie-Bäcklund symmetry vector field as defined in chapter 9. Consider the evolution equation (2).

Definition 13.7 *A function*

$$\sigma \left(\mathbf{x}, x_{m+1}, \mathbf{u}(\mathbf{x}), \frac{\partial \mathbf{u}}{\partial \mathbf{x}}, \frac{\partial^2 \mathbf{u}}{\partial \mathbf{x}^2}, \cdots \right)$$

*($\sigma = (\sigma_1, \ldots, \sigma_n)$) is called a **symmetry** of the evolution equation (2) if and only if σ satisfies the linearized equation*

$$\frac{\partial \sigma}{\partial x_{m+1}} = \mathbf{F}'(\mathbf{u})[\sigma].$$

$\mathbf{F}'$ *is the Gateaux derivative of* $\mathbf{F}$, *i.e.,*

$$\mathbf{F}'(\mathbf{u})[\sigma] = \frac{\partial}{\partial \epsilon} \mathbf{F}(\mathbf{u} + \epsilon \sigma)|_{\epsilon=0}.$$

The Lie-Bäcklund symmetry vector field of the evolution equation (2) is then given by

$$Z_B = \sum_{j=1}^{n} \sigma_j \frac{\partial}{\partial u_j}.$$

Remark: *From definition 12.7 it follows that a function σ is a symmetry of the evolution equation (2) if it leaves (2) invariant within order ϵ, i.e., the equation*

$$\frac{\partial}{\partial x_{m+1}} (\mathbf{u} + \epsilon \sigma) = \mathbf{F}(\mathbf{u} + \epsilon \sigma)$$

must be correct up to order ϵ.

Remark: *In literature a symmetry σ is also defined as a function that satisfies the equation*

$$\frac{\partial \sigma}{\partial x_{m+1}} = \mathbf{F}'[\sigma] - \sigma'[\mathbf{F}]$$

where $\partial \sigma / \partial x_{m+1}$ denotes that σ is only differentiated explicitly with respect to the variable x_{m+1}. Obviously this is only a notational difference.

Example: Let $m = 1$ and $n = 1$. We put $u_1 = u$ and consider the Korteweg-de Vries equation

$$\frac{\partial u}{\partial x_2} = 6u \frac{\partial u}{\partial x_1} + \frac{\partial^3 u}{\partial x_1^3}.$$

We show that

$$\sigma = 3x_2 \frac{\partial u}{\partial x_1} + \frac{1}{2}$$

is a symmetry for the Korteweg-de Vries equation where we have put $\sigma_1 = \sigma$.

Remark: *In this case the symmetry vector field, given by*

$$Z_V = \left(3x_2 \frac{\partial u}{\partial x_1} + \frac{1}{2} \right) \frac{\partial}{\partial u},$$

is the vertical Lie point symmetry vector field that can also be written in the form

$$Z = -3x_2 \frac{\partial}{\partial x_1} + \frac{1}{2} \frac{\partial}{\partial u}.$$

F is defined by

$$F := 6u \frac{\partial u}{\partial x_1} + \frac{\partial^3 u}{\partial x_1^3}$$

where we have put $F_1 = F$. First we calculate the Gateaux derivative of F, i.e.,

$$F'(u)[\sigma] = \frac{\partial}{\partial \epsilon} F(u + \epsilon \sigma)|_{\epsilon=0} = \frac{\partial}{\partial \epsilon} \left[6(u + \epsilon \sigma) \left(\frac{\partial (u + \epsilon \sigma)}{\partial x_1} \right) + \frac{\partial^3}{\partial x_1^3} (u + \epsilon \sigma) \right] \Bigg|_{\epsilon=0}.$$

Therefore

$$F'(u)[\sigma] = 6 \frac{\partial u}{\partial x_1} \sigma + 6u \frac{\partial \sigma}{\partial x_1} + \frac{\partial^3 \sigma}{\partial x_1^3}. \tag{4}$$

It follows that

$$\frac{\partial \sigma}{\partial x_2} = 3 \frac{\partial u}{\partial x_1} + 3x_2 \frac{\partial^2 u}{\partial x_1 \partial x_2}. \tag{5}$$

Inserting σ into (4) gives

$$F'(u)[\sigma] = 18x_2 u \frac{\partial^2 u}{\partial x_1^2} + 18x_2 \left(\frac{\partial u}{\partial x_1} \right)^2 + 3 \frac{\partial u}{\partial x_1} + 3x_2 \frac{\partial^4 u}{\partial x_1^4}. \tag{6}$$

From the Korteweg-de Vries equation we obtain

$$\frac{\partial^2 u}{\partial x_1 \partial x_2} = 6 \left(\frac{\partial u}{\partial x_1} \right)^2 + 6u \frac{\partial^2 u}{\partial x_1^2} + \frac{\partial^4 u}{\partial x_1^4}. \tag{7}$$

Inserting (7) into (5) leads to

$$\frac{\partial \sigma}{\partial x_2} = 3 \frac{\partial u}{\partial x_1} + 18x_2 \left(\frac{\partial u}{\partial x_1} \right)^2 + 18x_2 u \frac{\partial^2 u}{\partial x_1^2} + 3x_2 \frac{\partial^4 u}{\partial x_1^4}. \tag{8}$$

From (8) and (6) it follows that σ is a symmetry of the Korteweg-de Vries equation. ♣

13.2 Definition and Examples

Let us now define a recursion operator:

Definition 13.8 *An operator $R(\mathbf{u})$ is called a* **recursion operator** *if*

$$R'(\mathbf{u})[F(\mathbf{u})]\mathbf{v} = [F'(\mathbf{u})\mathbf{v}, R(\mathbf{u})\mathbf{v}]$$

whenever $\mathbf{u}$ *is a solution of (2), where*

$$R'(\mathbf{u})[F(\mathbf{u})]\mathbf{v} := \frac{\partial}{\partial \epsilon}(R(\mathbf{u} + \epsilon F(\mathbf{u}))\mathbf{v}|_{\epsilon=0}.$$

The commutator is defined as

$$[F'(\mathbf{u})\mathbf{v}, R(\mathbf{u})\mathbf{v}] := \frac{\partial}{\partial \epsilon}F'(\mathbf{u})(\mathbf{v} + \epsilon R(\mathbf{u})\mathbf{v})|_{\epsilon=0} - \frac{\partial}{\partial \epsilon}R(\mathbf{u})(\mathbf{v} + \epsilon F'(\mathbf{u})\mathbf{v})|_{\epsilon=0}.$$

Example: We consider Burgers' equation in the form

$$\frac{\partial u}{\partial x_2} = \frac{\partial^2 u}{\partial x_1^2} + 2u\frac{\partial u}{\partial x_1}$$

where $m = 1$ and $n = 1$. We put $u_1 = u$. F is given by

$$F(u) = \frac{\partial^2 u}{\partial x_1^2} + 2u\frac{\partial u}{\partial x_1}$$

where we have put $F_1 = F$. We show that

$$R(u) = \frac{\partial}{\partial x_1} + u + \frac{\partial u}{\partial x_1}D_{x_1}^{-1}$$

is a recursion operator for the given Burgers' equation, where

$$D_{x_1}^{-1}f(x_1) := \int^{x_1} f(s)ds.$$

First we calculate $R'(u)v$. When we apply the operator $R(u)$ to v we obtain

$$R(u)v = \frac{\partial v}{\partial x_1} + uv + \frac{\partial u}{\partial x_1}D_{x_1}^{-1}v.$$

It follows that

$$\begin{aligned}
R'(u)[F(u)]v &= \frac{\partial}{\partial \epsilon}(R(u + \epsilon F(u)v)|_{\epsilon=0} \\
&= \frac{\partial}{\partial \epsilon}\left(\frac{\partial}{\partial x_1} + u + \epsilon F(u) + \frac{\partial(u + \epsilon F(u))}{\partial x_1}D_{x_1}^{-1}\right)v\Big|_{\epsilon=0} \\
&= F(u)v + \frac{\partial F(u)}{\partial x_1}D_{x_1}^{-1}v \\
&= \left(\frac{\partial^2 u}{\partial x_1^2} + 2u\frac{\partial u}{\partial x_1}\right)v + \left(\frac{\partial^3 u}{\partial x_1^3} + 2\left(\frac{\partial u}{\partial x_1}\right)^2 + 2u\frac{\partial^2 u}{\partial x_1^2}\right)D_{x_1}^{-1}v.
\end{aligned}$$

To calculate the commutator $[F'(u)v,\ R(u)v]$ we first have to find the Gateaux derivative of F, i.e.

$$
\begin{aligned}
F'(u)[v] &= \left.\frac{\partial F(u + \epsilon v)}{\partial \epsilon}\right|_{\epsilon=0} \\
&= \left.\frac{\partial}{\partial \epsilon}\left(\frac{\partial^2}{\partial x_1^2}(u + \epsilon v) + 2(u + \epsilon v)\frac{\partial}{\partial x_1}(u + \epsilon v)\right)\right|_{\epsilon=0} \\
&= \frac{\partial^2 v}{\partial x_1^2} + 2u\frac{\partial v}{\partial x_1} + 2\frac{\partial u}{\partial x_1}v.
\end{aligned}
$$

Therefore

$$
F'(u)v = \frac{\partial^2 v}{\partial x_1^2} + 2u\frac{\partial v}{\partial x_1} + 2\frac{\partial u}{\partial x_1}v.
$$

For the first term of the commutator we obtain

$$
\begin{aligned}
\frac{\partial}{\partial \epsilon}F'(u)(v + \epsilon R(u)v)|_{\epsilon=0} &= \frac{\partial}{\partial \epsilon}\left[\frac{\partial^2}{\partial x_1^2}(v + \epsilon R(u)v) + 2u\frac{\partial}{\partial x_1}(v + \epsilon R(u)v)\right. \\
&\quad \left. + \ 2\frac{\partial u}{\partial x_1}(v + \epsilon R(u)v)\right]\Bigg|_{\epsilon=0} \\
&= \frac{\partial^3 v}{\partial x_1^3} + 3\frac{\partial^2 u}{\partial x_1^2}v + 5\frac{\partial u}{\partial x_1}\frac{\partial v}{\partial x_1} + 3u\frac{\partial^2 v}{\partial x_1^2} \\
&\quad + \ 6u\frac{\partial u}{\partial x_1}v + 2u^2\frac{\partial v}{\partial x_1} \\
&\quad + \ \left(\frac{\partial^3 u}{\partial x_1^3} + 2\left(\frac{\partial u}{\partial x_1}\right)^2 + 2u\frac{\partial^2 u}{\partial x_1^2}\right)D_{x_1}^{-1}v.
\end{aligned}
$$

For the second term of the commutator we obtain

$$
\begin{aligned}
\frac{\partial}{\partial \epsilon}R(u)(v + \epsilon F'(u)v)|_{\epsilon=0} &= \frac{\partial}{\partial \epsilon}\left[\frac{\partial}{\partial x_1}(v + \epsilon F'(u)v)\right. \\
&\quad + \ u(v + \epsilon F'(u)v) + \frac{\partial u}{\partial x_1}D_{x_1}^{-1}(v + \epsilon F'(u)v)\Bigg]\Bigg|_{\epsilon=0} \\
&= \frac{\partial F'(u)v}{\partial x_1} + uF'(u)v + \frac{\partial u}{\partial x_1}D_{x_1}^{-1}F'(u)v.
\end{aligned}
$$

Using the identity

$$
D_{x_1}^{-1}\left(v\frac{\partial u}{\partial x_1} + u\frac{\partial v}{\partial x_1}\right) \equiv uv
$$

and $F'(u)[v]$ as defined before, we find

$$\frac{\partial}{\partial \epsilon} R(u)(v + \epsilon F'(u)v)|_{\epsilon=0} = \frac{\partial^3 v}{\partial x_1^3} + 2\frac{\partial^2 u}{\partial x_1^2}v + 5\frac{\partial u}{\partial v}\frac{\partial v}{\partial x_1} + 3u\frac{\partial^2 v}{\partial x_1^2} + 4u\frac{\partial u}{\partial x_1}v + 2u^2\frac{\partial v}{\partial x_1}.$$

It follows that

$$[F'(u)v, R(u)v] = \left[\frac{\partial^2 u}{\partial x_1^2} + 2u\frac{\partial u}{\partial x_1} + \left(\frac{\partial^3 u}{\partial x_1^3} + 2\left(\frac{\partial u}{\partial x_1}\right)^2 + 2u\frac{\partial^2 u}{\partial x_1^2}\right)D_{x_1}^{-1}\right]v.$$

This proves that $R(u)$ is a recursion operator for Burgers' equation. ♣

An important property of the recursion operator is that we can generate symmetries for an evolution equation by considering

$$R(\mathbf{u})^p\sigma(\mathbf{u})$$

where σ is a known symmetry and p is some positive integer. The recursion operator then generates a hierarchy of evolution equations

$$\frac{\partial \mathbf{u}}{\partial x_{m+1}} = R(\mathbf{u})^p\mathbf{F}(\mathbf{u}).$$

Example: Let us consider the diffusion equation

$$\frac{\partial u}{\partial x_2} = \frac{\partial^2 u}{\partial x_1^2} - \frac{1}{2}\left(\frac{\partial u}{\partial x_1}\right)^2.$$

Two recursion operators are given by

$$R_1(u) = \frac{1}{2}\frac{\partial u}{\partial x_1} - D_{x_1}$$

$$R_2(u) = \frac{1}{2}\left(x_2\frac{\partial u}{\partial x_1} - x_1\right) - x_2 D_{x_1}.$$

It is clear that this evolution equation is invariant under translation in x_1 which corresponds to the symmetry

$$\sigma_1 = \frac{\partial u}{\partial x_1}.$$

Let us now make use of the given recursion operators in order to generate more symmetries for Burgers' equation. We find

$$R_1(u)\frac{\partial u}{\partial x_1} = \left(\frac{1}{2}\frac{\partial u}{\partial x_1} - D_{x_1}\right)\frac{\partial u}{\partial x_1} = \frac{1}{2}\left(\frac{\partial u}{\partial x_1}\right)^2 - \frac{\partial^2 u}{\partial x_1^2} = -\frac{\partial u}{\partial x_2}$$

so that

$$\sigma_2 = -\frac{\partial u}{\partial x_2}.$$

σ_2 corresponds to the translational symmetry in x_2. From the second recursion operator we find

$$
\begin{aligned}
R_2(u)\frac{\partial u}{\partial x_1} &= \left(\frac{1}{2}x_2\frac{\partial u}{\partial x_1} - \frac{1}{2}x_1 - x_2 D_{x_1}\right)\frac{\partial u}{\partial x_1} \\
&= \frac{1}{2}x_2\left(\frac{\partial u}{\partial x_1}\right)^2 - x_2\frac{\partial^2 u}{\partial x_1^2} - \frac{1}{2}x_1\frac{\partial u}{\partial x_1} \\
&= -x_2\frac{\partial u}{\partial x_2} - \frac{1}{2}x_1\frac{\partial u}{\partial x_1}
\end{aligned}
$$

so that

$$\sigma_3 = -x_2\frac{\partial u}{\partial x_2} - x_1\frac{\partial u}{\partial x_1}.$$

σ_3 corresponds to the scaling symmetry in x_1 and x_2. We now consider

$$
\begin{aligned}
(R(u))^3\frac{\partial u}{\partial x_1} &= \left(\frac{1}{2}\frac{\partial u}{\partial x_1} - D_{x_1}\right)^3\frac{\partial u}{\partial x_1} \\
&= \left(\frac{1}{8}\left(\frac{\partial u}{\partial x_1}\right)^3 - \frac{1}{4}\left(\frac{\partial u}{\partial x_1}\right)^2 D_{x_1} - \frac{1}{4}\frac{\partial u}{\partial x_1}\frac{\partial^2 u}{\partial x_1^2} + \frac{1}{2}\frac{\partial u}{\partial x_1}D_{x_1}^2\right. \\
&\quad \left. - \frac{1}{4}\frac{\partial u}{\partial x_1}\frac{\partial^2 u}{\partial x_1^2} + \frac{1}{2}\frac{\partial^2 u}{\partial x_1^2}D_{x_1} + \frac{1}{2}\frac{\partial u}{\partial x_1}D_{x_1}^2 + \frac{1}{2}\frac{\partial^3 u}{\partial x_1^3} - D_{x_1}^3\right)\frac{\partial u}{\partial x_1} \\
&= \frac{1}{8}\left(\frac{\partial u}{\partial x_1}\right)^4 + \frac{3}{2}\frac{\partial u}{\partial x_1}\frac{\partial^3 u}{\partial x_1^3} - \frac{1}{2}\left(\frac{\partial u}{\partial x_1}\right)^2\frac{\partial^2 u}{\partial x_1^2} + \frac{1}{2}\left(\frac{\partial^2 u}{\partial x_1^2}\right)^2 - \frac{\partial^4 u}{\partial x_1^4}
\end{aligned}
$$

so that

$$\sigma_4 = \frac{1}{8}\left(\frac{\partial u}{\partial x_1}\right)^4 + \frac{\partial u}{\partial x_1}\frac{\partial^3 u}{\partial x_1^3} - \frac{3}{4}\left(\frac{\partial u}{\partial x_1}\right)^2\frac{\partial^2 u}{\partial x_1^2} + \frac{1}{2}\left(\frac{\partial^2 u}{\partial x_1^2}\right)^2 - \frac{\partial^4 u}{\partial x_1^4}.$$

σ_4 corresponds to a Lie-Bäcklund symmetry vector field for the given Burgers' equation, i.e.

$$Z_B = \sigma_4\frac{\partial}{\partial u}.$$

This procedure can be continued to find an infinite hierarchy of Lie-Bäcklund symmetry vector fields. We can also construct a hierarchy of Burgers' equations from the given recursion operators. For example

$$\frac{\partial u}{\partial x_2} = R_2(u)\left(\frac{\partial^2 u}{\partial x_1^2} - \frac{1}{2}\left(\frac{\partial u}{\partial x_1}\right)^2\right)$$

$$= \frac{3}{2}x_2\frac{\partial u}{\partial x_1}\frac{\partial^2 u}{\partial x_1^2} - \frac{1}{4}x_2\left(\frac{\partial u}{\partial x_1}\right)^3 - x_2\frac{\partial^3 u}{\partial x_1^3} - \frac{1}{2}x_1\frac{\partial^2 u}{\partial x_1^2} + \frac{1}{4}x_1\left(\frac{\partial u}{\partial x_1}\right)^2.$$

♣

Remark: *One may well ask the question whether a recursion operator $R(u)$ for an evolution equation (2) is also a recursion operator for the hierarchy of evolution equations*

$$\frac{\partial \mathbf{u}}{\partial x_{m+1}} = R(\mathbf{u})^p \mathbf{F}(\mathbf{u}).$$

Such conditions on the recursion operator were investigated by Fokas and Fuchssteiner [48]). They have introduced the name **hereditary** *for those recursion operators. A consequence of the hereditary property of $R(\mathbf{u})$ is the fact that the symmetries*

$$R(\mathbf{u})^p \sigma(\mathbf{u})$$

form an abelian Lie algebra.

13.3 Computer Algebra Applications

In the program we evaluate

$$\frac{d}{d\epsilon} f(u(\mathbf{x}) + \epsilon v(\mathbf{x}))\Big|_{\epsilon=0}$$

where

$$f(u(\mathbf{x})) = u\frac{\partial u}{\partial x_1} + u^2 + \frac{\partial^2 u}{\partial x_2^2}$$

```
% Gateaux derivative
% gat.red

depend u, x1, x2;
depend v, x1, x2;

A := df(u,x2,2) + u*df(u,x1) + u*u;
B := sub(u=u+ep*v,A);
C := df(B,ep);
ep := 0;
C;
```

The output is

```
df(u,x1)*v + df(v,x1)*u + df(v,x2,2) + 2*u*v$
```

An alternative program is

```
% Gateaux derivative
% gat1.red

operator u, v;

A := df(u(x1,x2),x2,2) + u(x1,x2)*df(u(x1,x2),x1)
     + u(x1,x2)*u(x1,x2);

B := sub(u(x1,x2)=u(x1,x2)+ep*v(x1,x2),A);
C := df(B,ep);
ep := 0;
C;
```

Chapter 14

Bäcklund Transformations

14.1 Definitions

Bäcklund transformations play an important role in finding solutions of a certain class nonlinear partial differential equations. From a solution of a nonlinear partial differential equation, we can sometimes find a relationship that will generate the solution of a different partial differential equation, which is known as a Bäcklund transformation, or of the same partial differential equation where such a relation is then known as an auto-Bäcklund transformation. First we describe the Bäcklund transformation in the classical notation, Rogers and Shadwick [97] and it may be extended to incorporate both higher-order and higher-dimensional equations. In particular Lamb [74] developed Clairin's method for certain higher-order nonlinear evolution equations while Dodd and Bullough [25]) have considered Bäcklund transformations for higher-dimensional sine-Gordon equations. An underlying modern theory of Bäcklund transformations is provided also provided by Rogers and Shadwick [97] based on the jet bundle formalism whereby such extensions can be readily accommodated.

We describe first the problem in $\mathbf{R}^3$ (Rogers and Shadwick [97]).

Definition 14.1 *Let*

$$u = u(x_1, x_2) \tag{1a}$$

$$\tilde{u} = \tilde{u}(\tilde{x}_1, \tilde{x}_2) \tag{1b}$$

represent two smooth surfaces Λ and $\tilde{\Lambda}$ respectively, in $\mathbf{R}^3$. A set of four relations

$$B_i^* \left(x_1, x_2, \tilde{x}_1, \tilde{x}_2, u, \tilde{u}, \frac{\partial u}{\partial x_1}, \frac{\partial u}{\partial x_2}, \frac{\partial \tilde{u}}{\partial \tilde{x}_1}, \frac{\partial \tilde{u}}{\partial \tilde{x}_2} \right) = 0 \tag{2}$$

where $i = 1, \ldots, 4$, which connect the surface elements

$$\left\{ x_1, x_2, u, \frac{\partial u}{\partial x_1}, \frac{\partial u}{\partial x_2} \right\} \tag{3}$$

and

$$\left\{ \tilde{x}_1, \tilde{x}_2, \tilde{u}, \frac{\partial \tilde{u}}{\partial \tilde{x}_1}, \frac{\partial \tilde{u}}{\partial \tilde{x}_2} \right\} \tag{4}$$

of Λ and $\tilde{\Lambda}$ respectively, is called a **Bäcklund transformation**.

In particular we consider (1) to represent integral surfaces of partial differential equations. Consider the following explicit form of (2)

$$\frac{\partial \tilde{u}}{\partial \tilde{x}_i} = \tilde{B}_i \left(x_1, x_2, u, \tilde{u}, \frac{\partial u}{\partial x_1}, \frac{\partial u}{\partial x_2} \right) \tag{5}$$

and

$$\frac{\partial u}{\partial x_i} = B_i \left(\tilde{x}_1, \tilde{x}_2, u, \tilde{u}, \frac{\partial \tilde{u}}{\partial \tilde{x}_1}, \frac{\partial \tilde{u}}{\partial \tilde{x}_2} \right) \tag{6}$$

together with

$$\tilde{x}_i = X_i \left(x_1, x_2, u, \tilde{u}, \frac{\partial u}{\partial x_1}, \frac{\partial u}{\partial x_2} \right) \tag{7}$$

where $i = 1, 2$. In order that these relations transform a surface $u = u(x_1, x_2)$ with surface element (3) to a surface $\tilde{u} = \tilde{u}(\tilde{x}_1, \tilde{x}_2)$ with surface element (4) it is required that the relations

$$du - B_1 dx_1 - B_2 dx_2 = 0$$

$$d\tilde{u} - \tilde{B}_1 d\tilde{x}_1 - \tilde{B}_2 d\tilde{x}_2 = 0,$$

be integrable. Hence from

$$\frac{\partial^2 u}{\partial x_1 \partial x_2} = \frac{\partial^2 u}{\partial x_2 \partial x_1}$$

and

$$\frac{\partial^2 \tilde{u}}{\partial \tilde{x}_1 \partial \tilde{x}_2} = \frac{\partial^2 \tilde{u}}{\partial \tilde{x}_2 \partial \tilde{x}_1}$$

we obtain the conditions

$$\frac{\partial B_1}{\partial x_2} - \frac{\partial B_2}{\partial x_1} = 0 \tag{8}$$

$$\frac{\partial \tilde{B}_1}{\partial \tilde{x}_2} - \frac{\partial \tilde{B}_2}{\partial \tilde{x}_1} = 0 \tag{9}$$

respectively. Application of (9) to the Bäcklund relations (5) and (7) lead to a nonlinear equation of the form

$$\left(\frac{\partial^2 u}{\partial x_1^2}\frac{\partial^2 u}{\partial x_2^2} - \left(\frac{\partial^2 u}{\partial x_1 \partial x_2}\right)^2\right) f_1 + f_2 \frac{\partial^2 u}{\partial x_1^2} + 2f_3 \frac{\partial^2 u}{\partial x_2 \partial x_1} + f_4 \frac{\partial^2 u}{\partial x_2^2} + f_5 = 0 \qquad (10)$$

where

$$f_i\left(x_1, x_2, u, \tilde{u}, \frac{\partial u}{\partial x_1}, \frac{\partial u}{\partial x_2}\right)$$

with $i = 1, \ldots, 5$. Thus, if $\tilde{u}$ is absent in (10), the Monge-Ampére form is obtained. In a similar manner, application of (8) to the Bäcklund relations (6) and (7) leads to

$$\left(\frac{\partial^2 \tilde{u}}{\partial \tilde{x}_1^2}\frac{\partial^2 \tilde{u}}{\partial \tilde{x}_2^2} - \left(\frac{\partial^2 \tilde{u}}{\partial \tilde{x}_1 \partial \tilde{x}_2}\right)^2\right) \tilde{f}_1 + \tilde{f}_2 \frac{\partial^2 \tilde{u}}{\partial \tilde{x}_1^2} + 2\tilde{f}_3 \frac{\partial^2 \tilde{u}}{\partial \tilde{x}_2 \partial \tilde{x}_1} + \tilde{f}_4 \frac{\partial^2 \tilde{u}}{\partial \tilde{x}_2^2} + \tilde{f}_5 = 0 \qquad (11)$$

where

$$\tilde{f}_i\left(\tilde{x}_1, \tilde{x}_2, u, \tilde{u}, \frac{\partial \tilde{u}}{\partial \tilde{x}_1}, \frac{\partial \tilde{u}}{\partial \tilde{x}_2}\right).$$

Thus, if u is absent in (11), the Monge-Ampére form is obtained. In particular, if the equations for u and $\tilde{u}$ as derived in (10) and (11) are both of the Monge-Ampére form, then the Bäcklund transformation may be regarded as a mapping between their integral surfaces.

There are certain Monge-Ampére equations of significance which do possess Bäcklund transformations in the classical sense.

To extend the Bäcklund transformation to more than two independent variables and more than one dependent variable one introduces jet bundle transformations known as Bäcklund maps. Bäcklund maps have the feature of being maps of finite-dimensional spaces that admit a simple geometric characterization. For details we refer to Rogers and Shadwick [97].

14.2   Examples

Example: We show that an auto-Bäcklund transformation for the **sine-Gordon equation**

$$\frac{\partial^2 u}{\partial x_1 \partial x_2} = \sin u$$

is given by

$$\tilde{x}_1(x_1, x_2) \;=\; x_1$$

$$\tilde{x}_2(x_1, x_2) \;=\; x_2$$

$$\frac{\partial \tilde{u}}{\partial \tilde{x}_1}(\tilde{x}_1(x_1, x_2), \tilde{x}_2(x_1, x_2)) \;=\; \frac{\partial u}{\partial x_1} - 2\lambda \sin\left(\frac{u + \tilde{u}}{2}\right)$$

$$\frac{\partial \tilde{u}}{\partial \tilde{x}_2}(\tilde{x}_1(x_1, x_2), \tilde{x}_2(x_1, x_2)) \;=\; -\frac{\partial u}{\partial x_2} + \frac{2}{\lambda} \sin\left(\frac{u - \tilde{u}}{2}\right)$$

where λ is a nonzero parameter. Thus,

$$\tilde{B}_1\left(x_1, x_2, u, \tilde{u}, \frac{\partial u}{\partial x_1}, \frac{\partial u}{\partial x_2}\right) \;=\; \frac{\partial u}{\partial x_1} - 2\lambda \sin\left(\frac{u + \tilde{u}}{2}\right)$$

$$\tilde{B}_2\left(x_1, x_2, u, \tilde{u}, \frac{\partial u}{\partial x_1}, \frac{\partial u}{\partial x_2}\right) \;=\; -\frac{\partial u}{\partial x_2} + \frac{2}{\lambda} \sin\left(\frac{u - \tilde{u}}{2}\right)$$

$$B_1\left(\tilde{x}_1, \tilde{x}_2, u, \tilde{u}, \frac{\partial \tilde{u}}{\partial \tilde{x}_1}, \frac{\partial \tilde{u}}{\partial \tilde{x}_2}\right) \;=\; \frac{\partial \tilde{u}}{\partial \tilde{x}_1} + 2\lambda \sin\left(\frac{u + \tilde{u}}{2}\right)$$

$$B_2\left(\tilde{x}_1, \tilde{x}_2, u, \tilde{u}, \frac{\partial \tilde{u}}{\partial \tilde{x}_1}, \frac{\partial \tilde{u}}{\partial \tilde{x}_2}\right) \;=\; -\frac{\partial \tilde{u}}{\partial \tilde{x}_2} + \frac{2}{\lambda} \sin\left(\frac{u - \tilde{u}}{2}\right).$$

From

$$\frac{\partial B_1}{\partial x_2} - \frac{\partial B_2}{\partial x_1} = 0,$$

it follows that

$$\frac{\partial^2 \tilde{u}}{\partial \tilde{x}_1 \partial x_2} + \lambda \left(\frac{\partial u}{\partial x_2} + \frac{\partial \tilde{u}}{\partial x_2}\right) \cos\left(\frac{u + \tilde{u}}{2}\right)$$

$$- \left[-\frac{\partial^2 \tilde{u}}{\partial \tilde{x}_2 \partial x_1} + \frac{1}{\lambda}\left(\frac{\partial u}{\partial x_1} - \frac{\partial \tilde{u}}{\partial x_1}\right) \cos\left(\frac{u - \tilde{u}}{2}\right)\right] = 0.$$

By making use of the given transformation we find

$$2\frac{\partial^2 \tilde{u}}{\partial \tilde{x}_1 \partial \tilde{x}_2} + 2\sin\left(\frac{u - \tilde{u}}{2}\right) \cos\left(\frac{u + \tilde{u}}{2}\right) - 2\sin\left(\frac{u + \tilde{u}}{2}\right) \cos\left(\frac{u - \tilde{u}}{2}\right) = 0$$

so that

$$\frac{\partial^2 \tilde{u}}{\partial \tilde{x}_1 \partial \tilde{x}_2} = \sin \tilde{u}.$$

From the condition

$$\frac{\partial \tilde{B}_1}{\partial \tilde{x}_2} - \frac{\partial \tilde{B}_2}{\partial \tilde{x}_1} = 0,$$

it follows that

$$\frac{\partial^2 u}{\partial x_1 \partial x_2} = \sin u.$$

This proves that the given transformation is an auto-Bäcklund transformation of the sine-Gordon equation. ♣

Example: We again consider the sine-Gordon equation with the auto-Bäcklund transformation given in the previous example. Obviously $u(x_1, x_2) = 0$ is a solution of the sine-Gordon equation. This is known as the vacuum solution. We now make use of the auto-Bäcklund transformation to construct another solution of the sine-Gordon equation from the vacuum solution. Inserting $u(x_1, x_2) = 0$ in the given Bäcklund transformation results in

$$\frac{\partial \tilde{u}}{\partial x_1} = -2\lambda \sin \frac{\tilde{u}}{2}$$

$$\frac{\partial \tilde{u}}{\partial x_2} = \frac{2}{\lambda} \sin \left(-\frac{\tilde{u}}{2} \right).$$

Since

$$\int \frac{du}{\sin u/2} = 2 \ln \left(\tan \frac{u}{2} \right)$$

we obtain a new solution of the sine-Gordon equation, namely

$$\tilde{u}(x_1, x_2) = 4 \tan^{-1} \exp(\lambda x_1 + \lambda^{-1} x_2 + C)$$

where C is a constant of integration. This new solution may be used to determine another solution for the sine-Gordon equation and so on. ♣

Example: Consider Burgers' equation

$$\frac{\partial u}{\partial x_2} + u \frac{\partial u}{\partial x_1} = \sigma \frac{\partial^2 u}{\partial x_1^2}.$$

If $\phi(x_1, x_2)$ is defined to be the solution of the linear partial differential equation

$$\frac{\partial \phi}{\partial x_2} + u \frac{\partial \phi}{\partial x_1} = \sigma \frac{\partial^2 \phi}{\partial x_1^2}$$

and $\tilde{u}(x_1, x_2)$ is defined by

$$\tilde{u}(x_1, x_2) = -2\sigma \frac{1}{\phi} \frac{\partial \phi}{\partial x_1} + u \equiv -2\sigma \frac{\partial}{\partial x_1} \ln \phi + u$$

then $\tilde{u}(x_1, x_2)$ also satisfies Burgers' equation. Hence, one solution of Burgers' equation can be used to generate another. Note that, with $u(x_1, x_2) = 0$ (which is a solution of Burgers' equation) ϕ satisfies the linear diffusion equation

$$\frac{\partial \phi}{\partial x_2} = \sigma \frac{\partial^2 \phi}{\partial x_1^2}.$$

The Bäcklund transformation

$$\tilde{u} = -2\sigma \frac{1}{\phi} \frac{\partial \phi}{\partial x_1}$$

is known as the **Cole-Hopf transformation**.   ♣

Example: The Korteweg-de Vries equation

$$\frac{\partial u}{\partial x_2} + 6u \frac{\partial u}{\partial x_1} + \frac{\partial^3 u}{\partial x_1^3} = 0$$

and the modified Korteweg-de Vries equation

$$\frac{\partial \tilde{u}}{\partial \tilde{x}_2} - 6\tilde{u}^2 \frac{\partial \tilde{u}}{\partial \tilde{x}_1} + \frac{\partial^3 \tilde{u}}{\partial \tilde{x}_1^3} = 0$$

are related by the Bäcklund transformation

$$\tilde{x}_1(x_1, x_2) = x_1$$

$$\tilde{x}_2(x_1, x_2) = x_2$$

$$\frac{\partial \tilde{u}}{\partial \tilde{x}_1}(\tilde{x}_1(x_1, x_2), \tilde{x}_2(x_1, x_2)) = u + \tilde{u}^2$$

$$\frac{\partial \tilde{u}}{\partial \tilde{x}_2}(\tilde{x}_1(x_1, x_2), \tilde{x}_2(x_1, x_2)) = -\frac{\partial^2 u}{\partial x_1^2} - 2\left(\tilde{u} \frac{\partial u}{\partial x_1} + u \frac{\partial \tilde{u}}{\partial x_1}\right).$$

♣

Example: The **Liouville equation**

$$\frac{\partial^2 u}{\partial x_1 \partial x_2} = e^u$$

and the linear equation

$$\frac{\partial^2 \tilde{u}}{\partial \tilde{x}_1 \partial \tilde{x}_2} = 0$$

are related by the Bäcklund transformation

$$\tilde{x}_1(x_1, x_2) = x_1$$

$$\tilde{x}_2(x_1, x_2) = x_2$$

$$\frac{\partial \tilde{u}}{\partial \tilde{x}_1}(\tilde{x}_1(x_1, x_2), \tilde{x}_2(x_1, x_2)) = \frac{\partial u}{\partial x_1} + \lambda \exp\left(\frac{1}{2}(\tilde{u} + u)\right)$$

$$\frac{\partial \tilde{u}}{\partial \tilde{x}_2}(\tilde{x}_1(x_1, x_2), \tilde{x}_2(x_1, x_2)) = -\frac{\partial u}{\partial x_2} - \frac{2}{\lambda} \exp\left(\frac{1}{2}(u - \tilde{u})\right)$$

where λ is a nonzero constant. ♣

Example: Bäcklund transformations can also be found for certain ordinary differential equations. Let

$$\frac{d^2 u}{dx^2} = \sin u \tag{12}$$

and

$$\frac{d^2 \tilde{u}}{d\tilde{x}^2} = \sinh \tilde{u} \tag{13}$$

where u and $\tilde{u}$ are real valued functions. We show that

$$\tilde{x}(x) = x \tag{14a}$$

$$\frac{du(x)}{dx} - i\frac{d\tilde{u}(\tilde{x}(x))}{d\tilde{x}} = 2e^{i\lambda} \sin\left(\frac{1}{2}(u(x) + i\tilde{u}(\tilde{x}(x)))\right) \tag{14b}$$

defines a Bäcklund transformation, where λ is a real parameter. Taking the x-derivative of (11) gives

$$\frac{d^2 u}{dx^2} - i\frac{d^2 \tilde{u}}{d\tilde{x}^2} = e^{i\lambda} \cos\left(\frac{1}{2}(u + i\tilde{u})\right)\left(\frac{du}{dx} + i\frac{d\tilde{u}}{d\tilde{x}}\right). \tag{15}$$

The complex conjugate of (14) is

$$\frac{du}{dx} + i\frac{d\tilde{u}}{d\tilde{x}} = 2e^{-i\lambda} \sin\left(\frac{1}{2}(u - i\tilde{u})\right) \tag{16}$$

where we have used that $(\sin z)^* \equiv \sin(z^*)$. The * denotes complex conjugate. Inserting (16) into (15) yields

$$\frac{d^2 u}{dx^2} - i\frac{d^2 \tilde{u}}{d\tilde{x}^2} = 2\cos\left(\frac{1}{2}(u + i\tilde{u})\right)\sin\left(\frac{1}{2}(u - i\tilde{u})\right).$$

Using the identity

$$\cos\left(\frac{1}{2}(u + i\tilde{u})\right)\sin\left(\frac{1}{2}(u - i\tilde{u})\right) \equiv \frac{1}{2}\sin u + \frac{1}{2}\sin(-i\tilde{u}) \equiv \frac{1}{2}\sin u - \frac{1}{2}i \sinh \tilde{u}$$

we find that

$$\frac{d^2 u}{dx^2} - i\frac{d^2 \tilde{u}}{d\tilde{x}^2} = \sin u - i \sinh \tilde{u}.$$

Equations (12) and (13) follow. ♣

14.3   Computer Algebra Applications

To apply the Bäcklund transformation it is necessary to implement the chain rule. We show how this is done for the Bäcklund transformation of

$$\frac{d^2u}{dx^2} = \sin(u), \qquad \frac{d^2\tilde{u}}{d\tilde{x}^2} = \sinh(\tilde{u})$$

```
% B\"acklund transformation

depend ut, xt;
depend xt, x;
depend u, x;

let df(ut,x) = df(ut,xt)*df(xt,x);

res1 := df(u,x) - i*df(ut,xt) - 2*exp(i*lam)*sin((u + i*ut)/2);

res2 := df(res1,x);

let df(xt,x) = 1;

res2 := res2;
```

The output is

```
res2 :=  - e**(i*lam)*cos((i*ut + u)/2)*df(u,x)
  - e**(i*lam)*cos((i*ut + u)/2)*df(ut,xt)*i
  + df(u,x,2) - df(ut,xt,2)*i$
```

Chapter 15

Lax Representations

15.1 Definitions

In this section we consider the Lax representation of differential equations and give some examples. Most soliton equations admit a Lax representation. Also for a class of ordinary differential equations Lax representations can be found. The Lax representation for a soliton equation is the starting point of the inverse scattering method (Ablowitz and Segur [1]).

Definition 15.1 *Let L and M be two linear differential operators. The* **spectral problem** *is given by*

$$Lv = \lambda v \tag{1}$$

where

$$\frac{\partial v}{\partial x_{m+1}} = Mv \tag{2}$$

Here v is a smooth function depending on $x_1, \ldots, x_m, x_{m+1}$ and λ is a parameter.

We show that the operator equation

$$\frac{\partial L}{\partial x_{m+1}} = [M, L] \tag{3}$$

holds, where

$$[L, M] := LM - ML$$

is the commutator. Equation (3) is called the **Lax representation** which contains a nonlinear evolution equation if L and M are correctly chosen. Equations (1) and (2) are known as the **Lax equations**.

241

From (1) we obtain

$$\frac{\partial}{\partial x_{m+1}}(Lv) = \frac{\partial}{\partial x_{m+1}}(\lambda v).$$

It follows that

$$\frac{\partial L}{\partial x_{m+1}}v + L\frac{\partial v}{\partial x_{m+1}} = \lambda\frac{\partial v}{\partial x_{m+1}}. \tag{4}$$

Inserting (1) and (2) into (4) gives

$$\frac{\partial L}{\partial x_{m+1}}v + LMv = \lambda Mv. \tag{5}$$

From (1) we obtain

$$MLv = \lambda Mv.$$

Inserting this expression into (5) yields

$$\frac{\partial L}{\partial x_{m+1}}v = -LMv + MLv.$$

Consequently

$$\frac{\partial L}{\partial x_{m+1}}v = [M, L]v.$$

Since the smooth function v is arbitrary we obtain the operator equation (3).

The Lax representation can also be applied to discrete systems. Let

$$\psi_{m+1} = L_m\psi_m$$

and

$$\frac{d\psi_m}{dt} = M_m\psi_m.$$

Then we find the operator equation

$$\frac{dL_m}{dt} = M_{m+1}L_m - L_mM_m.$$

15.2 Examples

Example: We consider $n = 1$ and $m = 2$, where we put $u_1 = u$ and $v_1 = v$. Let

$$L = \frac{\partial^2}{\partial x_1^2} + u(x_1, x_2)$$

$$M = 4\frac{\partial^3}{\partial x_1^3} + 6u(x_1, x_2)\frac{\partial}{\partial x_1} + 3\frac{\partial u}{\partial x_1}.$$

We now calculate the evolution equation for u. Since

$$\frac{\partial L}{\partial x_2}v = \frac{\partial}{\partial x_2}(Lv) - L\left(\frac{\partial v}{\partial x_2}\right),$$

we obtain

$$\frac{\partial L}{\partial x_2}v = \frac{\partial}{\partial x_2}\left(\left(\frac{\partial^2}{\partial x_1^2} + u\right)v\right) - \left(\frac{\partial^2}{\partial x_1^2} + u\right)\frac{\partial v}{\partial x_2} = \frac{\partial u}{\partial x_2}v.$$

Now

$$
\begin{aligned}
[M, L]v &= (ML)v - (LM)v = M(Lv) - L(Mv) \\[2mm]
&= \left(4\frac{\partial^3}{\partial x_1^3} + 6u\frac{\partial}{\partial x_1} + 3\frac{\partial u}{\partial x_1}\right)\left(\frac{\partial^2 v}{\partial x_1^2} + uv\right) \\[2mm]
&\quad - \left(\frac{\partial^2}{\partial x_1^2} + u\right)\left(4\frac{\partial^3 v}{\partial x_1^3} + 6u\frac{\partial v}{\partial x_1} + 3\frac{\partial u}{\partial x_1}v\right) \\[2mm]
&= 6u\frac{\partial u}{\partial x_1}v + \frac{\partial^3 u}{\partial x_1^3}v.
\end{aligned}
$$

Since v is arbitrary, we obtain

$$\frac{\partial u}{\partial x_2} = 6u\frac{\partial u}{\partial x_1} + \frac{\partial^3 u}{\partial x_1^3}.$$

This is the Korteweg-de Vries equation. ♣

Example: We consider $n = 1$ and $m = 3$ where we put $u_1 = u$ and $v_1 = v$. Let

$$L = \frac{\partial^2}{\partial x_1^2} + bu(x_1, x_2, x_3) + \frac{\partial}{\partial x_2}$$

and

$$T = -4\frac{\partial^3}{\partial x_1^3} - 6bu(x_1, x_2, x_3)\frac{\partial}{\partial x_1} - 3b\frac{\partial u}{\partial x_1} - 3bD_{x_1}^{-1}\frac{\partial u}{\partial x_2} + \frac{\partial}{\partial x_3}$$

where $b \in \mathbf{R}$ and

$$D_{x_1}^{-1}f := \int_{-\infty}^{x_1} f(s)ds.$$

The nonlinear evolution equation which follows from the condition

$$[T, L]v = 0,$$

where v is a smooth function, is given by

$$\frac{\partial^2 u}{\partial x_3 \partial x_1} = \frac{\partial^4 u}{\partial x_1^4} + 6b \left(\frac{\partial u}{\partial x_1}\right)^2 + 6bu\frac{\partial^2 u}{\partial x_1^2} + 3\frac{\partial^2 u}{\partial x_2^2}.$$

This is the Kadomtsev-Petviashvili equation.   ♣

Consider a system of linear equations for an eigenfunction and an evolution equation, i.e.,

$$L(\mathbf{x}, D)\psi(\mathbf{x}, \lambda) = \lambda\psi(\mathbf{x}, \lambda) \tag{6}$$

$$\frac{\partial\psi(\mathbf{x}, \lambda)}{\partial x_k} = B_k(\mathbf{x}, D)\psi(\mathbf{x}, \lambda) \tag{7}$$

where $D \equiv \partial/\partial\mathbf{x}$. We now show that

$$\frac{\partial L}{\partial x_k} = [B_k, L] \equiv B_k L - L B_k$$

and

$$\frac{\partial B_l}{\partial x_k} - \frac{\partial B_k}{\partial x_l} = [B_k, B_l]$$

where $1 \le k, l \le m$. From (6) we obtain

$$\frac{\partial}{\partial x_k}(L\psi) = \frac{\partial L}{\partial x_k}\psi + L\frac{\partial\psi}{\partial x_k} = \frac{\partial}{\partial x_k}(\lambda\psi) = \lambda\frac{\partial\psi}{\partial x_k}. \tag{8}$$

Inserting (7) into (8) yields

$$\frac{\partial L}{\partial x_k}\psi + LB_k\psi = \lambda B_k\psi = B_k\lambda\psi$$

or

$$\frac{\partial L}{\partial x_k}\psi = -LB_k\psi + B_k L\psi = [B_k, L]\psi. \tag{9}$$

From (9) we obtain

$$\frac{\partial}{\partial x_k}(L\psi) - L\frac{\partial\psi}{\partial x_k} = B_k L\psi - LB_k\psi \tag{10}$$

and

$$\frac{\partial}{\partial x_l}(L\psi) - L\frac{\partial\psi}{\partial x_l} = B_l L\psi - LB_l\psi. \tag{11}$$

Taking the derivative of (10) with respect to x_l and of (11) with respect to x_k gives

$$\frac{\partial^2}{\partial x_l \partial x_k}(L\psi) - \frac{\partial}{\partial x_l}\left(L\frac{\partial \psi}{\partial x_k}\right) = \frac{\partial}{\partial x_l}(B_k L\psi) - \frac{\partial}{\partial x_l}(LB_k\psi) \tag{12}$$

$$\frac{\partial^2}{\partial x_k \partial x_l}(L\psi) - \frac{\partial}{\partial x_k}\left(L\frac{\partial \psi}{\partial x_l}\right) = \frac{\partial}{\partial x_k}(B_l L\psi) - \frac{\partial}{\partial x_k}(LB_l\psi). \tag{13}$$

Subtracting (13) from (12) we obtain

$$-\frac{\partial}{\partial x_l}\left(L\frac{\partial \psi}{\partial x_k}\right) + \frac{\partial}{\partial x_k}\left(L\frac{\partial \psi}{\partial x_l}\right) = \frac{\partial}{\partial x_l}(B_k L\psi) - \frac{\partial}{\partial x_l}(LB_k\psi) - \frac{\partial}{\partial x_k}(B_l L\psi) + \frac{\partial}{\partial x_k}(LB_l\psi). \tag{14}$$

Inserting

$$\frac{\partial \psi}{\partial x_k} = B_k\psi, \qquad \frac{\partial \psi}{\partial x_l} = B_l\psi \tag{15}$$

into (14) and taking into account (6) gives

$$\frac{\partial}{\partial x_l}(B_k\psi) - \frac{\partial}{\partial x_k}(B_l\psi) = 0. \tag{16}$$

From (16) we obtain

$$\frac{\partial B_k}{\partial x_l}\psi + B_k\frac{\partial \psi}{\partial x_l} - \frac{\partial B_l}{\partial x_k}\psi - B_l\frac{\partial \psi}{\partial x_k} = 0. \tag{17}$$

Inserting (15) into (17) gives

$$\frac{\partial B_k}{\partial x_l}\psi + B_k B_l\psi - \frac{\partial B_l}{\partial x_k}\psi - B_l B_k\psi = 0.$$

Since ψ is arbitrary, it follows that

$$\frac{\partial B_l}{\partial x_k} - \frac{\partial B_k}{\partial x_l} = [B_k, B_l].$$

Example: Let

$$L := D + u_2(\mathbf{x})D^{-1} + u_3(\mathbf{x})D^{-2} + u_4(\mathbf{x})D^{-3} + \cdots$$

where $\mathbf{x} = (x_1, x_2, x_3, \ldots)$,

$$D := \frac{\partial}{\partial x_1}$$

and

$$D^{-1}f(\mathbf{x}) = \int\limits^{x_1} f(s)ds.$$

We define $B_n(\mathbf{x}, D)$ as the differential part of $(L(\mathbf{x}, D))^n$ and show that

$$B_1 \;=\; D$$

$$B_2 \;=\; D^2 + 2u_2$$

$$B_3 \;=\; D^3 + 3u_2 D + 3u_3 + \frac{\partial u_2}{\partial x_1}$$

$$B_4 \;=\; D^4 + 4u_2 D^2 + \left(4u_3 + 6\frac{\partial u_2}{\partial x_1}\right) D + 4u_4 + 6\frac{\partial u_3}{\partial x_1} + 4\frac{\partial^2 u_2}{\partial x_1^2} + 6u_2^2.$$

From L it is obvious that $B_1 = D$. Let f be a smooth function. Then

$$L^2 f \;=\; L(Lf)$$

$$=\; L(Df + u_2 D^{-1} f + u_3 D^{-2} f + u_4 D^{-3} f + u_5 D^{-4} f + \ldots)$$

$$=\; D^2 f + u_2 f + (Du_2)D^{-1} f + u_3(D^{-1} f)$$

$$+\; (Du_3)D^{-2} f + u_4 D^{-2} f + (Du_4)D^{-3} f + \ldots$$

where

$$D^{-2} f := \int^{x_1} \left(\int^{s_1} f(s_2) ds_2\right) ds_1.$$

Since f is an arbitrary smooth function we have

$$B_2 = D^2 + 2u_2.$$

In the same manner we find B_3 and B_4. Let us now find the equation of motion for

$$\frac{\partial B_l}{\partial x_k} - \frac{\partial B_k}{\partial x_l} = [B_k, B_l]$$

with $k = 2$ and $l = 3$. Since

$$\frac{\partial B_3}{\partial x_2}\psi \;=\; \frac{\partial}{\partial x_2}(D^3\psi) + 3\frac{\partial u_2}{\partial x_2}(D\psi) + 3u_2 D\left(\frac{\partial \psi}{\partial x_2}\right) + 3\frac{\partial u_3}{\partial x_2}\psi + 3u_3\frac{\partial \psi}{\partial x_2}$$

$$+\; 3\frac{\partial^2 u_2}{\partial x_1 \partial x_2}\psi + 3\frac{\partial u_2}{\partial x_1}\frac{\partial \psi}{\partial x_2}$$

and

$$-\frac{\partial B_2}{\partial x_3}\psi = -D^2\frac{\partial \psi}{\partial x_3} - 2\frac{\partial u_2}{\partial x_3}\psi - 2u_2\frac{\partial \psi}{\partial x_3},$$

we find

$$\frac{\partial}{\partial x_1}\left(\frac{\partial u_2}{\partial x_3} - \frac{1}{4}\frac{\partial^3 u_2}{\partial x_1^3} - 3u_2\frac{\partial u_2}{\partial x_1}\right) - \frac{3}{4}\frac{\partial^2 u_2}{\partial x_2^2} = 0$$

which is the Kadomtsev-Petviashvili equation.   ♣

Example: The nonlinear partial differential equation

$$\frac{\partial^2 u}{\partial x_1 \partial x_2} + \alpha\frac{\partial u}{\partial x_1} + \beta\frac{\partial u}{\partial x_2} + \gamma\frac{\partial u}{\partial x_1}\frac{\partial u}{\partial x_2} = 0$$

is called the **Thomas equation**. Here α, β and γ are real constants with $\gamma \neq 0$. We show that

$$\frac{\partial \phi}{\partial x_1} = -\frac{\partial u}{\partial x_1} - \left(2\beta + \gamma\frac{\partial u}{\partial x_1}\right)\phi, \qquad \frac{\partial \phi}{\partial x_2} = \frac{\partial u}{\partial x_2} - \left(2\alpha + \gamma\frac{\partial u}{\partial x_2}\right)\phi \qquad (18)$$

provides a Lax representation for the Thomas equation. By taking the derivative of (18a) with respect to x_1 gives

$$\frac{\partial^2 \phi}{\partial x_1 \partial x_2} = \frac{\partial^2 u}{\partial x_1 \partial x_2} - 2\alpha\frac{\partial \phi}{\partial x_1} - \gamma\frac{\partial^2 u}{\partial x_1 \partial x_2}\phi - \gamma\frac{\partial u}{\partial x_2}\frac{\partial \phi}{\partial x_1} \qquad (19)$$

and taking the derivative of (18b) with respect to x_2 gives

$$\frac{\partial^2 \phi}{\partial x_2 \partial x_1} = -\frac{\partial^2 u}{\partial x_2 \partial x_1} - 2\beta\frac{\partial \phi}{\partial x_2} - \gamma\frac{\partial^2 u}{\partial x_2 \partial x_1}\phi - \gamma\frac{\partial u}{\partial x_1}\frac{\partial \phi}{\partial x_2}. \qquad (20)$$

Subtracting (19) and (20) leads to

$$0 = 2\frac{\partial^2 u}{\partial x_2 \partial x_1} - \left(2\alpha + \gamma\frac{\partial u}{\partial x_2}\right)\frac{\partial \phi}{\partial x_1} + \left(2\beta + \gamma\frac{\partial u}{\partial x_1}\right)\frac{\partial \phi}{\partial x_2}. \qquad (21)$$

Inserting (18a) and (18b) into (21) gives the Thomas equation.

Remark: The Thomas equation can be linearized by applying the transformation

$$u = \frac{1}{\gamma}\ln v.$$ ♣

15.3 Computer Algebra Applications

In our application we calculate the right-hand side of the Korteweg-de Vries equation

$$\frac{\partial u}{\partial t} = 6u\frac{\partial u}{\partial x} + \frac{\partial^3 u}{\partial x^3}$$

using the Lax pair

$$L := \frac{\partial^2}{\partial x^2} + u(x,t), \qquad M := 4\frac{\partial^3}{\partial x^3} + 6u(x,t)\frac{\partial}{\partial x} + 3\frac{\partial u}{\partial x}.$$

```
% Lax pair
% lax.red

depend u, x;
depend v, x;

L := df(v,x,2) + u*v;
M := 4*df(v,x,3) + 6*u*df(v,x) + 3*df(u,x)*v;

A := sub(v=L,M);
B := sub(v=M,L);
com := A - B;
```

The output is

```
com := v*(df(u,x,3) + 6*df(u,x)*u)$
```

Chapter 16

Conservation Laws

16.1 Basic Concepts

Conservation laws describe quantities that remain invariant during the evolution of a partial differential equation (Ablowitz [1], Zwillinger [149], Ablowitz et al [2], Calogero and Degasperis [10], Steeb *et al* [134]). This provides simple and efficient methods for the study of many qualitative properties of solutions, including stability, evolution of solitons, and decomposition into solitons, as well as the theoretical description of solution manifolds. A soliton equation is a partial differential equation with a wave-like solution, known as a solitary wave. A solitary wave is a localized, travelling wave and several nonlinear partial differential equations have a solution of this type. A soliton is a specific type of stable solitary wave which is best described in terms of its interaction with other solitary waves. Conservation laws also allow estimates of the accuracy of a numerical solution scheme. For a partial differential equation that is not written in conserved form there is a number of ways to attempt to write the equation in conserved form. These include a change of the dependent as well as the independent variables and applications of Noether's theorem (see Bluman et al [8] for the techniques). Bluman and Kumei [7] have introduced the so called 'potential symmetries' whereby conservation laws can be found. We also refer to chapter 20 for details on the computer algebra packages for determining conservation laws. We also discuss Noethers theorem in detail using the jet bundle formalism (Shadwick [114]).

Let us first consider the case of two independent variables x_1 and x_2 with one dependent variable u. Here x_2 plays the role of the time variable. We consider an evolution equation of the form

$$\frac{\partial u}{\partial x_2} = F\left(u, \frac{\partial u}{\partial x_1}, \frac{\partial^2 u}{\partial x_1^2}, \ldots, \frac{\partial^r u}{\partial x_1^r}\right). \tag{1}$$

Definition 16.1 *A* **conservation law** *for (1) is a partial differential equation of the form*

$$\frac{\partial}{\partial x_2} T\left(u(x_1, x_2)\right) + \frac{\partial}{\partial x_1} X\left(u(x_1, x_2)\right) = 0 \tag{2}$$

which is satisfied by all solutions of (1). T is defined to be the **conserved density** *and X the* **conserved flux** *or flow. Using differential forms (2) can also be written as*

$$d(T(u(x_1, x_2)) dx_1 - X(u(x_1, x_2)) dx_2) = 0.$$

Definition 16.2 *The functional*

$$I(u) \equiv \int\limits_{-\infty}^{\infty} T\left(u(x_1, x_2)\right) dx_1 \tag{3}$$

is a **constant of the motion** *since*

$$\frac{d}{dx_2} I(u) = 0 \tag{4}$$

for all solutions of (1), provided the integral (3) exists and the integrand satisfies the appropriate boundary conditions at $x_1 = \pm\infty$.

A standard procedure is to determine a set of conservation laws and then use (3) to obtain constants of the motion.

Example: The **Korteweg-de Vries equation**

$$\frac{\partial u}{\partial x_2} = u\frac{\partial u}{\partial x_1} + \frac{\partial^3 u}{\partial x_1^3}$$

has an infinite set of conservation laws. The first few, in order of increasing rank, have the conserved densities

$$T_1(u) = u$$
$$T_2(u) = u^2$$
$$T_3(u) = u^3 - 3\left(\frac{\partial u}{\partial x_1}\right)^2$$
$$T_4(u) = 5u^4 - 60u\left(\frac{\partial u}{\partial x_1}\right)^2 - 36\frac{\partial u}{\partial x_1}\frac{\partial^3 u}{\partial x_1^3}$$
$$\vdots$$

Since the Korteweg-de Vries equation can be written as a conservation law

$$\frac{\partial u}{\partial x_2} + \frac{\partial}{\partial x_1}\left(-\frac{u^2}{2} - \frac{\partial^2 u}{\partial x_1^2}\right) = 0,$$

it follows that $T(u) = u$ is a conserved density and the conserved flux is given by

$$X(u) = -\frac{u^2}{2} - \frac{\partial^2 u}{\partial x_1^2}.$$

To demonstrate that $T(u) = u^2$ is a conserved density, we consider

$$\frac{\partial T}{\partial x_2} = \frac{\partial(u^2)}{\partial x_2} = 2u\frac{\partial u}{\partial x_2} = 2u\frac{\partial^3 u}{\partial x_1^3} + 2u^2\frac{\partial u}{\partial x_1}$$

where we make use of the given Korteweg-de Vries equation to replace the $\partial u/\partial x_2$ term. The flux X such that (2) is satisfied is given by

$$X(u) = \left(\frac{\partial u}{\partial x_1}\right)^2 - 2u\frac{\partial^2 u}{\partial x_1^2} - \frac{2}{3}u^3. \qquad \clubsuit$$

Example: The Schrödinger equation in one space dimension

$$i\frac{\partial u}{\partial x_2} = -\frac{\partial^2 u}{\partial x_1^2} + V(x_1)u$$

can be expressed in the conserved form (2) where

$$T(u) = ig(x_1)u, \qquad X(u) = g(x_1)\frac{\partial u}{\partial x_1} - \frac{dg}{dx_1}u$$

and $g(x_1)$ is defined by

$$\frac{d^2 g}{dx_1^2} - V(x_1)g(x_1) = 0. \qquad \clubsuit$$

Example: The sine-Gordon equation

$$\frac{\partial^2 u}{\partial x_1 \partial x_2} = \sin u$$

can be expressed in the conserved form (2) where

$$T(u) = \frac{1}{2}\left(\frac{\partial u}{\partial x_1}\right)^2, \qquad X(u) = \cos u. \qquad \clubsuit$$

Example: We consider the one-dimensional diffusion equation

$$\frac{\partial u}{\partial x_2} = \frac{\partial^2 u}{\partial x_1^2}.$$

Let us derive a constant of motion from a given conserved density. Consider

$$\frac{\partial}{\partial x_2}\left(-x_1\frac{\partial u}{\partial x_1} - 2x_2\frac{\partial u}{\partial x_2}\right) + \frac{\partial}{\partial x_1}(X(u)) = 0.$$

Taking into account the diffusion equation and integrating we find the conservation law

$$\frac{\partial}{\partial x_2}\left(-x_1\frac{\partial u}{\partial x_1} - 2x_2\frac{\partial u}{\partial x_2}\right) + \frac{\partial}{\partial x_1}\left(\frac{\partial u}{\partial x_1} + x_1\frac{\partial^2 u}{\partial x_1^2} + 2x_2\frac{\partial^3 u}{\partial x_1^3}\right) = 0.$$

This leads to the constant of motion

$$I(u) = \int\limits_{-\infty}^{+\infty}\left(-x_1\frac{\partial u}{\partial x_1} - 2x_2\frac{\partial u}{\partial x_2}\right)dx_1 = \int\limits_{-\infty}^{+\infty} u\,dx_1 = C.$$

Since u is the concentration of the diffusing substance, the quantity C is the total amount of the diffusing substance. ♣

We now demonstrate with the help of an example that, in general, one cannot derive a constant of motion from a conservation law, even if one assumes that the dependent variable u and all its derivatives with respect to the space coordinates vanish rapidly as the space coordinates tend to infinity. In the following we let x_2 be fixed but arbitrary. We assume that the dependent function u is an element of the Schwartz space $S(\mathbf{R})$ with respect to the space coordinate x_1.

Definition 16.3 *The functions in the* **Schwartz space** *$S(\mathbf{R})$ are those functions which, together with their derivatives fall off more quickly than the inverse of any polynomial.*

Example: We consider the one-dimensional diffusion equation

$$\frac{\partial u}{\partial x_2} = \frac{\partial^2 u}{\partial x_1^2}.$$

There exist solutions of the diffusion equation which for a fixed time x_2 are elements of the Schwartz space $S(\mathbf{R})$. We assume that the conserved density T has the form

$$T(u) = u\ln u.$$

Now the equation

$$\frac{\partial}{\partial x_2}(u\ln u) + \frac{\partial}{\partial x_1}X(u) = 0$$

can easily be solved by integration and we find the conservation law

$$\frac{\partial}{\partial x_2}(u\ln u) + \frac{\partial}{\partial x_1}\left(-\frac{\partial^2 u}{\partial x_1^2} - \int\limits_{-\infty}^{x_1}\frac{\partial^2 u(\eta)}{\partial \eta^2}\ln u(\eta, x_2)\,d\eta\right) = 0$$

where the integral in the second bracket exists. But one cannot conclude that the integral

$$\int\limits_{-\infty}^{+\infty} u(x_1, x_2)\ln(u(x_1, x_2))\,dx_1 \tag{5}$$

which exists, does not depend on x_2, i.e.

$$\frac{d}{dx_2} \int_{-\infty}^{+\infty} u(x_1, u_2) \ln(u(x_1, x_2)) \, dx_1 \neq 0.$$

This can be seen when we insert the following solution of the diffusion equation ($x_2 > 0$)

$$u(x_1, x_2) = \frac{1}{\sqrt{x_2}} \exp\left(-\frac{x_1^2}{4x_2}\right)$$

into(5). This behaviour is due to the fact that the function f which is given by

$$f(x_1, x_2) = \int_{-\infty}^{x_1} \left(\frac{\partial^2 u(\eta, x_2)}{\partial \eta^2} \ln u(\eta, x_2) \, d\eta\right)$$

does not vanish as $|x_1| \to \infty$. For the given solution of the diffusion equation we find

$$\frac{d}{dx_2} \int_{-\infty}^{+\infty} u(x_1, x_2) \ln u(x_1, x_2) \, dx_1 \leq 0$$

for $x_2 > 0$ and for sufficiently large x_2 we find

$$\left| \frac{d}{dx_2} \int_{-\infty}^{+\infty} u(x_1, x_2) \ln(u(x_1, x_2)) \, dx_1 \right| \leq \frac{2\sqrt{2}}{\exp(x_2)}. \qquad \clubsuit$$

We now consider the case of m independent variables $\mathbf{x} = (x_1, \ldots, x_m)$ and one dependent variable u, where x_m plays the role of the time variable.

Definition 16.4 *A partial differential equation with m independent variables and one dependent variable is in conservation form if it can be written in the form*

$$\frac{\partial}{\partial x_m} T(u) + div\mathbf{X}(u) = 0, \qquad div\mathbf{X} := \frac{\partial X_1}{\partial x_1} + \cdots + \frac{\partial X_{m-1}}{\partial x_{m-1}} \qquad (6)$$

where $\mathbf{X} = (X_1, \ldots, X_{m-1})$.

Example: We consider the three dimensional diffusion equation

$$\frac{\partial u}{\partial x_4} = \frac{\partial^2 u}{\partial x_1^2} + \frac{\partial^2 u}{\partial x_2^2} + \frac{\partial^2 u}{\partial x_3^2}.$$

Let $T(u) = u \ln u$. Then we find, by integration, the conservation law

$$\frac{\partial}{\partial x_4}(u \ln u) - divV(u) = 0$$

where

$$V(u(\mathbf{x}, x_4)) = -\frac{1}{4\pi} \int\limits_{-\infty}^{+\infty} \left(\frac{\partial^2 u}{\partial \eta_1^2} + \frac{\partial^2 u}{\partial \eta_2^2} + \frac{\partial^2 u}{\partial \eta_3^2} \right) \ln(u+1) \frac{\mathbf{x} - \boldsymbol{\eta}}{|\mathbf{x} - \boldsymbol{\eta}|} \, d\eta_1 \, d\eta_2 \, d\eta_3$$

with $\mathbf{x} = (x_1, x_2, x_3)$ and $\boldsymbol{\eta} = (\eta_1, \eta_2, \eta_3)$ and $u = u(\boldsymbol{\eta}, x_4)$ on the right-hand side of the equation.   ♣

Definition 16.5 *A conservation law is called trivial if* $T(u)$ *is itself the* $\mathbf{x}$ *derivative of some expression.*

Example: We again consider the one-dimensional diffusion equation

$$\frac{\partial u}{\partial x_2} = \frac{\partial^2 u}{\partial x_1^2}.$$

An example of a trivial conservation law is

$$\frac{\partial}{\partial x_2} \left(-g(u) \frac{\partial u}{\partial x_1} \right) + \frac{\partial}{\partial x_1} \left(g(u) \frac{\partial u}{\partial x_2} \right) = 0$$

where g is a smooth function of u. This is an identity. Let g be a polynomial in u. Then integrating by parts we obtain

$$\int\limits_{-\infty}^{+\infty} g(u) \frac{\partial u}{\partial x_1} dx_1 = 0.$$

The trivial conservation law has no physical meaning.   ♣

From the definition of conservation laws a definition of integrability can be proposed:

Definition 16.6 *If an evolution equation has an infinite sequence of nontrivial conservation laws, the equation is called formally integrable, sometimes called* **integrable**.

From our previous example of the Korteweg-de Vries equation it follows from the given definition of integrability that the Korteweg-de Vries equation is integrable since it has an infinite set of conservation laws. Other examples of integrable evolution equations are the modified Korteweg-de Vries equation, the Boussinesq equation, the Calogero-Degasperis-Fokas equation and the one-dimensional nonlinear Schrödinger equation. These are examples of soliton equations that can be solved by the inverse scattering transform. It must be noted that Burgers' equation (which is not a soliton equation) does not have an infinite number of conservation laws, but it has an infinite number of Lie-Bäcklund symmetry vector fields.

16.2 Exterior Differential Systems

In this section the correspondence between conservation laws and Lie symmetry vector fields is studied. A definition of higher order conservation laws (hierarchy of conservation laws) is presented in terms of an exterior differential system.

Let M be an oriented differentiable manifold of dimension m with local coordinates $\mathbf{x} = (x_1, \ldots, x_m)$ and volume m-form ω given in these coordinates by

$$\omega = dx_1 \wedge dx_2 \wedge \ldots \wedge dx_m.$$

Let N be an n-dimensional manifold with local coordinates $\mathbf{u} = (u_1, u_2, \ldots, u_n)$ and let (E, π, M) be a fibre bundle with fibre N and $E = M \times N$. The k-jet bundle of local sections of (E, π, M) is denoted by $J^k(E)$ and $J(E)$ denotes the infinite jet bundle. The canonical projections from $J^k(E)$ to M and from $J(E)$ to M are denoted by π^k_M and π_M respectively, and π_k is the canonical projection from $J(E)$ to $J^k(E)$. The k-th order and infinite-order contact modules are denoted by Ω^k and Ω respectively, with standard bases

$$\left\{ \theta_j, \theta_{j,i_1}, \ldots, \theta_{j,i_1 \ldots i_{k-1}} \right\}$$

for Ω^k, and $\{ \theta_j, \theta_{j,i_1}, \ldots \}$ for Ω . The contact forms are given by

$$\theta_{j,i_1 \ldots i_l} = du_{j,i_1 \ldots i_l} - \sum_{b=1}^{m} u_{j,i_1 \ldots i_l b} dx_b$$

We consider a system of partial differential equations of order k

$$F_\nu \left(x_i, u_j, \frac{\partial u_j}{\partial x_i}, \ldots, \frac{\partial^k u_j}{\partial x_{i_1} \partial x_{i_2} \ldots \partial x_{i_k}} \right) = 0 \tag{7}$$

where $\nu = 1, \ldots, q$ and $i = 1, \ldots, m$, $j = 1, \ldots, n$. Within the jet bundle formalism the submanifold R^k, given by the constrained equations

$$F_\nu(x_i, u_j, u_{j,i}, \ldots, u_{j,i_1 \ldots i_k}) = 0 \tag{8}$$

is then considered.

The modules of contact m-forms $\Omega^k_{(m)}$ and $\Omega_{(m)}$ are the modules generated by $\Omega^k \wedge \{\omega_i\}$ and $\Omega \wedge \{\omega_i\}$ respectively, where ω_i is the $(m-1)$ form defined by

$$\omega_i := \frac{\partial}{\partial x_i} \,\lrcorner\, \omega.$$

If s is a local section of (E, π, M) then $j^k s$ and js are the k-jet and infinite jet extensions of s, respectively. Obviously

$$j^k s^* \Omega^k_{(m)} = 0, \qquad j s^* \Omega_{(m)} = 0.$$

Let $\mathcal{A}$ denote the Lie algebra of vector fields on $J(E)$, defined by

$$\mathcal{A} := \{\ Z \in \Xi(J(E))\ :\ L_Z \Omega \subset \Omega \text{ and } \pi_{M^*} Z = 0\ \}$$

where $\Xi(J(E))$ denotes the vector fields on $J(E)$ and Z is a Lie symmetry vector field. Clearly the vector fields in $\mathcal{A}$ preserve the contact m-forms on $J(E)$ as well as preserving Ω, that is, if $Z \in \mathcal{A}$ then

$$L_Z \Omega_{(m)} \subset \Omega_{(m)}.$$

The vector field D_i, defined on $J(N)$, is referred to as the operator of total differentiation, and is given by

$$D_i := \frac{\partial}{\partial x_i} + \sum_{j=1}^{m} u_{j,i} \frac{\partial}{\partial u_j} + \cdots + \sum_{j=1}^{m} \sum_{i_1,\ldots,i_r=1}^{n} u_{j,ii_1\ldots i_k} \frac{\partial}{\partial u_{j,i_1\ldots i_k}}, \qquad i_1 \le \ldots \le i_r.$$

We now consider conservation laws in connection with exterior differential forms. The space of sections of a fibred manifold is denoted by $S(E)$.

Definition 16.7 *Let (E, π, M) be a fibred manifold and let Σ^{k-1} be a finitely generated ideal of homogeneous differential forms on $J^{k-1}(E)$ where $k \ge 1$. If $d\Sigma^{k-1}$ is contained in Σ^{k-1} then Σ^{k-1} is an **exterior differential system** or an **exterior ideal**.   A solution of Σ^{k-1} is a section $s \in S(E)$ such that*

$$j^{k-1} s^* \sigma^{k-1} = 0.$$

Example: The ideal generated by the contact modules $\{\theta_{j,i_1\ldots i_k},\ d\theta_{j,i_1\ldots i_l}\ :\ l \le k-1\}$ is an exterior differential system on $J^{k-1}(E)$ whose solutions are the sections $s \in S(E)$.
♣

We denote by R^k the differential equation associated with Σ^{k-1}.

Definition 16.8 *Let Σ^{k-1} be an exterior differential system on $J^{k-1}(E)$. A **conserved current** for Σ^{k-1} is an $(m-1)$-form ζ such that*

$$d\zeta \in \Sigma^{k-1}.$$

Thus for each solution s of Σ^{k-1} the $(m-1)$-form $j^{k-1} s^* \zeta$ is closed, i.e.,

$$d\left(j^{k-1} s^* \zeta\right) = 0. \tag{9}$$

*Equation (9) is called a **conservation law** for Σ^{k-1}. The local coordinate expression of this conservation law is*

$$\sum_{i=1}^{m} \frac{\partial}{\partial x_i} (j^k s^* f_i) = 0$$

where the functions f_i are given by

$$\pi_{k-1}^{k}{}^* \zeta = \sum_{i_1,\ldots,i_{m-1}=1}^{m} f_{i_1\ldots i_{m-1}} dx_{i_1} \wedge \cdots \wedge dx_{i_{m-1}} \qquad mod\ contact\ forms \quad \Omega^k.$$

It is a consequence of Stokes theorem that a conserved current determines a conserved function on the space of solutions of Σ^{k-1} that can be written as

$$s \mapsto \int j^{k-1}s^*\zeta.$$

The integration is understood to be carried out over a suitable $(m-1)$-dimensional submanifold of M.

Example: Let $M = \mathbf{R}^2$ and $(E, \pi, M) = (M \times \mathbf{R}, pr_1, M)$ with (x_1, x_2, u) local coordinates on E and (x_1, x_2, u, u_1, u_2) local coordinates on the jet bundle $J^2(E)$. pr_1 is the projection map $pr_1 : \mathbf{R}^2 \times \mathbf{R} \to \mathbf{R}^2$. The exterior differential system Σ^1 generated by

$$du \wedge dx_2 - u_1 \, dx_1 \wedge dx_2$$

$$du \wedge dx_1 + u_2 \, dx_1 \wedge dx_2$$

$$du_1 \wedge dx_1 + \sin u \, dx_1 \wedge dx_2$$

$$du_2 \wedge dx_2 - \sin u \, dx_1 \wedge dx_2$$

has as its associated equation the sine-Gordon equation

$$\frac{\partial^2 u}{\partial x_1 \partial x_2} = \sin u.$$

The one-form

$$\zeta := \frac{1}{2}u_1^2 dx_1 - \cos u \, dx_2$$

is a conserved current for Σ^1. We find

$$d\zeta = u_1 du_1 \wedge dx_1 + \sin u \, du \wedge dx_2.$$

Let s be a solution of Σ^1. Then s satisfies the conservation law

$$\frac{1}{2}\frac{\partial}{\partial x_2}\left(\frac{\partial s}{\partial x_1}\right)^2 + \frac{\partial}{\partial x_1}(\cos s) = 0.$$

Consider the class of solutions of the sine-Gordon equation which has compact support and let M_c be the strip in $\mathbf{R}^2$ bounded by $x_2 = c$ and $x_2 = c'$. It follows from Stokes theorem that

$$\int_{\partial M_c} j^1 s^*\zeta = 0.$$

Since

$$\int_{\partial M_c} j^1 s^*\zeta = \frac{1}{2}\int_{-\infty}^{+\infty}\left(\frac{\partial s}{\partial x_1}(x_1, c)\right)^2 dx_1 - \frac{1}{2}\int_{-\infty}^{+\infty}\left(\frac{\partial s}{\partial x_1}(x_1, c')\right)^2 dx_1$$

it follows that the functional defined by

$$s \mapsto \frac{1}{2} \int\limits_{-\infty}^{+\infty} \left(\frac{\partial s}{\partial x_1} \right)^2 dx_1$$

is independent of x_2 and is thus conserved.   ♣

It is sometimes possible to enlarge the class of conserved functionals by considering $(m-1)$-forms which depend on derivatives of higher order than those which appear on $J^{k-1}(E)$. Thus it is convenient to extend the definitions to the prolongation of Σ^{k-1} on the infinite jet bundle $J(E)$.

Definition 16.9 *If ζ is a differentiable $(m-1)$-form on $J(E)$ and $d\zeta \in \Sigma^\infty$, then ζ is called a conserved current for Σ^∞. It follows from the definition of smooth forms that ζ is a conserved current for some Σ^l, i.e.*

$$\zeta = \pi_l^* \zeta_l$$

*and $d\zeta_l \in \Sigma^l$. If ζ does not factor through $J^{k-1}(E)$, then it is called a **higher order conserved current** for Σ^{k-1}. If ζ is a conserved current for Σ^∞ and*

$$j s^* \zeta = 0$$

for every solution s of Σ^∞, then ζ is called trivial, as the conserved functional determined by ζ is vacuous. The trivial conserved current is denoted by ζ_T.

Example: The sine-Gordon equation

$$\frac{\partial^2 u}{\partial x_1 \partial x_2} = \sin u$$

has an infinite hierarchy of higher order conserved currents. The first three which are non-trivial are

$$\zeta^{(1)} \;=\; \frac{1}{2} u_1^2 \, dx_1 - \cos u \, dx_2$$

$$\zeta^{(3)} \;=\; \left(\frac{1}{4} u_1^4 + u_1 u_{111} \right) dx_1 + u_{11} \sin u \, dx_2$$

$$\zeta^{(5)} \;=\; \left(\frac{1}{6} u_1^6 + \frac{11}{3} u_1^2 u_{11} + \frac{7}{3} u_1^3 u_{111} + \frac{4}{3} u_1 u_{11111} \right) dx_1 + \left(\frac{4}{3} u_{1111} + \frac{5}{3} u_1^2 u_{11} \right) \sin u \, dx_2.$$

♣

We now state the following

Theorem 16.1 *Assume that the system of partial differential equations (7) is invariant under the vector field $\bar{Z}$. Let ζ be a conserved current of (7). Then $L_{\bar{Z}}\zeta$ is also a conserved current of (7). Since*

$$d(js^*\zeta) = 0$$

is a conservation law, it follows that

$$d(js^* L_{\bar{Z}}\zeta) = 0$$

is also a conservation law of (7).

Proof: First we note that

$$d(js^*(\cdot)) = js^*(d(\cdot)).$$

Let ζ be a conservation law. Since solutions $s : M \to N$ are mapped into solutions by the Lie transformation group which is generated by $\bar{Z}$, we obtain

$$(j \exp(\varepsilon \bar{Z})s)^* d\zeta = 0.$$

Owing to the identity

$$\frac{d}{d\varepsilon}(j \exp(\varepsilon \bar{Z})s)^* \zeta \equiv (j \exp(\varepsilon \bar{Z})s)^* L_{\bar{Z}}\zeta$$

we find

$$0 = \frac{d}{d\varepsilon}(j \exp(\varepsilon \bar{Z})s)^* d\zeta = (j \exp(\varepsilon \bar{Z})s)^* d(L_{\bar{Z}}\zeta).$$

In the last step we used the fact that the Lie derivative and the exterior derivative commute. Setting $\varepsilon = 0$ it follows that

$$(js)^* d(L_{\bar{Z}}\zeta) = 0$$

which completes the proof. ♠

We give two examples to illustrate this theorem.

Example: Consider the diffusion equation

$$\frac{\partial u}{\partial x_2} = \frac{\partial^2 u}{\partial x_1^2}$$

where we have put $u_1 = u$. Consequently the submanifold is given by

$$F \equiv u_2 - u_{11} = 0.$$

. This diffusion equation admits the vertical vector field

$$Z_V = (x_1 u_1 + 2x_2 u_2)\frac{\partial}{\partial u_1}.$$

Let D_1 and D_2 be the total derivative vector fields. The vector field Z_V is associated with the (x_1, x_2)-scale change. We find that

$$L_{\tilde{V}}F = 2F + x_1(D_1 F) + x_2(D_2 F).$$

A conserved current of the diffusion equation can be given at once, namely

$$\zeta = u dx_1 + u_1 dx_2$$

since

$$(js)^* d\zeta = 0$$

is nothing more than the diffusion equation. Straightforward calculation shows that

$$L_{\tilde{Z}_V}\zeta = (x_1 u_1 + 2x_2 u_2)dx_1 + (u_1 + x_1 u_{11} + 2x_2 u_{12})dx_2$$

where $\tilde{Z}_V$ is the prolonged vertical vector field. According to the theorem the conservation law follows from $js^* d(L_{\tilde{Z}}\zeta) = 0$. Repeated application of $\tilde{Z}_V$ leads to a hierarchy of conservation laws.   ♣

Example: Consider the nonlinear diffusion equation

$$\frac{\partial u}{\partial x_2} = \left(\frac{\partial u}{\partial x_1}\right)^2 + \frac{\partial^2 u}{\partial x_1^2},$$

Since $n = 1$ we have put $u_1 = u$. The submanifold is given by

$$F \equiv u_2 - u_1^2 - u_{11} = 0.$$

A conserved current is given by

$$\zeta = x_1 \exp(u)dx_1 + (x_1 u_1 \exp(u) - \exp u)dx_2$$

and $js^* d\zeta = 0$, where

$$d\zeta = x_1 e^u du \wedge dx_1 + x_1 e^u du_1 \wedge dx_2 + u_1 e^u dx_1 \wedge dx_2 + x_1 u_1 e^u du \wedge dx_2 - e^u du \wedge dx_2.$$

The nonlinear equation admits the vertical vector field

$$Z_V = (x_1 u_1 + 2x_2 u_2)\frac{\partial}{\partial u}.$$

Again $js^* d(L_{\tilde{Z}_V}\zeta) = 0$ is a conservation law.   ♣

16.3 Cartan Fundamental Form

A version of Noether's theorem appropriate in the context of the Hamilton-Cartan formalism is given where Cartan's fundamental form plays a central role. We study conservation laws for which we let $\mathcal{L}$ be a first order **Lagrangian**,

$$\mathcal{L} : J^1(E) \to \mathbf{R}. \tag{10}$$

Definition 16.10 *The* **Cartan fundamental form** *associated with $\mathcal{L}$ is the m-form defined in local coordinates by*

$$\Theta := \left(\mathcal{L} - \sum_{j=1}^{n} \sum_{i=1}^{m} \frac{\partial \mathcal{L}}{\partial u_{j,i}} u_{j,i} \right) \omega + \sum_{j=1}^{n} \sum_{i=1}^{m} \frac{\partial \mathcal{L}}{\partial u_{j,i}} du_j \wedge \left(\frac{\partial}{\partial x_i} \lrcorner \omega \right) \tag{10}$$

where $\omega = dx_1 \wedge \cdots \wedge dx_m$.

We now show that the exterior differential system Σ^1 generated by $\{Z_V \lrcorner d\Theta\}$, is equivalent to the **Euler-Lagrange equation** for $\mathcal{L}$. Here Z_V denotes the vertical Lie symmetry vector fields on $J^1(E)$. Let

$$Z_V = \sum_{l=1}^{n} \xi_l(\mathbf{x}, \mathbf{u}) \frac{\partial}{\partial u_l}$$

where $\mathbf{x} = (x_1, \ldots, x_m)$ and $\mathbf{u} = (u_1, \ldots, u_n)$. The **Euler-Lagrange equation** is then given by

$$js^*(Z_V \lrcorner d\Theta) = 0. \tag{11}$$

From Θ we obtain

$$
\begin{aligned}
d\Theta &= \left[d\left(\mathcal{L} - \sum_{j=1}^{n} \sum_{i=1}^{m} \frac{\partial \mathcal{L}}{\partial u_{j,i}} u_{j,i} \right) \right] \omega + \sum_{j=1}^{n} \sum_{i=1}^{m} d\left(\frac{\partial \mathcal{L}}{\partial u_{j,i}} \right) u_j \wedge \left(\frac{\partial}{\partial x_i} \lrcorner \omega \right) \\
&= \left(\sum_{k=1}^{n} \frac{\partial \mathcal{L}}{\partial u_k} du_k + \sum_{k=1}^{n} \sum_{l=1}^{m} \frac{\partial \mathcal{L}}{\partial u_{k,l}} du_{k,l} \right) \wedge \omega \\
&\quad - \left[\sum_{j=1}^{n} \sum_{i=1}^{m} \left(\sum_{k=1}^{n} \frac{\partial^2 \mathcal{L}}{\partial u_k \partial u_{j,i}} du_k + \sum_{k=1}^{n} \sum_{l=1}^{m} \frac{\partial^2 \mathcal{L}}{\partial u_{k,l} \partial u_{j,i}} du_{k,l} \right) u_{j,i} \right] \omega \\
&\quad - \left(\sum_{j=1}^{n} \sum_{i=1}^{m} \frac{\partial \mathcal{L}}{\partial u_{j,i}} du_{j,i} \right) \wedge \omega + \sum_{j=1}^{n} \sum_{i=1}^{m} \left(\sum_{k=1}^{n} \frac{\partial^2 \mathcal{L}}{\partial u_k \partial u_{j,i}} du_k \right. \\
&\quad \left. + \sum_{k=1}^{n} \sum_{l=1}^{m} \frac{\partial^2 \mathcal{L}}{\partial u_{j,i} \partial u_{k,l}} du_{k,l} \right) \wedge du_j \wedge \left(\frac{\partial}{\partial x_i} \lrcorner \omega \right).
\end{aligned}
$$

It then follows that

$$
\begin{aligned}
Z_V \lrcorner d\Theta &= \left[\sum_{k=1}^{n} \sum_{l=1}^{n} \frac{\partial \mathcal{L}}{\partial u_k} \xi_l \delta_{lk} - \sum_{j=1}^{n} \sum_{i=1}^{m} \left(u_{j,i} \sum_{k=1}^{n} \frac{\partial^2 \mathcal{L}}{\partial u_k \partial u_{j,i}} \sum_{l=1}^{n} \delta_{lk} \xi_l \right) \right] \omega \\
&\quad + \sum_{j=1}^{n} \sum_{i=1}^{m} \left(\sum_{k=1}^{n} \frac{\partial^2 \mathcal{L}}{\partial u_k \partial u_{j,i}} \sum_{l=1}^{n} \delta_{lk} \xi_l \right) du_j \wedge \left(\frac{\partial}{\partial x_i} \lrcorner \omega \right) \\
&\quad - \sum_{j=1}^{n} \sum_{i=1}^{m} \left(\sum_{k=1}^{n} \frac{\partial^2 \mathcal{L}}{\partial u_k \partial u_{j,i}} du_k \right) \wedge \left(\sum_{l=1}^{n} \delta_{lj} \xi_l \right) \wedge \left(\frac{\partial}{\partial x_i} \lrcorner \omega \right) \\
&\quad - \sum_{j=1}^{n} \sum_{i=1}^{m} \sum_{l=1}^{n} \delta_{lj} \xi_l \left(\sum_{k=1}^{n} \sum_{p=1}^{m} \frac{\partial^2 \mathcal{L}}{\partial u_{j,i} \partial u_{k,p}} du_{k,p} \right) \wedge \left(\frac{\partial}{\partial x_i} \lrcorner \omega \right) \\
&= \left[\sum_{k=1}^{n} \xi_k \left(\frac{\partial \mathcal{L}}{\partial u_k} - \sum_{j=1}^{n} \sum_{i=1}^{m} \frac{\partial^2 \mathcal{L}}{\partial u_k \partial u_{j,i}} u_{j,i} \right) \right] \omega \\
&\quad + \sum_{k=1}^{n} \xi_k \left(\sum_{j=1}^{n} \sum_{i=1}^{m} \frac{\partial^2 \mathcal{L}}{\partial u_k \partial u_{j,i}} du_j \wedge \left(\frac{\partial}{\partial x_i} \lrcorner \omega \right) \right) \\
&\quad - \sum_{j=1}^{n} \xi_j \left(\sum_{k=1}^{n} \sum_{i=1}^{m} \frac{\partial^2 \mathcal{L}}{\partial u_k \partial u_{j,i}} du_k \wedge \left(\frac{\partial}{\partial x_i} \lrcorner \omega \right) \right) \\
&\quad - \sum_{j=1}^{n} \xi_j \left(\sum_{i=1}^{n} \sum_{k=1}^{n} \sum_{l=1}^{m} \frac{\partial^2 \mathcal{L}}{\partial u_{j,i} \partial u_{k,l}} du_{k,l} \wedge \left(\frac{\partial}{\partial x_i} \lrcorner \omega \right) \right)
\end{aligned}
$$

where we have used the identity ($\alpha \ : \ r$-form)

$$
Z_V \lrcorner (\alpha \wedge \beta) \equiv (Z_V \lrcorner \alpha) \wedge \beta + (-1)^r \alpha \wedge (Z_V \lrcorner \beta).
$$

From $js^*(Z_V \lrcorner d\Theta) = 0$ we obtain

$$
\begin{aligned}
0 &= \sum_{k=1}^{n} \xi_k \left(\frac{\partial \mathcal{L}}{\partial u_k} - \sum_{j=1}^{n} \sum_{i=1}^{m} \frac{\partial^2 \mathcal{L}}{\partial u_k \partial u_{j,i}} \frac{\partial u_j}{\partial x_i} \right) \omega \\
&\quad + \sum_{k=1}^{n} \xi_k \left(\sum_{j=1}^{n} \sum_{i=1}^{m} \frac{\partial^2 \mathcal{L}}{\partial u_k \partial u_{j,i}} \frac{\partial u_j}{\partial x_i} dx_i \wedge \left(\frac{\partial}{\partial x_i} \lrcorner \omega \right) \right) \\
&\quad - \sum_{j=1}^{n} \xi_j \left(\sum_{k=1}^{n} \sum_{i=1}^{m} \frac{\partial^2 \mathcal{L}}{\partial u_k \partial u_{j,i}} \frac{\partial u_k}{\partial x_i} dx_i \wedge \left(\frac{\partial}{\partial x_i} \lrcorner \omega \right) \right) \\
&\quad - \sum_{j=1}^{n} \xi_j \left(\sum_{i=1}^{m} \sum_{k=1}^{n} \sum_{l=1}^{m} \frac{\partial^2 \mathcal{L}}{\partial u_{j,i} \partial u_{k,l}} \frac{\partial^2 u_k}{\partial x_i \partial x_l} \right) dx_i \wedge \left(\frac{\partial}{\partial x_i} \lrcorner \omega \right)
\end{aligned}
$$

and it follows that

$$
0 = \left[\sum_{k=1}^{n} \xi_k \left(\frac{\partial \mathcal{L}}{\partial u_k} - \sum_{j=1}^{n} \sum_{i=1}^{m} \frac{\partial^2 \mathcal{L}}{\partial u_k \partial u_{j,i}} \frac{\partial u_j}{\partial x_i} \right) \right.
$$

$$+ \sum_{k=1}^{n} \xi_k \left(\sum_{j=1}^{n} \sum_{i=1}^{m} \frac{\partial^2 \mathcal{L}}{\partial u_k \partial u_{j,i}} \frac{\partial u_j}{\partial x_i} \right)$$

$$- \sum_{j=1}^{n} \xi_j \left(\sum_{k=1}^{n} \sum_{i=1}^{m} \frac{\partial^2 \mathcal{L}}{\partial u_k \partial u_{j,i}} \frac{\partial u_k}{\partial x_i} \right)$$

$$- \sum_{j=1}^{n} \xi_j \left(\sum_{i=1}^{m} \sum_{k=1}^{n} \sum_{l=1}^{m} \frac{\partial^2 \mathcal{L}}{\partial u_{j,i} \partial u_{k,l}} \frac{\partial^2 u_k}{\partial x_i \partial x_l} \right) \Bigg] \omega.$$

Finally

$$0 = \sum_{k=1}^{n} \xi_k \left(\frac{\partial \mathcal{L}}{\partial u_k} - \sum_{j=1}^{n} \sum_{i=1}^{m} \frac{\partial^2 \mathcal{L}}{\partial u_j \partial u_{k,i}} \frac{\partial u_j}{\partial x_i} - \sum_{j=1}^{n} \sum_{i=1}^{m} \sum_{l=1}^{m} \frac{\partial^2 \mathcal{L}}{\partial u_{k,i} \partial u_{j,l}} \frac{\partial^2 u_j}{\partial x_i \partial x_l} \right).$$

Consequently, the **Euler-Lagrange equation** follows

$$\frac{\partial \mathcal{L}}{\partial u_k} - \sum_{j=1}^{n} \sum_{i=1}^{m} \frac{\partial^2 \mathcal{L}}{\partial u_j \partial u_{k,i}} \frac{\partial u_j}{\partial x_i} - \sum_{j=1}^{n} \sum_{i=1}^{m} \sum_{l=1}^{m} \frac{\partial^2 \mathcal{L}}{\partial u_{k,i} \partial u_{j,l}} \frac{\partial^2 u_j}{\partial x_i \partial x_l} = 0 \qquad (12)$$

where $k = 1, \ldots, n$.

Definition 16.11 $\mathcal{L}$ *is called* **regular** *if the* $mn \times mn$ *matrix*

$$\frac{\partial^2 \mathcal{L}}{\partial u_{j,i} \partial u_{k,l}}$$

is non-singular.

We assume here that $\mathcal{L}$ is regular.

Example: Consider the sine-Gordon equation

$$\frac{\partial^2 u}{\partial x_1 \partial x_2} = \sin u.$$

Here $n = 1$ and $m = 2$ and we have put $u_1 = u$. The sine-Gordon equation has the Lagrangian

$$\mathcal{L} = \frac{1}{2} u_1 u_2 - \cos u.$$

From the Euler-Lagrange equation we obtain

$$\frac{\partial \mathcal{L}}{\partial u} - \sum_{i=1}^{2} \frac{\partial^2 \mathcal{L}}{\partial u \partial u_i} \frac{\partial u}{\partial x_i} - \sum_{i=1}^{2} \sum_{l=1}^{2} \frac{\partial^2 \mathcal{L}}{\partial u_i \partial u_l} \frac{\partial^2 u}{\partial x_i \partial x_l} = 0$$

so that the sine-Gordon equation follows. The Cartan fundamental form (10) for the sine-Gordon equation is given by

$$
\begin{aligned}
\Theta &= \left(\mathcal{L} - \frac{\partial \mathcal{L}}{\partial u_1} u_1 - \frac{\partial \mathcal{L}}{\partial u_2} u_2 \right) dx_1 \wedge dx_2 + \frac{\partial \mathcal{L}}{\partial u_1} du \wedge dx_2 + \frac{\partial \mathcal{L}}{\partial u_2} du \wedge dx_1 \\
&= \left(-\frac{1}{2} u_1 u_2 - \cos u_1 \right) dx_1 \wedge dx_2 + \frac{1}{2} u_2 du \wedge dx_2 + \frac{1}{2} u_1 du \wedge dx_1 .
\end{aligned}
$$

♣

Theorem 16.2 *If ζ is a higher-order conserved current for Σ^1 then there is a vector field Z in the Lie algebra L and a trivial conserved current ζ_T such that*

$$
d\zeta \equiv Z \lrcorner d\Theta + d\zeta_T \; mod \; \pi_l^* \Omega_{(m)}^l
$$

for some integer l. The vector field Z satisfies

$$
L_Z \Theta \equiv d\chi \; mod \; \pi_l^* \Omega_{(m)}^l
$$

where χ is the $(m-1)$-form defined by

$$
\chi := \zeta_T + Z \lrcorner \Theta
$$

In addition, Z is a symmetry vector field of Σ^∞.

This theorem is the analogue of the converse of the usual Noether theorem which associates a conserved current to each infinitesimal symmetry of the Cartan form. The proof of the theorem is left as an exercise.

For higher-order conserved currents we have the following formulation of Noethers's theorem.

Theorem 16.3 *If Z is a vector field in the Lie algebra L and there is an $(m-1)$-form χ such that*

$$
L_Z \Theta \equiv d\chi \; mod \; \pi_l^* \Omega_{(m)}^l
$$

for some integer l, then the $(m-1)$-form ζ defined by

$$
\zeta := \chi - Z \lrcorner \Theta
$$

is a higher-order conserved current for Σ^1.

Proof: Obviously

$$
d\zeta = d\chi - d(Z \lrcorner \Theta)
$$

so

$$d\zeta \equiv L_Z\Theta - d(Z\lrcorner\Theta) \bmod \pi_l^*\Omega_{(m)}^l$$

by the hypothesis. Since

$$L_Z\Theta = Z\lrcorner d\Theta + d(Z\lrcorner\Theta)$$

we obtain

$$d\zeta \equiv Z\lrcorner d\Theta \bmod \pi_l^*\Omega_{(m)}^l.$$

Consequently $d\zeta \in \pi_l^*\Sigma^l$. ♠

A vector field that satisfies the hypothesis of the above theorem is called an **infinitesimal Noether symmetry**.

Example: Consider the following nonlinear Dirac equation that consists of a system of eight coupled partial differential equations

$$\lambda\frac{\partial v_4}{\partial x_1} + \lambda\frac{\partial v_1}{\partial x_4} + u_1[1 + \lambda\epsilon K] = 0$$

$$\lambda\frac{\partial v_3}{\partial x_1} + \lambda\frac{\partial v_2}{\partial x_4} + u_2[1 + \lambda\epsilon K] = 0$$

$$-\lambda\frac{\partial v_2}{\partial x_1} - \lambda\frac{\partial v_3}{\partial x_4} + u_3[1 + \lambda\epsilon K] = 0$$

$$-\lambda\frac{\partial v_1}{\partial x_1} - \lambda\frac{\partial v_4}{\partial x_4} + u_4[1 + \lambda\epsilon K] = 0$$

$$-\lambda\frac{\partial u_4}{\partial x_1} - \lambda\frac{\partial u_1}{\partial x_4} + v_1[1 + \lambda\epsilon K] = 0$$

$$-\lambda\frac{\partial u_3}{\partial x_1} - \lambda\frac{\partial u_2}{\partial x_4} + v_2[1 + \lambda\epsilon K] = 0$$

$$\lambda\frac{\partial u_2}{\partial x_1} + \lambda\frac{\partial u_3}{\partial x_4} + v_3[1 + \lambda\epsilon K] = 0$$

$$\lambda\frac{\partial u_1}{\partial x_1} + \lambda\frac{\partial u_4}{\partial x_4} + v_4[1 + \lambda\epsilon K] = 0 \tag{13}$$

where

$$K(\mathbf{u}, \mathbf{v}) \equiv \sum_{j=1}^{2}(u_j^2 + v_j^2) - \sum_{j=3}^{4}(u_j^2 + v_j^2)$$

and $u_i = u_i(x_1, x_4)$, $v_i = v_i(x_1, x_4)$ with $\mathbf{u} = (u_1, u_2, u_3, u_4)$ and $\mathbf{v} = (v_1, v_2, v_3, v_4)$. System (13) is obtained from the nonlinear Dirac equation of the form

$$\lambda\frac{\partial}{\partial x_1}(\gamma_1\psi) - \lambda i\frac{\partial}{\partial x_4}(\gamma_4\psi) + \psi(1 + \lambda\epsilon\bar\psi\psi) = 0$$

where we consider the case with one space dimension x_1 and $x_4 \equiv ct$ (t is time and c speed of light). $\lambda := \hbar/m_0 c$ where $m_0 > 0$ is the rest mass and ϵ is a real parameter (coupling constant). $\psi = (\psi_1, \psi_2, \psi_3, \psi_4)^T$ (T means transpose), $\bar{\psi} \equiv (\psi_1^*, \psi_2^*, -\psi_3^*, -\psi_4^*)$ and γ_1 and γ_4 are the following 4×4 matrices

$$\gamma_1 = \begin{pmatrix} 0 & 0 & 0 & -i \\ 0 & 0 & -i & 0 \\ 0 & i & 0 & 0 \\ i & 0 & 0 & 0 \end{pmatrix}, \qquad \gamma_4 = \begin{pmatrix} 1 & 0 & 0 & 0 \\ 0 & 1 & 0 & 0 \\ 0 & 0 & -1 & 0 \\ 0 & 0 & 0 & -1 \end{pmatrix}.$$

Now we have to put $\psi_j(\mathbf{x}) \equiv u_j(\mathbf{x}) + iv_j(\mathbf{x})$ where $j = 1, \ldots, 4$ and $\mathbf{x} = (x_1, \ldots, x_4)$. $\mathbf{u}$ and $\mathbf{v}$ are real fields. This leads to system (13). In the jet bundle we have

$$F_1(u_1, \ldots, v_4, u_{1,1}, \ldots, v_{4,4}) \equiv \lambda(v_{4,1} + v_{1,4}) + u_1(1 + \lambda\epsilon K(\mathbf{u}, \mathbf{v})) = 0$$

$$F_2(u_1, \ldots, v_4, u_{1,1}, \ldots, v_{4,4}) \equiv \lambda(v_{3,1} + v_{2,4}) + u_2(1 + \lambda\epsilon K(\mathbf{u}, \mathbf{v})) = 0$$

$$F_3(u_1, \ldots, v_4, u_{1,1}, \ldots, v_{4,4}) \equiv \lambda(-v_{2,1} - v_{3,4}) + u_3(1 + \lambda\epsilon K(\mathbf{u}, \mathbf{v})) = 0$$

$$F_4(u_1, \ldots, v_4, u_{1,1}, \ldots, v_{4,4}) \equiv \lambda(-v_{1,1} - v_{4,4}) + u_4(1 + \lambda\epsilon K(\mathbf{u}, \mathbf{v})) = 0$$

$$F_5(u_1, \ldots, v_4, u_{1,1}, \ldots, v_{4,4}) \equiv \lambda(-u_{4,1} - u_{1,4}) + v_1(1 + \lambda\epsilon K(\mathbf{u}, \mathbf{v})) = 0$$

$$F_6(u_1, \ldots, v_4, u_{1,1}, \ldots, v_{4,4}) \equiv \lambda(-u_{0,1} - u_{0,1}) + v_0(1 + \lambda\epsilon K(\mathbf{u}, \mathbf{v})) = 0$$

$$F_7(u_1, \ldots, v_4, u_{1,1}, \ldots, v_{4,4}) \equiv \lambda(u_{2,1} + u_{3,4}) + v_3(1 + \lambda\epsilon K(\mathbf{u}, \mathbf{v})) = 0$$

$$F_8(u_1, \ldots, v_4, u_{1,1}, \ldots, v_{4,4}) \equiv \lambda(u_{1,1} + u_{4,4}) + v_4(1 + \lambda\epsilon K(\mathbf{u}, \mathbf{v})) = 0 \qquad (14)$$

with the contact forms

$$\alpha_i \equiv du_i - u_{i,1}dx_1 - u_{i,4}dx_4, \qquad \beta_i \equiv dv_i - v_{i,1}dx_1 - v_{i,4}dx_4$$

where $i = 1, \ldots, 4$.

We consider the Lie symmetry vector fields

$$Z = \sum_{k=1}^{4} \left(u_k \frac{\partial}{\partial v_k} - v_k \frac{\partial}{\partial u_k} \right) \qquad (15)$$

$$X = \frac{\partial}{\partial x_1}, \qquad T = \frac{\partial}{\partial x_4} \qquad (16)$$

as well as the vertical vector fields

$$U = -\sum_{i=1}^{4} \left(u_{i1} \frac{\partial}{\partial u_i} + v_{i1} \frac{\partial}{\partial v_i} \right), \qquad V = -\sum_{i=1}^{4} \left(u_{i2} \frac{\partial}{\partial u_i} + v_{i2} \frac{\partial}{\partial v_i} \right). \qquad (17)$$

We now calculate the conservation laws which are associated with the symmetries described by the vector fields (15) and (16). For system (14) the Lagrangian density $\mathcal{L}$ takes the form

$$\mathcal{L} = \lambda(-u_4v_{1,1} + v_4u_{1,1} - u_3v_{2,1} + v_3u_{2,1} - u_2v_{3,1} + v_2u_{3,1} - u_1v_{4,1} + v_1u_{4,1}$$

$$-u_1v_{1,4} + v_1u_{1,4} - u_2v_{2,4} + v_2u_{2,4} - u_3v_{3,4} + v_3u_{3,4} - u_4v_{4,4} + v_4u_{4,4}) - K(1 + \lambda\epsilon K).$$

First we calculate the Cartan fundamental form. From (10) we obtain

$$\left(\mathcal{L} - \sum_{j=1}^{4}\left(\frac{\partial\mathcal{L}}{\partial u_{j,1}}u_{j,1} + \frac{\partial\mathcal{L}}{\partial u_{j,4}}u_{j,4} + \frac{\partial\mathcal{L}}{\partial v_{j,1}}v_{j,1} + \frac{\partial\mathcal{L}}{\partial v_{j,4}}v_{j,4}\right)\right)^{\cdot} = -K(1 + \lambda\epsilon K)$$

and

$$\sum_{j=1}^{4}\left(\frac{\partial\mathcal{L}}{\partial u_{j,1}}du_j \wedge dx_4 - \frac{\partial\mathcal{L}}{\partial u_{j,4}}du_j \wedge dx_1 + \frac{\partial\mathcal{L}}{\partial v_{j,1}}dv_j \wedge dx_4 - \frac{\partial\mathcal{L}}{\partial v_{j,4}}dv_j \wedge dx_1\right)$$

$$= \lambda(v_4du_1 \wedge dx_4 + v_3du_2 \wedge dx_4 + v_2du_3 \wedge dx_4 + v_1du_4 \wedge dx_4$$

$$-v_1du_1 \wedge dx_1 - v_2du_2 \wedge dx_1 - v_3du_3 \wedge dx_1 - v_4du_4 \wedge dx_1$$

$$-u_4dv_1 \wedge dx_4 - u_3dv_2 \wedge dx_4 - u_2dv_3 \wedge dx_4 - u_1dv_4 \wedge dx_4$$

$$-u_1dv_1 \wedge dx_1 + u_2dv_2 \wedge dx_1 + u_3dv_3 \wedge dx_1 + u_4dv_4 \wedge dx_1).$$

Consequently Θ is given by the two-form

$$\Theta = -K(1 + \lambda\epsilon K)dx_1 \wedge dx_4 + \lambda(v_4du_1 - u_4dv_1 + v_3du_2$$

$$-u_3dv_2 + v_2du_3 - u_2dv_3 + v_1du_4 - u_1dv_4) \wedge dx_4$$

$$+\lambda(u_1dv_1 - v_1du_1 + u_2dv_2 - v_2du_2 + u_3dv_3 - v_3du_3 + u_4dv_4 - v_4du_4) \wedge dx_1.$$

Since

$$L_X\Theta = 0, \qquad L_T\Theta = 0, \qquad L_Z\Theta = 0$$

we find that $\chi = 0$. The conserved currents are thus given by

$$X\lrcorner\Theta = -K(1 + \lambda\epsilon K)dx_4 - \lambda\sum_{i=1}^{4}(u_idv_i - v_idu_i)$$

$$T\lrcorner\Theta = K(1 + \lambda\epsilon K)dx_1 - \lambda\sum_{i=1}^{4}(v_{5-i}du_i - u_{5-i}dv_i)$$

$$Z \lrcorner \Theta = -2\lambda(u_1 u_4 + u_2 u_3 + v_1 v_4 + v_2 v_3) dx_4 + \lambda \left(\sum_{i=1}^{4} (u_i^2 + v_i^2) \right) dx_1. \qquad (18)$$

It follows that

$$\begin{aligned}
-js^*(X \lrcorner \Theta) &= \left[\lambda \sum_{i=1}^{4} \left(u_i \frac{\partial v_i}{\partial x_4} - v_i \frac{\partial u_i}{\partial x_4} \right) \right] dx_1 \\
&\quad + \left[\lambda \sum_{i=1}^{4} \left(u_i \frac{\partial v_i}{\partial x_4} - v_i \frac{\partial u_i}{\partial x_4} \right) + K(\mathbf{u}, \mathbf{v})(1 + \lambda \epsilon K(\mathbf{u}, \mathbf{v})) \right] dx_4 \\
-js^*(T \lrcorner \Theta) &= \left[\lambda \sum_{i=1}^{4} \left(v_{5-i} \frac{\partial u_i}{\partial x_1} - u_{5-i} \frac{\partial v_i}{\partial x_1} \right) - K(\mathbf{u}, \mathbf{v})(1 + \lambda \epsilon K(\mathbf{u}, \mathbf{v})) \right] dx_1 \\
&\quad + \left[\lambda \sum_{i=1}^{4} \left(v_{5-i} \frac{\partial u_i}{\partial x_4} - u_{5-i} \frac{\partial v_i}{\partial x_4} \right) \right] dx_4
\end{aligned}$$

and

$$-js^*(Z \lrcorner \Theta) = -\lambda \left[\sum_{i=1}^{4} (u_i^2 + v_i^2) \right] dx_1 + 2\lambda \left[u_1 u_4 + u_2 u_3 + v_1 v_4 + v_2 v_3 \right] dx_4.$$

We are interested in solutions u_i, v_i ($i = 1, 2, 3, 4$) which vanish at infinity, i.e., we are looking for localized solutions. Then we have

$$\frac{d}{dx_4} \int_{-\infty}^{+\infty} \sum_{i=1}^{4} \left(u_i \frac{\partial v_i}{\partial x_4} - v_i \frac{\partial u_i}{\partial x_4} \right) dx_1 = 0$$

$$\frac{d}{dx_4} \int_{-\infty}^{+\infty} \left[\lambda \sum_{i=1}^{4} \left(v_{5-i} \frac{\partial u_i}{\partial x_1} - u_{5-i} \frac{\partial v_i}{\partial x_1} \right) - K(\mathbf{u}, \mathbf{v})(1 + \lambda \epsilon K(\mathbf{u}, \mathbf{v})) \right] dx_1 = 0,$$

$$\frac{d}{dx_4} \int_{-\infty}^{+\infty} \left[\sum_{i=1}^{4} (u_i^2 + v_i^2) \right] dx_1 = 0.$$

We now find conserved currents with the help of the vector field U and V given by (16). By straightforward calculation we find

$$\begin{aligned}
L_U \Theta &= \sum_{i=1}^{4} u_{i,1} \left[\frac{\partial}{\partial u_i} (K(1 + \lambda \epsilon K)) \right] dx_1 \wedge dx_4 \\
&\quad + \sum_{i=1}^{4} v_{i,1} \left[\frac{\partial}{\partial v_i} (K(1 + \lambda \epsilon K)) \right] dx_1 \wedge dx_4 \\
&\quad + \lambda \sum_{i=1}^{4} (v_i du_{i,1} \wedge dx_1 - u_{i,1} dv_i \wedge dx_1)
\end{aligned}$$

$$+ \ \lambda \sum_{i=1}^{4} (u_{i,1} dv_{5-i} \wedge dx_4 - v_{5-i} du_{i,1} \wedge dx_4)$$

$$+ \ \lambda \sum_{i=1}^{4} (-u_{i,1} dv_{i,1} \wedge dx_1 + v_{i,1} du_i \wedge dx_1)$$

$$+ \ \lambda \sum_{i=1}^{4} (-v_{i,1} du_{5-i} \wedge dx_4 + u_{5-i} dv_{i,1} \wedge dx_4).$$

Owing to the identities

$$\sum_{i=1}^{4} \left(\frac{\partial}{\partial u_i} K(1 + \lambda \epsilon K) \right) \alpha_i \wedge dx_4 \equiv d(K(1 + \lambda \epsilon K) dx_4)$$

$$- \sum_{i=1}^{4} u_{i,1} \left(\frac{\partial}{\partial u_i} K(1 + \lambda \epsilon K) \right) dx_1 \wedge dx_4$$

$$u_{i,1} dx_1 \equiv -\alpha_i + du_i - u_{i,4} dx_4$$

$$du_{i,1} \wedge dx_1 \equiv -d\alpha_i - du_{i,4} \wedge dx_4$$

$$du_i \wedge dx_1 \simeq \alpha_i \wedge dx_1 - u_{i,4} dx_1 \wedge dx_4,$$

and so on, we obtain

$$L_U \theta = d\chi \ \text{mod (contact forms } \alpha_i, \beta_i)$$

where

$$\chi \ = \ K(1 + \lambda \epsilon K) dx_4 + \sum_{i=1}^{4} (-uv_{i,1} dx_1 + v_i u_{i,1} dx_1 + u_{5-i} v_{i,1} dx_4 - v_{5-i} u_{i,1} dx_4)$$

$$+ \ \lambda \sum_{i=1}^{4} (u_i dv_i - v_i du_i).$$

The conserved current is given by $\chi - U \lrcorner \theta$ and we find the result given by (17).   ♣

Example: Consider the nonlinear wave equation

$$\frac{\partial^2 u}{\partial x_1^2} + \frac{\partial^2 u}{\partial x_2^2} + \frac{\partial^2 u}{\partial x_3^2} - \frac{\partial^2 u}{\partial x_4^2} = u^3.$$

The Lagrangian for this nonlinear wave equation is given by

$$\mathcal{L} = \frac{1}{2}(-u_1^2 - u_2^2 - u_3^2 + u_4^2) - \frac{1}{4} u^4$$

so that the Cartan fundamental form is

$$\Theta \;=\; \left[-\frac{1}{2}\left(u_1^2 + u_2^2 + u_3^2 - u_4^2\right) + \frac{1}{4}u^4\right]\omega + u_1 du \wedge dx_2 \wedge dx_3 \wedge dx_4$$

$$-u_2 du \wedge dx_1 \wedge dx_3 \wedge dx_4 + u_3 du \wedge dx_1 \wedge dx_2 \wedge dx_4 + u_4 du \wedge dx_1 \wedge dx_2 \wedge dx_3.$$

Here we have one dependent variable, where we have put $u_1 = u$, and four independent variables $(x_1, \ldots, x_4)$. $\omega = dx_1 \wedge \ldots \wedge dx_4$ is the volume form. As an example we calculate the conserved currents associated with the Lie symmetry vector field

$$T = \frac{\partial}{\partial x_4}.$$

We find that

$$L_T \Theta = 0$$

so that $\chi = 0$. A conserved current is thus given by

$$\frac{\partial}{\partial x_4}\rfloor\Theta \;=\; \left(\frac{1}{2}(u_1^2 + u_2^2 + u_3^2 - u_4^2) + \frac{1}{4}u^4\right) dx_1 \wedge dx_2 \wedge dx_3$$

$$u_1 du \wedge dx_2 \wedge dx_3 + u_2 du \wedge dx_1 \wedge dx_3 + u_3 du \wedge dx_1 \wedge dx_2.$$

♣

Example: Let us again consider the nonlinear Dirac equation. We show that the conserved current

$$Z\rfloor\Theta = -2\lambda(u_1 u_4 + u_2 u_3 + v_1 v_4 + v_2 v_3)dx_4 + \lambda\left(\sum_{i=1}^{4}(u_i^2 + v_1^2)\right)dx_1$$

can be derived by the use of exterior differential systems. We make the ansatz

$$\zeta = f_1(\mathbf{u}, \mathbf{v})dx_1 + f_2(\mathbf{u}, \mathbf{v})dx_4,$$

where f_1 and f_2 are smooth functions. Let

$$J \equiv \langle F_1, \ldots, F_8,\; dF_1, \ldots, dF_8,\; \alpha_1, \ldots, \beta_4,\; d\alpha_1, \ldots, d\beta_4\rangle$$

denote the ideal generated by $F_1, \ldots, d\beta_4$. We recall that the condition for obtaining the conservation laws is as follows: If

$$d\zeta \in \langle F_1, \ldots, F_8,\; dF_1, \ldots, dF_8,\; \alpha_1, \ldots, \beta_4,\; d\alpha_1, \ldots, d\beta_4\rangle,$$

then $(js)^*(d\zeta) = 0$. We construct a convenient two-form which is an element of the ideal generated by $F_1, \ldots, d\beta_4$. Let

$$\sigma_1 = \beta_1 \wedge dx_1 - \beta_4 \wedge dx_4, \qquad \sigma_2 = \beta_2 \wedge dx_1 - \beta_3 \wedge dx_4$$

$$\sigma_3 = -\beta_3 \wedge dx_1 + \beta_2 \wedge dx_4, \qquad \sigma_4 = -\beta_4 \wedge dx_1 + \beta_1 \wedge dx_4$$

$$\sigma_5 = -\alpha_1 \wedge dx_1 + \alpha_4 \wedge dx_4, \qquad \sigma_6 = -\alpha_2 \wedge dx_1 + \alpha_3 \wedge dx_4$$

$$\sigma_7 = \alpha_3 \wedge dx_1 - \alpha_2 \wedge dx_4, \qquad \sigma_8 = \alpha_4 \wedge dx_1 - \alpha_1 \wedge dx_4.$$

It is obvious that the two-forms σ_i $(i = 1, \ldots, 8)$ are elements of the ideal. Since both f_1 and f_2 do not depend on $u_{j,i}$ and $v_{j,i}$, we have to eliminate the terms which contain $u_{j,i}$ and $v_{j,i}$. Therefore we consider the two-forms

$$\tau_j \equiv \lambda \sigma_j - F_j dx_1 \wedge dx_4$$

which are elements of the ideal. We find

$$\tau_1 = \lambda(dv_1 \wedge dx_1 - dv_4 \wedge dx_4) - u_1(1 + \lambda \epsilon K(\mathbf{u}, \mathbf{v})) dx_1 \wedge dx_4$$

$$\tau_2 = \lambda(dv_2 \wedge dx_1 - dv_3 \wedge dx_4) - u_2(1 + \lambda \epsilon K(\mathbf{u}, \mathbf{v})) dx_1 \wedge dx_4$$

$$\tau_3 = \lambda(-dv_3 \wedge dx_1 + dv_2 \wedge dx_4) - u_3(1 + \lambda \epsilon K(\mathbf{u}, \mathbf{v})) dx_1 \wedge dx_4$$

$$\tau_4 = \lambda(-dv_4 \wedge dx_1 + dv_1 \wedge dx_4) - u_4(1 + \lambda \epsilon K(\mathbf{u}, \mathbf{v})) dx_1 \wedge dx_4$$

$$\tau_5 = \lambda(-du_1 \wedge dx_1 + du_4 \wedge dx_4) - v_1(1 + \lambda \epsilon K(\mathbf{u}, \mathbf{v})) dx_1 \wedge dx_4$$

$$\tau_6 = \lambda(-du_2 \wedge dx_1 + du_3 \wedge dx_4) - v_2(1 + \lambda \epsilon K(\mathbf{u}, \mathbf{v})) dx_1 \wedge dx_4$$

$$\tau_7 = \lambda(du_3 \wedge dx_1 - du_2 \wedge dx_4) - v_3(1 + \lambda \epsilon K(\mathbf{u}, \mathbf{v})) dx_1 \wedge dx_4$$

$$\tau_8 = \lambda(du_4 \wedge dx_1 - du_1 \wedge dx_4) - v_4(1 + \lambda \epsilon K(\mathbf{u}, \mathbf{v})) dx_1 \wedge dx_4. \qquad (19)$$

We note that the conditions

$$js^* \tau_j = 0$$

$(j = 1, \ldots, 8)$ lead to the given nonlinear Dirac equation.

We consider now the two-form

$$\tau = \lambda(v_1 \tau_1 + v_2 \tau_2 - v_3 \tau_3 - v_4 \tau_4 - u_1 \tau_5 - u_2 \tau_6 + u_3 \tau_7 + u_4 \tau_8)$$

which is again an element of the ideal. It follows that

$$\tau = 2\lambda \left(\sum_{i=1}^{4} (v_i dv_i + u_i du_i) \wedge dx_1 \right)$$

$$+ 2\lambda(-v_1 dv_4 - v_2 dv_3 - v_3 dv_2 - v_4 dv_1 - u_1 du_4 - u_2 du_3 - u_3 du_2 - u_4 du_4) \wedge dx_4.$$

Now the two-form τ can be represented as the exterior derivative of the one-form ζ, i.e.,

$$\tau = d\zeta$$

where

$$f_1(\mathbf{u}, \mathbf{v}) = \lambda \sum_{i=1}^{4} (u_i^2 + v_i^2),$$

$$f_2(\mathbf{u}, \mathbf{v}) = -2\lambda(u_1 u_4 + u_2 u_3 + v_1 v_4 + v_2 v_3)$$

and the conserved current (18) follows.

Remark: *The Cartan fundamental form Θ for the given nonlinear Dirac equation can be expressed with the help of the two-forms τ_i $(i = 1, \ldots, 8)$, namely*

$$\Theta = -u_1\tau_1 - u_2\tau_2 + u_3\tau_3 + u_4\tau_4 - v_1\tau_5 - v_2\tau_6 + v_3\tau_7 + v_4\tau_8.$$

It follows that $\Theta \in \langle F_1, \ldots, d\beta_4 \rangle$, but $d\Theta \neq 0$ and therefore Θ cannot be obtained as the exterior derivative of a one-form. For field equations which can be derived from a Lagrangian density it is obvious that $\Theta \in \langle F_1, \ldots, d\beta_4 \rangle$. ♣

Example: Thus far we have derived the conserved current (18) for the nonlinear Dirac equation (13) by applying two approaches. In the first approach we have taken into account the Cartan fundamental form Θ which contains the Lagrangian density and the Lie symmetry vector field Z. In the second approach we have only considered the differential forms which are equivalent to the nonlinear Dirac equation. Now we describe a third approach for obtaining the conserved current given by (18), where we take into account the Lie symmetry vector field Z and the differential forms which are equivalent to the nonlinear Dirac equation. Hence we consider the differential forms given by (19). We consider the two-form

$$\alpha = \sum_{j=1}^{8} h_j(u, v)\tau_j$$

where the two-forms τ_j are given by (19). h_j are smooth functions. Let Z be the symmetry generator given by (15) and let $\bar{J}$ be the ideal given by (19). If $Z\lrcorner d\alpha \in \bar{J}$, then taking into account $L_Z(\bar{J}) \subset \bar{J}$ and the definition of the Lie derivative

$$L_Z\alpha \equiv Z\lrcorner d\alpha + d(Z\lrcorner \alpha)$$

it follows that the one-form $Z\lrcorner \alpha$ is a conserved current since $d(Z\lrcorner \chi) \in \bar{J}$. Now we have to determine the unknown functions $h_j(\mathbf{u}, \mathbf{v})$ with the help of the equation $Z\lrcorner d\alpha \in \bar{J}$. The condition gives the solution

$$h_1(\mathbf{u}, \mathbf{v}) = u_1, \qquad h_2(\mathbf{u}, \mathbf{v}) = u_2, \qquad h_3(\mathbf{u}, \mathbf{v}) = -u_3, \qquad h_4(\mathbf{u}, \mathbf{v}) = -u_4$$

$$h_5(\mathbf{u}, \mathbf{v}) = v_1, \qquad h_6(\mathbf{u}, \mathbf{v}) = v_2, \qquad h_7(\mathbf{u}, \mathbf{v}) = -v_3, \qquad h_8(\mathbf{u}, \mathbf{v}) = v_4.$$

Consequently, we find $\alpha = \Theta$, where Θ is the Cartan fundamental form for (13). ♣

16.4 Computer Algebra Applications

We show that the sine-Gordon equation

$$\frac{\partial^2 u}{\partial x_1 \partial x_2} = \sin(u)$$

admits the conservation law

$$\frac{\partial}{\partial x_2} \frac{1}{2} \left(\frac{\partial u}{\partial x_1} \right)^2 + \frac{\partial}{\partial x_1} \cos(u) = 0.$$

```
% conservation law
% conser.red

depend u, x1, x2;

for all x1, x2 let df(u,x1,x2) = sin(u);

S := df(u,x2);
X := df(u,x1);

A := sub(u=(df(u,x1))^2/2,S);
B := sub(u=cos(u),X);
CL := A + B;
```

The output is

```
A := df(u,x1)*sin(u)$
B := - df(u,x1)*sin(u)$
CL := 0$
```

We show that the equation

$$\frac{d^2u_1}{dt^2} = 2\omega\frac{du_2}{dt}, \qquad \frac{d^2u_2}{dt^2} = -2\omega\frac{du_1}{dt}$$

can be derived from the Lagrange function

$$L = \frac{1}{2}(\dot{u}_1^2 + \dot{u}_2^2) + \omega(u_1\dot{u}_2 - \dot{u}_1 u_2) + \omega(u_1\dot{u}_1 + u_2\dot{u}_2)\tan(\omega t) + \frac{1}{2}\omega(u_1^2 + u_2^2)\sec^2(\omega t).$$

The system describes the two-dimensional motion of a charged particle in a plane perpendicular to the direction of a constant magnetic field.

```
% Euler-Lagrange equation
% lag.red

depend L, ts, u1, u2, u1d, u2d;

L  := (1/2)*(u1d^2 + u2d^2)
       + om*(u1*u2d - u1d*u2)
       + om*(u1*u1d + u2*u2d)*tan(om*ts)
       + om^2*(u1^2 + u2^2)*(sec(om*ts))^2/2;

LUD1 := df(L,u1d);
res1 := df(LUD1,ts)
           + u1d*df(LUD1,u1)
           + u1dd*df(LUD1,u1d) - df(L,u1);

LUD2 := df(L,u2d);
res2 := df(LUD2,ts)
           + u2d*df(LUD2,u2)
           + u2dd*df(LUD2,u2d) - df(L,u2);

for all q let (sec(q))^2 = 1 + (tan(q))^2;

res1;
res2;

operator v1, v2;
eq1 := sub({u1dd=df(v1(ts),ts,2),u2d=df(v2(ts),ts)},res1);
eq2 := sub({u2dd=df(v2(ts),ts,2),u1d=df(v1(ts),ts)},res2);

eq1;
eq2;
```

Chapter 17

Symmetries and Painlevé Test

17.1 Introduction

One of the most fundamental, important and fascinating problems in the investigation of nonlinear dynamical systems is to find a general criterion which describes integrability. The Painlevé test provides an approach for the study of integrability (see Steeb and Euler [126]). The basic idea goes back to Kowalevski [69], [68]. In two celebrated papers she showed that the only algebraic completely integrable systems among the rigid body motions are Euler's rigid body, Lagrange's top and Kowalevski's top. Painlevé and coworkers determined whether or not equations of the form

$$\frac{d^2 w}{dz^2} = F(z, w, dw/dz)$$

exist, where F is rational in dw/dz, algebraic in w, and analytic in z, which have their critical points (that is their branch points and essential simngularities) fixed Ince [60], Davis [24]. In this chapter we give examples of some applications of the Painlevé test. In particular we are interested in the connection between the Painlevé test and Lie symmetry vector fields. We consider nonlinear ordinary and partial differential equations. The Painlevé test for partial differemtial equations has been introduced by Weiss et al [147], (see also Weiss [146]). A definition of the Painlevé property for partial differential equations has been proposed by Ward [144], [145]. Of interest is also the relation between the Painlevé test for a partial differential equation and the ordinary differential equation obtained by a similarity reduction (group theoretical reduction). A large number of authors investigated the Painlevé propertry for ordinary and partial differential equations (Steeb and Euler [126], Steeb [121], Strampp [139]). The Painlevé test has been applied to Nambu mechanics (Steeb and Euler [126]), Discrete Boltzmann equations (Euler and Steeb [39], Energy level motion (Steeb and Euler [128]), Inviscid Burgers' equation (Steeb and Euler [128]) as well as a class of scale invariant partial differential equations (Steeb and Euler [125]). For more details on the Painlevé test for ordinary and partial differential equations see Steeb and Euler [126] and references therein. We also discuss the different concepts of integrability in this chapter.

17.2   Ordinary Differential Equations

In classical mechanics (i.e., ordinary differential equations) the so-called **Painlevé test** (also called singular point analysis) can serve in a certain sense to decide between integrable and nonintegrable dynamical systems. The differential equation is considered in the complex domain (complex time-plane) and the structure of the singularities of the solution of the ordinary differential equation is then studied.

Definition 17.1 *A system of ordinary differential equations considered in the complex domain is said to have the **Painlevé property** when every solution is single valued, except at the fixed singularities of the coefficients. This means that the Painlevé property requires that all its solutions be free of moving critical points.*

A necessary condition that an n-th order ordinary differential system of the form

$$\frac{d\mathbf{w}}{dz} = \mathbf{g}(\mathbf{w}) \tag{1}$$

where $\mathbf{g}$ is rational in $\mathbf{w}$, pass the Painlevé test is that there is a **Laurent expansion**

$$w_k(z) = (z - z_1)^m \sum_{j=0}^{\infty} a_{kj}(z - z_1)^j \tag{2}$$

with $n - 1$ arbitrary expansion coefficients, besides the pole position z_1 which is arbitrary. More than one branch may arise. In these sub-branches the number of arbitrary expansion coefficients can be smaller than $n - 1$.

We demonstrate now with the help of an example how the Painlevé test is performed.

Example: The **semiclassical Jaynes-Cummings model** is given by

$$\frac{dS_1}{dt} = -S_2, \qquad \frac{dS_2}{dt} = S_1 + S_3 E, \qquad \frac{dS_3}{dt} = -S_2 E \tag{3a}$$

$$\frac{d^2 E}{dt^2} + \mu^2 E = \alpha S_1. \tag{3b}$$

Here t (time) is the independent variable.

Remark: *μ is the dimensionless parameter given by $\mu = \omega/\omega_0$, and the coupling constant $\alpha = N\mu(2\lambda/\omega_0)^2$. Time is scaled with the atomic transition frequency ω_0 and is therefore dimensionless. S_1, S_2 and S_3 are components of the Bloch vector and represent atomic polarization and inversion, whereas the electric field $E = -2(\lambda/\omega_0)A$. There is numerical evidence that the system shows chaotic behaviour for certain parameter values and initial conditions.*

To perform the Painlevé test we consider system (3) in the complex domain. For the sake of simplicity we do not change our notation. First we look for the dominant behaviour. Inserting the ansatz

$$S_1(t) \propto S_{10}(t - t_1)^{m_1}, \qquad S_2(t) \propto S_{20}(t - t_1)^{m_2}, \qquad S_3(t) \propto S_{30}(t - t_1)^{m_3} \tag{4a}$$

$$E(t) \propto E_0(t - t_1)^{m_4} \tag{4b}$$

into the system (3a) and (3b) we find that the dominant terms given by the system

$$\frac{dS_1}{dt} = -S_2, \qquad \frac{dS_2}{dt} = S_3E, \qquad \frac{dS_3}{dt} = -S_2E \tag{5a}$$

$$\frac{d^2E}{dt^2} = \alpha S_1 \tag{5b}$$

where $m_1 = -3$, $m_2 = -4$, $m_3 = -4$ and $m_4 = -1$. For the expansion coefficients we obtain

$$S_{10} = \frac{8i}{\alpha}, \qquad S_{20} = \frac{24i}{\alpha}, \qquad S_{30} = -\frac{24}{\alpha}, \qquad E_0 = 4i.$$

From the dominant behaviour we conclude that the system of the dominant terms (5a) and (5b) is scale invariant under

$$t \to \varepsilon^{-1}t \qquad S_1 \to \varepsilon^3 S_1 \qquad S_2 \to \varepsilon^4 S_2 \qquad S_3 \to \varepsilon^4 S_3 \qquad E \to \varepsilon E.$$

Next we determine the resonances. To find the resonances we insert the ansatz

$$S_1(t) = S_{10}(t - t_1)^{-3} + A(t - t_1)^{-3+r}$$

$$S_2(t) = S_{20}(t - t_1)^{-4} + B(t - t_1)^{-4+r}$$

$$S_3(t) = S_{30}(t - t_1)^{-4} + C(t - t_1)^{-4+r}$$

$$E(t) = E_0(t - t_1)^{-1} + D(t - t_1)^{-1+r}$$

into the system with the dominant terms, where A, B, C ,D are arbitrary constants. Taking into account terms linear in A, B, C, D we obtain

$$A(r - 3)(t - t_1)^{r-4} = -B(t - t_1)^{r-4}$$

$$DS_{30}(t - t_1)^{r-5} + CE_0(t - t_1)^{r-5} = B(r - 4)(t - t_1)^{r-5}$$

$$C(r - 4)(t - t_1)^{r-5} = -DS_{20}(t - t_1)^{r-5} - BE_0(t - t_1)^{r-5}$$

$$D(r - 1)(r - 2)(t - t_1)^{r-3} = \alpha A(t - t_1)^{r-3}.$$

The above system can be written in matrix form

$$\begin{pmatrix} (r-3) & 1 & 0 & 0 \\ 0 & -(r-4) & E_0 & S_{30} \\ 0 & E_0 & (r-4) & S_{20} \\ -\alpha & 0 & 0 & (r-1)(r-2) \end{pmatrix} \begin{pmatrix} A \\ B \\ C \\ D \end{pmatrix} = \begin{pmatrix} 0 \\ 0 \\ 0 \\ 0 \end{pmatrix}.$$

Let

$$Q(r) := \begin{pmatrix} (r-3) & 1 & 0 & 0 \\ 0 & -(r-4) & E_0 & S_{30} \\ 0 & E_0 & (r-4) & S_{20} \\ -\alpha & 0 & 0 & (r-1)(r-2) \end{pmatrix}.$$

The resonances are the values of r such that

$$\det Q(r) = 0.$$

It follows that

$$(r-3)[-(r-4)^2(r-1)(r-2) - E_0^2(r-1)(r-2)] - [-S_{20}E_0\alpha + S_{30}(r-4)\alpha] = 0$$

so that the resonances are

$$\left\{ -1, \quad 4, \quad 8, \quad \frac{3}{2} \pm i\frac{\sqrt{15}}{2} \right\}.$$

The Jaynes-Cummings model does not pass the Painlevé test since there exist complex resonances. The **Kowalevski exponents** can be found from the variational equation of system (5a) and (5b). We find that the resonances and the Kowalevski exponents coincide. The two Kowalevski exponents 4 and 8 can be related to first integrals of system (5a) and (5b). We obtain

$$\begin{aligned}
I_1(S_1, S_2, S_3, E) &= S_2^2 + S_3^2 \\
I_2(S_1, S_2, S_3, E) &= \alpha S_3 - \alpha S_1 E + \frac{1}{2}\left(\frac{dE}{dt}\right)^2
\end{aligned}$$

since

$$\begin{aligned}
I_1(\varepsilon^3 S_1, \varepsilon^4 S_2, \varepsilon^4 S_3, \varepsilon^1 E) &= \varepsilon^8(S_2^2 + S_3^2) \\
I_2(\varepsilon^3 S_1, \varepsilon^4 S_2, \varepsilon^4 S_3, \varepsilon^1 E) &= \varepsilon^4\left(\alpha S_3 - \alpha S_1 E + \frac{1}{2}\left(\frac{dE}{dt}\right)^2\right).
\end{aligned}$$

Using the first integrals for the dominant system (5a) and (5b) we find by inspection that the first integrals for system (3a) and (3b) are given by

$$\begin{aligned}
I_1(S_1, S_2, S_3, E) &= S_1^2 + S_2^2 + S_3^2 \\
I_2(S_1, S_2, S_3, E) &= \alpha S_3 - \alpha S_1 E + \frac{1}{2}\mu^2 E^2 + \frac{1}{2}\left(\frac{dE}{dt}\right)^2.
\end{aligned}$$

From the Painlevé test we find that the system (3) admits a Laurent expansion of the form

$$S_1(t) = \sum_{j=0}^{\infty} S_{1j}(t - t_1)^{j-3} \tag{6a}$$

$$S_2(t) = \sum_{j=0}^{\infty} S_{2j}(t - t_1)^{j-4} \tag{6b}$$

$$S_3(t) = \sum_{j=0}^{\infty} S_{3j}(t - t_1)^{j-4} \qquad (6c)$$

$$E(t) = \sum_{j=0}^{\infty} E_j(t - t_1)^{j-1} \qquad (6d)$$

with three arbitrary constants (including t_1). The expansion coefficients are determined by a recursion relations

$$\sum_{j=0}^{\infty} S_{1j}(j-3)(t-t_1)^{j-4} = -\sum_{j=0}^{\infty} S_{2j}(t-t_1)^{j-4}$$

$$\sum_{j=0}^{\infty} S_{2j}(j-4)(t-t_1)^{j-5} = \sum_{j=0}^{\infty} S_{1j}(t-t_1)^{j-3} + \sum_{i=0}^{\infty}\sum_{j=0}^{\infty} S_{3j}E_i(t-t_1)(t-t_1)^{i+j-5}$$

$$\sum_{j=0}^{\infty} S_{3j}(j-4)(t-t_1)^{j-5} = -\sum_{i=0}^{\infty}\sum_{j=0}^{\infty} S_{2j}E_i(t-t_1)^{i+j-5}$$

$$\sum_{j=0}^{\infty} E_j(j-1)(j-2)(t-t_1)^{j-3} = -\mu^2\sum_{j=0}^{\infty} E_j(t-t_1)^{j-1} + \alpha\sum_{j=0}^{\infty} S_{1j}(t-t_1)^{j-3}.$$

In particular we find for the first expansion coefficients $S_{11} = S_{21} = S_{31} = E_1 = 0$. Obviously, this local expansion is not the general solution (owing to the complex resonances) which requires five arbitrary constants.

Let us now demonstrate how we can find the exact solution from the Laurent expansions (6). Let k be the modulus of the **elliptic functions** $\mathrm{sn}(z,k)$, $\mathrm{cn}(z,k)$ and $\mathrm{dn}(z,k)$. We define

$$K'(k) := \int_0^1 (1-t^2)^{-1/2}(1-k'^2 t^2)^{-1/2} dt$$

where $k'^2 := 1 - k^2$. By the addition-theorem of the Jacobi elliptic functions, we have

$$\mathrm{sn}(z + iK', k) = \frac{1}{k\,\mathrm{sn}(z,k)}.$$

Similarly

$$\mathrm{cn}(z + iK', k) = -\frac{i}{k}\frac{\mathrm{dn}(z,k)}{\mathrm{sn}(z,k)}, \qquad \mathrm{dn}(z + iK', k) = -i\frac{\mathrm{cn}(z,k)}{\mathrm{sn}(z,k)}.$$

For points in the neighbourhood of the point $z = 0$, the function $\mathrm{sn}(z,k)$ can be expanded by Taylor's theorem in the form

$$\mathrm{sn}(z,k) = \mathrm{sn}(0,k) + z\,\mathrm{sn}'(0,k) + \frac{1}{2}z^2\mathrm{sn}''(0,k) + \frac{1}{3!}z^3\mathrm{sn}'''(0,k) + \cdots$$

where accents denote derivatives. Since $sn(0, k) = 0$, $sn'(0, k) = 1$, $sn''(0, k) = 0$, $sn'''(0, k) = -(1 + k^2)$ etc. the expansion becomes

$$sn(z, k) = z - \frac{1}{6}(1 + k^2)z^3 + \cdots .$$

Therefore

$$cn(z, k) = (1 - sn^2 z)^{1/2} = 1 - \frac{1}{2}z^2 + \cdots$$

and

$$dn(z, k) = (1 - k^2 sn^2 z)^{1/2} = 1 - \frac{1}{2}k^2 z^2 + \cdots .$$

Consequently

$$sn(z + iK', k) = \frac{1}{k sn(z, k)} = \frac{1}{kz}\left(1 - \frac{1}{6}(1 + k^2)z^2 + \cdots\right)^{-1}$$

$$= \frac{1}{kz} + \frac{1 + k^2}{6k}z + \frac{1}{6^2 k}(1 + k^2)^2 z^3 + \cdots .$$

Similarly, we have

$$cn(z + iK', k) = \frac{-i}{kz} + \frac{2k^2 - 1}{6k}iz + \frac{(1 + k^2)(2k^2 - 1)}{36k}iz^3 +$$

and

$$dn(z + iK', k) = -\frac{i}{z} + \frac{2 - k^2}{6}iz + \frac{(1 + k^2)(2 - k^2)}{36}iz^3 + \cdots .$$

It follows that at the point $z = iK'$ the functions $sn(z, k)$, $cn(z, k)$ and $dn(z, k)$ have simple poles, with the residues

$$\frac{1}{k}, \qquad -\frac{i}{k}, \qquad -i,$$

respectively. We can focus our attention on the quantity E and (3b) since S_1 can be derived from (3b). Then the quantity S_2 and S_3 can be found from (3).

Comparing the Laurent expansion given by (6) and the expansion of the eliptic functions we find that E admits the particular solution (now considered again in the real domain)

$$E(t) = E_0 \, dn(\Omega t, k)$$

where $E_0^2 = 16\Omega^2$ and

$$k = 2\left(1 + \frac{1}{c}(1 - \sqrt{1 + c})\right), \qquad \Omega^2 = \frac{c(\mu^2 - \frac{1}{3})}{4(\sqrt{1 + c} - 1)}$$

$$c = -(\mu^2 - \frac{1}{3})^{-2}\left(\frac{4}{3}\left[\alpha^2 - 4(\mu^2 - \frac{1}{9})^3\right]^{1/2} + (\mu^2 - \frac{1}{9})(\mu^2 - \frac{17}{9})\right).$$

The quantities S_1, S_2 and S_3 can now easily be found from system (3a) and (3b).   ♣

17.3 Invertible Point Transformation

In this section we discuss how the techniques of an invertible point transformation and the Painlevé test can be used to construct integrable ordinary differential equations or equations that are related to the Painlevé transcendents (Steeb [114]). We compare both techniques for the second Painlevé transcendent.

The most well known second-order ordinary differential equations which have the Painlevé property are the so-called six Painlevé transcendents PI-PVI. One finds the classification of all equations of the form $d^2w/dz^2 = F(dw/dz, w, z)$ which have the Painlevé property, where F is rational in dw/dz, algebraic in w and locally analytic in z. Within a Möbius transformation, one finds that there exist fifty such equations. Distinguished among these fifty equations are PI-PVI. Any other of the fifty equations can either be integrated in terms of known functions or can be reduced to one of these six equations. Although PI-PVI were first discovered from strictly mathematical considerations, they have recently appeared in several physical applications. The six **Painlevé transcendents** are given by

$$\text{PI}: \quad \frac{d^2w}{dz^2} = 6w^2 + az$$

$$\text{PII}: \quad \frac{d^2w}{dz^2} = 2w^3 + zw + a$$

$$\text{PIII}: \quad \frac{d^2w}{dz^2} = \frac{1}{w}\left(\frac{dw}{dz}\right)^2 - \frac{1}{z}\frac{dw}{dz} + \frac{1}{z}(aw^2 + b) + cw^3 + \frac{d}{w}$$

$$\text{PIV}: \quad \frac{d^2w}{dz^2} = \frac{1}{2w}\left(\frac{dw}{dz}\right)^2 + \frac{3}{2}w^3 + 4zw^2 + 2(z^2 - a)w + \frac{b}{w}$$

$$\text{PV}: \quad \frac{d^2w}{dz^2} = \left(\frac{1}{2w} + \frac{1}{w-1}\right)\left(\frac{dw}{dz}\right)^2 - \frac{1}{z}\frac{dw}{dz} + \frac{(w-1)^2}{z^2}\left(aw + \frac{b}{w}\right)$$
$$+ c\frac{w}{z} + d\frac{w(w+1)}{w-1}$$

$$\text{PVI}: \quad \frac{d^2w}{dz^2} = \frac{1}{2}\left(\frac{1}{w} + \frac{1}{w-1} + \frac{1}{w-z}\right)\left(\frac{dw}{dz}\right)^2 - \left(\frac{1}{z} + \frac{1}{z-1} + \frac{1}{w-z}\right)\frac{dw}{dz}$$
$$+ \frac{w(w-1)(w-z)}{z^2(z-1)^2}\left(a + \frac{bz}{w^2} + c\frac{(z-1)}{(w-1)^2} + d\frac{z(z-1)}{(w-z)^2}\right).$$

For nonlinear ordinary and partial differential equations the general solution usually cannot be given explicitly. It is desirable to have an approach to find out whether a given nonlinear differential equation can explicitly be solved or reduced to one of the six Painlevé transcendents. For ordinary differential equations the Painlevé test and the invertible point transformation can be used to construct integrable nonlinear equations or equations which are related to one of the six Painlevé transcendents.

We consider the second Painlevé transendent

$$\frac{d^2w}{dz^2} = 2w^3 + zw + a \tag{7}$$

(a is an arbitrary constant) and perform an invertible point transformation to find the anharmonic oscillator

$$\frac{d^2u}{dt^2} + f_1(t)\frac{du}{dt} + f_2(t)u + f_3(t)u^3 = 0. \tag{8}$$

This provides a condition on f_1, f_2 and f_3 such that the anharmonic oscillator (8) can be transformed to the second Painlevé transcendent. We also perform the Painlevé test for (8). This gives also conditions on $f_1(t)$, $f_2(t)$ and $f_3(t)$ and it is shown that the two conditions are the same for the two approaches.

Let us first discuss the Painlevé test. We study the anharmonic oscillator (8), where f_1, f_2 and f_3 are smooth functions of t with the help of the Painlevé test. We assume that $f_3 \neq 0$. For arbitrary functions f_1, f_2 and f_3 the nonlinear equation (8) cannot explicity be solved.

Remark: *A remark is in order for applying the Painlevé test for non-autonomous systems. The coefficients that depend on the independent variable must themselves be expanded in terms of t. If non-autonomous terms enter the equation at lower order than the dominant balance the above mentioned expansion turns out to be unnecessary whereas if the non-autonomous terms are at dominant balance level they must be expanded with respect to t. Obviously f_1, f_2 and f_3 do not enter the expansion at dominant level.*

Inserting the Laurent expansion

$$u(t) = \sum_{j=0}^{\infty} a_j(t - t_1)^{j-1}$$

into (8) we find at the resonance $r = 4$, the condition

$$9f_3^{(4)}f_3^3 - 54f_3^{(3)}f_3'f_3^2 + 18f_3^{(3)}f_3^3f_1 - 36(f_3'')^2f_3^2 + 192f_3''(f_3')^2f_3 - 78f_3''f_3'f_3^2f_1$$

$$+36f_3''f_3^3f_2 + 3f_3''f_3^3f_1^2 - 112(f_3')^4 + 64(f_3')^3f_3f_1 + 6(f_3')^2f_1'f_3^2 - 72(f_3')^2f_3^2f_2$$

$$+90f_3'f_2'f_3^3 - 27f_3'f_1''f_3^3 - 57f_3'f_1'f_3^3f_1 + 72f_3'f_3^3f_2f_1 - 14f_3'f_3^3f_1^3 - 54f_2''f_3^4$$

$$-90f_2'f_3^4f_1 + 18f_1^{(3)}f_3^4 + 54f_1''f_3^4f_1 + 36(f_1')^2f_3^4 - 36f_1'f_3^4f_2 + 60f_1'f_3^4f_1^2$$

$$-36f_3^4f_2f_1^2 + 8f_3^4f_1^4 = 0 \tag{9}$$

where $f' \equiv df/dt$ and $f^{(4)} \equiv f'''' \equiv d^4f/dt^4$. This means that if this condition is satisfied then the expansion coefficient a_4 is arbitrary.

Now we ask whether the equation derived above can be found from (7) with the help of the invertible point transformation. We consider the invertible point transformation

$$z(u(t),t)) = G(u(t),t) \qquad w(z(u(t),t)) = F(u(t),t)$$

where

$$\Delta \equiv \frac{\partial G}{\partial t}\frac{\partial F}{\partial u} - \frac{\partial G}{\partial u}\frac{\partial F}{\partial t} \neq 0.$$

Since

$$\frac{dw}{dt} = \frac{dw}{dz}\frac{dz}{dt} = \frac{dw}{dz}\left(\frac{\partial G}{\partial u}\frac{du}{dt} + \frac{\partial G}{\partial t}\right) = \frac{\partial F}{\partial u}\frac{du}{dt} + \frac{\partial F}{\partial t}$$

and

$$
\begin{aligned}
\frac{d^2w}{dt^2} &= \frac{d^2w}{dz^2}\frac{dz}{dt}\left(\frac{\partial G}{\partial u}\frac{du}{dt} + \frac{\partial G}{\partial t}\right) \\
&\quad + \frac{dw}{dz}\left(\frac{\partial^2 G}{\partial u\partial t}\frac{du}{dt} + \frac{\partial^2 G}{\partial u^2}\left(\frac{du}{dt}\right)^2 + \frac{\partial G}{\partial u}\frac{d^2u}{dt^2} + \frac{\partial^2 G}{\partial t^2} + \frac{\partial^2 G}{\partial t\partial u}\frac{du}{dt}\right) \\
&= \frac{\partial^2 F}{\partial u\partial t}\frac{du}{dt} + \frac{\partial^2 F}{\partial u^2}\left(\frac{du}{dt}\right)^2 + \frac{\partial F}{\partial u}\frac{d^2u}{dt^2} + \frac{\partial^2 F}{\partial u\partial t}\frac{du}{dt} + \frac{\partial^2 F}{\partial t^2}
\end{aligned}
$$

we obtain

$$\frac{d^2u}{dt^2} + \Lambda_3\left(\frac{du}{dt}\right)^3 + \Lambda_2\left(\frac{du}{dt}\right)^2 + \Lambda_1\frac{du}{dt} + \Lambda_0 = 0$$

where

$$\Lambda_3 = \left(-\frac{\partial F}{\partial u}\frac{\partial^2 G}{\partial u^2} + \frac{\partial^2 F}{\partial u^2}\frac{\partial G}{\partial u} - 2F^3\left(\frac{\partial G}{\partial u}\right)^3 - FG\left(\frac{\partial G}{\partial u}\right)^3 - a\left(\frac{\partial G}{\partial u}\right)^3\right)\Delta^{-1}$$

$$
\begin{aligned}
\Lambda_2 = &\left(-2\frac{\partial F}{\partial u}\frac{\partial^2 G}{\partial u\partial t} - \frac{\partial F}{\partial t}\frac{\partial^2 G}{\partial u^2} + \frac{\partial^2 F}{\partial u^2}\frac{\partial G}{\partial t} + 2\frac{\partial^2 F}{\partial x\partial t}\frac{\partial G}{\partial u} - 6F^3\left(\frac{\partial G}{\partial u}\right)^2\frac{\partial G}{\partial t}\right. \\
&\left. -3FG\left(\frac{\partial G}{\partial u}\right)^2\frac{\partial G}{\partial t} - 3a\left(\frac{\partial G}{\partial u}\right)^2\frac{\partial G}{\partial t}\right)\Delta^{-1}
\end{aligned}
$$

$$
\begin{aligned}
\Lambda_1 = &\left(-\frac{\partial F}{\partial u}\frac{\partial^2 G}{\partial t^2} - 2\frac{\partial F}{\partial t}\frac{\partial^2 G}{\partial u\partial t} + 2\frac{\partial^2 F}{\partial u\partial t}\frac{\partial G}{\partial t} + \frac{\partial^2 F}{\partial t^2}\frac{\partial G}{\partial u} - 6F^3\frac{\partial G}{\partial u}\left(\frac{\partial G}{\partial t}\right)^2\right. \\
&\left. -3FG\frac{\partial G}{\partial u}\left(\frac{\partial G}{\partial t}\right)^2 - 3a\frac{\partial G}{\partial u}\left(\frac{\partial G}{\partial t}\right)^2\right)\Delta^{-1}
\end{aligned}
$$

$$\Lambda_0 = \left(-\frac{\partial F}{\partial t}\frac{\partial^2 G}{\partial t^2} + \frac{\partial^2 F}{\partial t^2}\frac{\partial G}{\partial t} - 2F^3\left(\frac{\partial G}{\partial t}\right)^3 - FG\left(\frac{\partial G}{\partial t}\right)^3 - a\left(\frac{\partial G}{\partial t}\right)^3\right)\Delta^{-1}.$$

We now make a particular choice for F and G, namely

$$F(u(t),t) = f(t)u(t), \qquad G(u(t),t) = g(t)$$

where f and g are arbitrary functions of t. With this special ansatz we find that

$$\Lambda_3 = \Lambda_2 = 0$$

and

$$\Lambda_1 = \frac{2\dot{f}\dot{g} - f\ddot{g}}{\dot{g}f}, \qquad \Lambda_0 = \frac{\left(\ddot{f}\dot{g} - \dot{f}\ddot{g} - fg\dot{g}^3\right)u - 2f^3\dot{g}^3u^3 - a\dot{g}^3}{\dot{g}f}$$

where $\dot{f} = df/dt$ etc. For $a = 0$ it follows that

$$\frac{d^2u}{dt^2} + f_1(t)\frac{du}{dt} + f_2(t)u + f_3(t)u^3 = 0$$

where

$$f_1(t) = \frac{2\dot{f}\dot{g} - f\ddot{g}}{\dot{g}f} \tag{10}$$

$$f_2(t) = \frac{\dot{g}\ddot{f} - \dot{f}\ddot{g} - fg\dot{g}^3}{\dot{g}f} \tag{11}$$

$$f_3(t) = -2(f\dot{g})^2. \tag{12}$$

For the case $a \neq 0$ one obtains the driven anharmonic oscillator. This case is not discussed here. We are now able to eliminate f and g from system (10), (11) and (12). We obtain

$$f(t) = Cf_3^{1/6}\exp\left(\int^t \frac{f_1(s)ds}{3}\right). \tag{13}$$

By inserting (13) and (12) into (11) we find

$$g(t) = \frac{C^2\exp(2\int(\frac{1}{3}f_1dt))}{18f_3^{8/3}}\left(-6f_3f_3'' + 2\left(f_3'\right)^2 - 2f_1f_3f_3' - 12f_1f_3^2 + 36f_2f_3^2 - 8f_1^2f_3^2\right).$$

Inserting the derived f and g in (10) we find, exactly, condition (9).

To summarize: The condition (9) that the anharmonic oscillator (8) pass the Painlevé test is identical to the condition that the anharmonic oscillator be transformable to the second Painlevé transcendent. Constraint (9) can be solved in such a way that f_2 can be given as a function of f_1 and f_3. The above approach can also be applied to the nonlinear ordinary differential equation (Steeb [121])

$$\frac{d^2u}{dt^2} + f_1(t)\frac{du}{dt} + f_2(t)u + f_3(t)u^n = 0$$

$(n \neq 0, 1)$.

17.4 Partial Differential Equations

Ward [144], [145] discussed the Painlevé property for partial differential equations in terms of meromorphic solutions on $\mathbb{C}^m$. The major difference between analytic functions of one complex variable and several complex variables is that, in general, the singularities of a function of several complex variables cannot be isolated. If f is a meromorphic function of m complex variables ($2n$ real variables), the singularities of f occur along analytic manifolds of (real) dimension $2n - 2$. These manifolds are known as **singularity manifolds** and are determined by conditions of the form

$$\phi(z_1, \ldots, z_n) = 0$$

where ϕ is an analyic function of $(z_1, \ldots, z_n)$ in a neighbourhood of the manifold.

Suppose that there are m independent variables and that the system of partial differential equations has coefficients that are holomorphic on $\mathbb{C}^m$. We cannot simply require that all the solutions of this system be holomorphic on $\mathbb{C}^m$ since arbitrarily 'nasty' singularities can occur along characteristic hypersurfaces. The following definition of the Painlevé property avoids this problem.

Definition 17.2 *If S is a holomorphic non-characteristic hypersurface in $\mathbb{C}^m$, then every solution that is holomorphic on $\mathbb{C}^m \backslash S$ extends to a meromorphic solution on $\mathbb{C}^m$.*

In other words, if a solution has a singularity on a non-characteristic hypersurface, that singularity is a pole and nothing worse.

A weaker form of the Painlevé property for partial differential equations, known as the **Painlevé test** was formulated by Weiss *et al* [147]. It involves looking for solutions ϕ of the system of partial differential equations in the form

$$u = \phi^{-\alpha} \sum_{n=0}^{\infty} u_n \phi^n$$

where ϕ is a holomorphic function whose vanishing defines a noncharacteristic hypersurface. Substituting this series into the partial differential equations yields conditions on the number α and recursion relations for the functions u_n. The requirements are that α should turn out to be a non-negative integer, the recursion relation should be consistent and the series expansion should contain the correct number of arbitrary functions (counting ϕ as one of them). If this weaker form of the Painlevé property is satisfied, we say that the partial differential equation passes the Painlevé test. In the following we apply this weaker form of the Painlevé test.

There is a controversy about this definition in the literature. With reference to the above, we say that a partial differential equation has the Painlevé property when the solutions of the partial differential equation are single valued about the movable singularity manifold. For partial differential equations we require that the solution be a single valued functional of the data, i.e., arbitrary functions. This is a formal property and not a restriction on the data itself. The Painlevé property requires all movable singularity manifolds to be single valued, whether characteristic or not. The Painlevé property is a statement of how the solutions behave as functionals of the data in a neighbourhood of a singularity manifold and not a statement about the data itself.

Example: Consider Burgers' equation

$$\frac{\partial u}{\partial x_2} + u\frac{\partial u}{\partial x_1} = \sigma\frac{\partial^2 u}{\partial x_1^2}$$

where σ is a constant. Inserting $u \propto u_0\phi^n$ we find that $n = -1$ and $u_0 = -2\sigma\partial\phi/\partial x$. The equation with the dominant terms takes the form

$$u\frac{\partial u}{\partial x_1} = \sigma\frac{\partial^2 u}{\partial x_1^2}.$$

The resonances are given by -1 and 2. The Kowalevski exponents are the same. That $r = 2$ is a Kowalevski exponent can be seen as follows: The equation with the dominant behaviour is invariant under $u \to \varepsilon u, x_1 \to \varepsilon^{-1}x_1$. Thus it can be written as $\partial I(u)/\partial x_1 = 0$, where $I = u^2/2 - \sigma\partial u/\partial x_1$. Therefore

$$I\left(\varepsilon u, \frac{\partial \varepsilon u}{\partial \varepsilon^{-1}x_1}\right) - \varepsilon^2 I\left(u, \frac{\partial u}{\partial x_1}\right)$$

and $r = 2$ has to be a Kowalevski exponent. Inserting the ansatz

$$u = \phi^{-1}\sum_{j=0}^{\infty} u_j\phi^j \tag{14}$$

into Burgers equation it follows that

$$j = 0: \quad u_0 = -2\sigma\frac{\partial\phi}{\partial x_1} \tag{15}$$

$$j = 1: \quad \frac{\partial\phi}{\partial x_2} + u_1\frac{\partial\phi}{\partial x_1} - \sigma\frac{\partial^2\phi}{\partial x_1^2} = 0 \tag{16}$$

$$j = 2: \quad \frac{\partial F}{\partial x_1} = 0 \tag{17}$$

where F is given by the left hand side of (16). By (16) the compatability condition (17) at the resonance $j = r = 2$ is satisfied identically. Consequently, Burgers' equation passes the Painlevé test. In the series expansion (14) the functions ϕ and u_2 can be chosen arbitrarily.

A Bäcklund transformation can now be found as follows: Let $u_j = 0$ for $j \geq 2$. Then we find the auto-Bäcklund transformation

$$u = -2\sigma\phi^{-1}\frac{\partial\phi}{\partial x_1} + u_1 \equiv -2\sigma\frac{\partial(\ln\phi)}{\partial x_1} + u_1$$

where u and u_1 satisfy Burgers' equation and

$$\frac{\partial\phi}{\partial x_2} + u_1\frac{\partial\phi}{\partial x_1} = \sigma\frac{\partial^2\phi}{\partial x_1^2}. \tag{18}$$

The trivial solution to the Burgers' equation is $u_1 = 0$. Then (18) simplifies to

$$\frac{\partial\phi}{\partial x_2} = \sigma\frac{\partial^2\phi}{\partial x_1^2} \tag{19}$$

(linear diffusion equation) and

$$u = -2\sigma\phi^{-1}\frac{\partial\phi}{\partial x_1}. \tag{20}$$

This is the **Cole-Hopf transformation**. Equation (20) can be solved with respect to ϕ. Let $\sigma = 1$. We find

$$\phi(x_1, x_2) = \exp(-\frac{1}{2}\int\limits_{-\infty}^{x_1} u(x_1', x_2)\, dx_1'). \tag{21}$$

Inserting (21) into (18) we arrive at

$$\int\limits_{-\infty}^{x_1} \frac{\partial u(x_1', x_2)}{\partial x_2}\, dx_1' = -\frac{1}{2}u^2 + \frac{\partial u}{\partial x_1}.$$

Taking the derivative with respect to x_1 yields Burgers equation ($\sigma = 1$). This enables us to construct a hierarchy of nonlinear partial differential equations which can be linearized with the help of $u = -2\phi^{-1}\partial\phi/\partial x_1$ and pass the Painlevé test. This means that the process of starting from $\partial\phi/\partial x_2 = \partial^3\phi/\partial x_1^3$, $\partial\phi/\partial x_2 = \partial^4\phi/\partial x_1^4$ and so on and inserting (21) yields a hierarchy of linearizable equations. All these equations pass the Painlevé test. From

$$\frac{\partial\phi}{\partial x_2} = \frac{\partial^3\phi}{\partial x_1^3}$$

we find

$$\frac{\partial u}{\partial x_2} = \frac{3}{4}u^2\frac{\partial u}{\partial x_1} - \frac{3}{2}\left(\frac{\partial u}{\partial x_1}\right)^2 - \frac{3}{2}u\frac{\partial^2 u}{\partial x_1^2} + \frac{\partial^3 u}{\partial x_1^3}. \qquad \clubsuit$$

17.5   Symmetries by Truncated Expansions

In this section we consider partial differential equations which pass the Painlevé test. In particular we study the following evolution equation

$$\frac{\partial u}{\partial x_2} = F(u(x_1, x_2)) \tag{22}$$

where F is a polynomial in u and

$$\frac{\partial u}{\partial x_1}, \frac{\partial^2 u}{\partial x_1^2}, \cdots, \frac{\partial^n u}{\partial x_1^n}.$$

Let us assume that we find a solution of (22) of the form

$$u = \phi^{-\alpha} \sum_{j=0}^{\infty} u_j \phi^j. \tag{23}$$

Let us recall the definition for an equation of the form (22) to pass the Painlevé test:
(i) There exist integers $\alpha > 0, \beta > 0$ such that by insertion of (23) into (22) yields

$$\sum_{j=0}^{\infty} G_j(u_j, u_{j-1}, \cdots, u_0, \phi)\phi^{-\alpha-\beta+j} = 0 \tag{24}$$

where G_j are differential polynomials.

(ii) The functions u_j can be determined recursively from the equations

$$G_j(u_j, u_{j-1}, \cdots, u_0, \phi) = 0, \qquad j \geq 0 \tag{25}$$

by purely algebraic manipulations. The functions u_j are determined uniquely with the exception of $\beta = n - 1$ resonance functions which can be chosen arbitrarily.

The series can be truncated if

(iii) By setting the resoncance functions equal to zero, we can achieve that $u_j = 0, j > \alpha$, in addition to (25).

(iv) Series (23), truncated at the constant level term, yields a solution to (22) whenever $u_0, \cdots, u_\alpha, \phi$ can be determined to satisfy (25) for $j = 0, \cdots, \alpha + \beta$.

We recall that the linearized equation (variational equation) of (22) about a solution u

$$\frac{\partial v}{\partial x_2} = F'(u)[v] \equiv \left.\frac{\partial F(u + \epsilon v)}{\partial \epsilon}\right|_{\epsilon=0}. \tag{26}$$

Any solution of (26) yields a symmetry or an infinitesimal transformation about u, i.e., the transformation $u + \epsilon v$ leaves (34) form-invariant.

Proposition 1: *Assume that (i) ~ (iv) hold. Then $u_{\alpha-1}$ will be an infinitesimal transformation about u_α.*

Proof: Write the truncated series as

$$u = u_\alpha + \sigma,$$

where

$$\sigma = u_0 \phi^{-\alpha} + \cdots + u_{\alpha-1} \phi^{-1}.$$

Insertion into (22) yields

$$\frac{\partial u_\alpha}{\partial x_2} + \frac{\partial \sigma}{\partial x_2} = F(u_\alpha) + \frac{\partial F(u_\alpha)}{\partial u} \sigma + \frac{\partial F(u_\alpha)}{\partial u_{x_1}} \frac{\partial \sigma}{\partial x_1} + \cdots + \frac{\partial F(u_\alpha)}{\partial u_{x_{1_1} \cdots x_{1_r}}} \frac{\partial^r \sigma}{\partial x_1^r} + p \qquad (27)$$

where

$$u_{x_{1_1} \cdots x_{1_r}} \equiv \frac{\partial^r u}{\partial x_1^r}$$

and p stands for higher terms occurring in the Taylor expansion of F about

$$\left(u_\alpha, \frac{\partial u_\alpha}{\partial x_1}, \ldots, \frac{\partial^r u_\alpha}{\partial x_1^r} \right).$$

Evaluating derivatives of σ gives

$$\frac{\partial \sigma}{\partial x_2} = \frac{\partial u_{\alpha-1}}{\partial x_2} \phi^{-1} + \cdots \text{ terms in } \phi^{-c} \qquad (28a)$$

$$\frac{\partial \sigma}{\partial x_1} = \frac{\partial u_{\alpha-1}}{\partial x_1} \phi^{-1} + \cdots \text{ terms in } \phi^{-c} \qquad (28b)$$

$$\frac{\partial^2 \sigma}{\partial x_1^2} = \frac{\partial^2 u_{\alpha-1}}{\partial x_1^2} \phi^{-1} + \cdots \text{ terms in } \phi^{-c} \qquad (28c)$$

where $c \geq 2$. Now, by inserting (28) into (27) we obtain

$$0 = \left(F(u_\alpha) - \frac{\partial u_\alpha}{\partial x_2} \right) \phi^0 + \left(F'(u_\alpha) u_{\alpha-1} - \frac{\partial u_{\alpha-1}}{\partial x_2} \right) \phi^{-1} + \cdots \text{ terms in } \phi^{-c},$$

with $c \geq 2$. By assumptions (i) to (iv) it follows that

$$G_\alpha = F(u_\alpha) - \frac{\partial u_\alpha}{\partial x_2} = 0$$

$$G_{\alpha-1} = F'(u_\alpha) u_{\alpha-1} - \frac{\partial u_{\alpha-1}}{\partial x_2} = 0. \qquad \spadesuit$$

Example: Consider the Korteweg-de Vries equation

$$\frac{\partial u}{\partial x_2} + u\frac{\partial u}{\partial x_1} + \frac{\partial^3 u}{\partial x_1^3} = 0$$

possessing the Painlevé expansion

$$u = \phi^{-2}\sum_{j=0}^{\infty} u_j\phi^j \ .$$

The resonances are $j = -1, \ 4, \ 6$ and we can achieve $u_j = 0, \ j > 2$, by setting

$$u_4 = u_6 = 0.$$

The truncated expansion leads to the following conditions $G_j = 0, \ j = 0, \ \ldots, \ 5$,

$$j = 0 \ (\phi^{-5}): \quad u_0 = -12\left(\frac{\partial\phi}{\partial x_1}\right)^2 \tag{29}$$

$$j = 1 \ (\phi^{-4}): \quad u_1 = 12\frac{\partial^2\phi}{\partial x_1^2} \tag{30}$$

$$j = 2 \ (\phi^{-3}): \quad \frac{\partial\phi}{\partial x_1}\frac{\partial\phi}{\partial x_2} + \left(\frac{\partial\phi}{\partial x_1}\right)^2 u_2 + 4\frac{\partial\phi}{\partial x_1}\frac{\partial^3\phi}{\partial x_1^3} - 3\left(\frac{\partial^2\phi}{\partial x_1^2}\right)^2 = 0 \tag{31}$$

$$j = 3 \ (\phi^{-2}): \quad \frac{\partial^2\phi}{\partial x_1\partial x_2} + \frac{\partial^2\phi}{\partial x_1^2}u_2 + \frac{\partial^4\phi}{\partial x_1^4} = 0 \tag{32}$$

$$j = 4 \ (\phi^{-1}): \quad \frac{\partial}{\partial x_1}\left(\frac{\partial^2\phi}{\partial x_1\partial x_2} + \frac{\partial^2\phi}{\partial x_1^2}u_2 + \frac{\partial^4\phi}{\partial x_1^4}\right) = 0 \tag{33}$$

$$j = 5 \ (\phi^0): \quad \frac{\partial u_2}{\partial x_2} + u_2\frac{\partial u_2}{\partial x_1} + \frac{\partial^3 u_2}{\partial x_1^3} = 0. \tag{34}$$

Obviously (34) means that u_2 solves the Korteweg-de Vries equation while (33), according to proposition 1, states that u_1 is a solution of the Korteweg-de Vries equation linearized about u_2. Thus by any solution ϕ of (31) and (32) we obtain an infinitesimal transformation about u_2.   ♣

Remark: *Proposition 1 can be generalized to hold for the following cases: equations involving higher order time derivatives, systems of equations, equations in more than one spatial variable and ordinary differential equations.*

17.6 Painlevé Test and Recursion Operators

In this section we ask how recursion operators can be obtained from Painlevé expansions. We recall that a condition for an operator $R(u)$ to be a recursion operator for the evolution equation (22) is that

$$\frac{\partial R(u)}{\partial x_2} = [F'(u),\, R(u)] \tag{35}$$

whenever u solves (22).

The following method can be applied for obtaining recursion operators.

Proposition 2: *Let*

$$L(u)v = \lambda v \tag{36}$$

$$\frac{\partial v}{\partial x_2} = F(u)v \tag{37}$$

be a Lax equations for (22). Let E_λ be the eigenspace belonging to the eigenvalue λ. Assume that the following holds for any solution u of (22)

$$\frac{\partial}{\partial x_2} T(v_1,\, \ldots,\, v_l,\, \lambda) = F'(u)T(v_1,\, \ldots,\, v_l,\, \lambda) \tag{38}$$

$$R(u)T(v_1,\, \ldots,\, v_l,\, \lambda) = f(\lambda)T(v_1,\, \ldots,\, v_l,\, \lambda) \tag{39}$$

where $R(u)$ is an operator, f a function and T a transformation defined on the l-fold product

$$E_\lambda \times \ldots \times E_\lambda \ .$$

Then it holds that

$$\left(\frac{\partial R(u)}{\partial x_2} - [F'(u),\, R(u)]\right) T(v_1, \ldots, v_l, \lambda) = 0 \ . \tag{40}$$

Proof: The proof is obvious from the fact that λ is time-independent. ♠

In order to find recursion operators we now proceed as follows:

(1) We must find the transformation $T(v_1,\, \ldots,\, v_l,\, \lambda)$ mapping $E_\lambda \times \cdots \times E_\lambda$ into solutions of the linearized equation.

(2) We have to derive an eigenequation (39) for the functions $T(v_1,\, \ldots,\, v_l,\, \lambda)$ from the

eigenequation $L(u)v_j = \lambda v_j$.

(3) By the above proposition 2 the condition (40) holds.

(4) Suppose we have enough functions $T(v_1, \ldots, v_l, \lambda)$ for concluding from (40) that (35) holds, then $R(u)$ will be a recursion operator.

The most important step is to find the transformation T. The Painlevé property can be connected with the Lax equations via the Schwarzian derivative.

Example: Consider the Korteweg-de Vries equation

$$\frac{\partial u}{\partial x_2} + u\frac{\partial u}{\partial x_1} + \frac{\partial^3 u}{\partial x_1^3} = 0 .$$

The **Schwarzian derivative** of ϕ is defined as

$$\{\phi;\, x_1\} := \frac{\partial}{\partial x_1}\left(\frac{\partial^2 \phi}{\partial x_1^2}\left(\frac{\partial \phi}{\partial x_1}\right)^{-1}\right) - \frac{1}{2}\left(\frac{\partial^2 \phi}{\partial x_1^2}\left(\frac{\partial \phi}{\partial x_1}\right)^{-1}\right)^2 .$$

The resulting equation in terms of the Schwarzian derivative takes the form

$$\frac{\partial}{\partial x_1}\left(\frac{\partial \phi}{\partial x_2}\left(\frac{\partial \phi}{\partial x_1}\right)^{-1} + \{\phi;\, x_1\}\right) = \lambda$$

where λ=constant. This equation is solved by setting $\phi = v_1/v_2$, where the function v_j satisfy the Lax equations for the Korteweg-de Vries equation

$$\frac{\partial^2 v}{\partial x_1^2} + \frac{1}{6}(u_2 + \lambda)v = 0$$

$$\frac{\partial v}{\partial x_2} = \left(-\frac{u_2}{3} + \frac{2}{3}\lambda\right)\frac{\partial v}{\partial x_1} + \frac{u_2}{6}v .$$

Therefore we obtain

$$T(v_1,\, v_2,\, \lambda) = \frac{\partial^2}{\partial x_1^2}\left(\frac{v_1}{v_2}\right) .$$

Transformation T then leads to the recursion operator

$$R(u) = D^2 + \frac{2}{3}u + \frac{1}{3}\frac{\partial u}{\partial x_1}D^{-1}$$

for the Korteweg-de Vries equation where $D := \partial/\partial x_1$.   ♣

17.7 Singular Manifold and Similarity Variables

In this section the nonintegrability of the nonlinear partial differential equation

$$\frac{\partial^2 v}{\partial \eta \partial \xi} = v^3 \tag{41}$$

is studied with the help of the Painlevé test. The condition at the resonance is discussed in detail. Particular solutions are given and the connection of the singular manifold at the resonance and the similarity variable is investigated.

It is well known that in one and more space dimensions polynomial field equations such as the **nonlinear Klein-Gordon equation**

$$\frac{\partial^2 u}{\partial x_1^2} - \frac{\partial^2 u}{\partial x_2^2} + m^2 u + \lambda u^3 = 0 \tag{42}$$

cannot be solved exactly even for the case $m = 0$.

Let us consider (42). For the sake of simplicity we assume that $m = 0$. We introduce **light-cone coordinates**

$$\xi(x_1, x_2) := \frac{1}{2}(x_1 - x_2)$$

$$\eta(x_1, x_2) := \frac{1}{2}(x_1 + x_2)$$

$$v(\eta(x_1, x_2), \xi(x_1, x_2)) := u(x_1, x_2)$$

and put $\lambda = 1$. Then we arrive at (41). It is well known that (42) (and therefore (41)) can be derived from a Lagrangian density and Hamiltonian density. Equation (41) is considered in the complex domain. For the sake of simplicity we do not change our notation. We focus our attention on three points. First we investigate whether (41) passes the Painlevé test. In particular we give the condition for the singular manifold at the resonance. This means we insert the expansion

$$v = \sum_{j=0}^{\infty} v_j \phi^{j-1} \tag{43}$$

where ϕ and u_j are locally analytic functions of η and ξ. The resonances are given by $r_1 = -1$ and $r_2 = 4$. The Kowalevski exponents are the same. The Kowalevski exponent $r_2 = 4$ is related to the Hamiltonian density. Secondly we give the Lie symmetry vector fields of (41), construct similarity ansätze via the similarity variable ς and perform group theoretical reductions of the partial differential equation (41) to ordinary differential equations. The connection of the similarity variables ς and the condition on ϕ at the resonance is discussed for every group theoretical reduction. Furthermore, the Painlevé test is performed for these ordinary differential equations. Finally we discuss the truncated expansion

$$v = \phi^{-1}v_0 + v_1 \tag{44}$$

where v_1 satisfies (41). The truncated expansion can be considered as an auto Bäcklund transformation, where ϕ satisfies certain conditions. This condition, i.e., the partial differential equation for ϕ, is compared with the condition for ϕ at the resonance.

For this section we make use of the notation

$$\frac{\partial \phi}{\partial \eta} \equiv \phi_\eta, \quad \frac{\partial^2 \phi}{\partial \xi \partial \eta} \equiv \phi_{\xi\eta}, \quad \text{etc.}$$

Inserting expansion (43) into (41) gives for the first four expansion coefficients

$$v_0^2 = 2\phi_\eta \phi_\xi \tag{45a}$$

$$-\phi_{\xi\eta}v_0 - \phi_\xi v_{0\eta} - \phi_\eta v_{0\xi} = 3v_0^2 v_1 \tag{45b}$$

$$v_{0\xi\eta} = 3v_0^2 v_2 + 3v_0 v_1^2 \tag{45c}$$

$$2\phi_\xi \phi_\eta v_3 + \phi_{\xi\eta}v_2 + \phi_\xi v_{2\eta} + \phi_\eta v_{2\xi} + v_{1\xi\eta} = 2v_0^2 v_3 + 6v_0 v_1 v_2. \tag{45d}$$

At the resonance $r = 4$ we obtain

$$2\phi_{\xi\eta}v_3 + 2\phi_\eta v_{3\xi} + 2\phi_\xi v_{3\eta} + v_{2\eta\xi} = 6v_0 v_1 v_3 + 3v_1^2 v_2 + 3v_0 v_2^2. \tag{46}$$

Inserting system (45) into (46) yields the following condition on ϕ

$$3\phi_\xi^2 \phi_\eta^2 (-\phi_{\xi\xi\xi\xi}\phi_\eta^4 - \phi_{\eta\eta\eta\eta}\phi_\xi^4 + 4\phi_{\xi\eta\eta\eta}\phi_\xi^3\phi_\eta + 4\phi_{\xi\xi\xi\eta}\phi_\xi\phi_\eta^3 + 6\phi_{\xi\xi\eta\eta}\phi_\xi^2\phi_\eta^2)$$

$$\times (2\phi_{\xi\eta}\phi_\xi\phi_\eta - \phi_{\xi\xi}\phi_\eta^2 - \phi_{\eta\eta}\phi_\xi^2) - 27\phi_\xi^4\phi_\eta^4(2\phi_{\xi\eta\eta}\phi_{\xi\xi\eta}\phi_\xi\phi_\eta - \phi_{\xi\eta\eta}^2\phi_\xi^2 - \phi_{\xi\xi\eta}^2\phi_\eta^2)$$

$$+66\phi_{\xi\eta}\phi_\xi^2\phi_\eta^2(\phi_{\xi\xi\xi}\phi_{\xi\xi}\phi_\eta^5 + \phi_{\eta\eta\eta}\phi_{\eta\eta}\phi_\xi^5) + 6\phi_{\xi\eta}\phi_\xi^4\phi_\eta^4(\phi_{\xi\xi\xi}\phi_{\eta\eta}\phi_\eta + \phi_{\eta\eta\eta}\phi_{\xi\xi}\phi_\xi)$$

$$-126\phi_{\xi\eta}\phi_\xi^3\phi_\eta^3(\phi_{\xi\eta\eta}\phi_{\eta\eta}\phi_t^3 + \phi_{\xi\xi\eta}\phi_{\xi\xi}\phi_\eta^3) - 36\phi_\eta^3\phi_\xi^3(\phi_{\xi\eta\eta}\phi_{\xi\xi}^2\phi_\eta^3 + \phi_{\xi\xi\eta}\phi_{\eta\eta}^2\phi_\xi^3)$$

$$+18\phi_\xi^4\phi_\eta^4(\phi_{\xi\eta\eta}\phi_{\xi\xi\xi}\phi_\eta^2 + \phi_{\xi\xi\eta}\phi_{\eta\eta\eta}\phi_\xi^2) - 18\phi_\xi^4\phi_\eta^4(\phi_{\xi\eta\eta}\phi_{\eta\eta\eta}\phi_\xi^2 + \phi_{\xi\xi\eta}\phi_{\xi\xi\xi}\phi_\eta^2)$$

$$+54\phi_{\xi\eta}\phi_\xi^4\phi_\eta^4(\phi_{\xi\eta\eta}\phi_{\xi\xi}\phi_\eta + \phi_{\xi\xi\eta}\phi_{\eta\eta}\phi_\xi) + 54\phi_\xi^2\phi_\eta^2(\phi_{\xi\eta\eta}\phi_{\eta\eta}^2\phi_\xi^5 + \phi_{\xi\xi\eta}\phi_{\xi\xi}^2\phi_\eta^5)$$

$$-24\phi_\xi\phi_\eta(\phi_{\xi\xi\xi}\phi_{\xi\xi}^2\phi_\eta^7 + \phi_{\eta\eta\eta}\phi_{\eta\eta}^2\phi_\xi^7) - 6\phi_\xi^4\phi_\eta^4(\phi_{\xi\xi\xi}\phi_{\eta\eta\eta}\phi_\eta\phi_\xi + \phi_{\xi\xi}^2\phi_{\eta\eta}^2)$$

$$-88\phi_{\xi\eta}\phi_\xi\phi_\eta(\phi_{\eta\eta}^3\phi_\xi^6 + \phi_{\xi\xi}^3\phi_\eta^6) + 3\phi_\xi^2\phi_\eta^2(\phi_{\xi\xi\xi}^2\phi_\eta^6 + \phi_{\eta\eta\eta}^2\phi_\xi^6)$$

$$+6\phi_\xi^4\phi_\eta^4(\phi_{\xi\xi\xi}\phi_{\eta\eta}^2\phi_\xi + \phi_{\eta\eta\eta}\phi_{\xi\xi}^2\phi_\eta) + 20(\phi_{\xi\xi}^4\phi_\eta^8 + \phi_{\eta\eta}^4\phi_\xi^8)$$

$$-4\phi_\xi^3\phi_\eta^3\phi_{\xi\eta}(3\phi_{\xi\xi}\phi_{\eta\eta} + 7\phi_{\eta\xi}^2)(\phi_{\xi\xi}\phi_\eta^2 + \phi_{\eta\eta}\phi_\xi^2)$$

$$+18\phi_\xi^4\phi_\eta^4(\phi_{\xi\xi}\phi_{\eta\eta} + 2\phi_{\xi\eta}^2)(\phi_{\xi\eta\eta}\phi_\xi + \phi_{\xi\xi\eta}\phi_\eta)$$

$$-18\phi_\xi^3\phi_\eta^3(\phi_{\xi\xi}\phi_{\eta\eta} + 2\phi_{\xi\eta}^2)(\phi_{\xi\xi\xi}\phi_\eta^3 + \phi_{\eta\eta\eta}\phi_\xi^3)$$

$$+\phi_\xi^2\phi_\eta^2(17\phi_{\xi\xi}\phi_{\eta\eta} + 111\phi_{\xi\eta}^2)(\phi_{\xi\xi}^2\phi_\eta^4 + \phi_{\eta\eta}^2\phi_\xi^4) - 2\phi_{\xi\eta}^2\phi_\xi^4\phi_\eta^4(2\phi_{\xi\eta}^2 + 15\phi_{\xi\xi}\phi_{\eta\eta}) = 0 \ . \quad (47)$$

If condition (47) is satisfied, then the expansion coefficient $v_4(\xi, \eta)$ is arbitrary. For example (47) cannot be satisfied when we set $\phi(\xi, \eta) = \xi\eta$. Consequently we conclude that (41) does not pass the Painlevé test. If we assume that $\phi_{\xi\eta} = 0$ it follows that

$$3\phi_\xi^2\phi_\eta^2(\phi_{\xi\xi\xi\xi}\phi_\eta^4 + \phi_{\eta\eta\eta\eta}\phi_\xi^4)(\phi_{\xi\xi}\phi_\eta^2 + \phi_{\eta\eta}\phi_\xi^2)$$

$$-24\phi_\xi\phi_\eta(\phi_{\xi\xi\xi}\phi_{\xi\xi}^2\phi_\eta^7 + \phi_{\eta\eta\eta}\phi_{\eta\eta}^2\phi_\xi^7) - 6\phi_\xi^4\phi_\eta^4(\phi_{\xi\xi\xi}\phi_{\eta\eta\eta}\phi_\eta\phi_\xi + \phi_{\xi\xi}^2\phi_{\eta\eta}^2)$$

$$+3\phi_\xi^2\phi_\eta^2(\phi_{\xi\xi\xi}^2\phi_\eta^6 + \phi_{\eta\eta\eta}^2\phi_\xi^6) + 6\phi_\xi^4\phi_\eta^4(\phi_{\xi\xi\xi}\phi_{\eta\eta}^2\phi_\xi + \phi_{\eta\eta\eta}\phi_{\xi\xi}^2\phi_\eta)$$

$$+20(\phi_{\xi\xi}^4\phi_\eta^8 + \phi_{\eta\eta}^4\phi_\xi^8) - 18\phi_{\xi\xi}\phi_{\eta\eta}\phi_\xi^3\phi_\eta^3(\phi_{\xi\xi\xi}\phi_\eta^3 + \phi_{\eta\eta\eta}\phi_\xi^3)$$

$$+17\phi_{\xi\xi}\phi_{\eta\eta}\phi_\xi^2\phi_\eta^2(\phi_{\xi\xi}^2\phi_\eta^4 + \phi_{\eta\eta}^2\phi_\xi^4) = 0. \tag{48}$$

We now consider the reduced singularity manifold $\phi(\eta, \xi) = \eta - g(\xi) = 0$. Then $\phi_{\xi\eta} = 0$ and $\phi_{\eta\eta} = 0$. Equation (41) reduces to a condition on $g(\xi)$ given by

$$3(g')^2 g'' g^{(4)} + 3(g')^2(g^{(3)})^2 - 24g'(g'')^2 g^{(3)} + 20(g'')^4 = 0 \tag{49}$$

where $g' \equiv dg/d\xi$ and $g^{(4)} \equiv d^4g/d\xi^4$. For $A = (\ln g')'$ equation (47) reduces to

$$3AA'' + 3(A')^2 - 9A^2A' + 2A^4 = 0 . \tag{50}$$

We now perform a Painlevé test on (50). Inserting the ansatz $A(\xi) \propto A_0\xi^n$ we find $n = -1$ and A_0 admits two solutions, namely $A_0 = -3/2$ and $A_0 = -3$. All terms are dominant. Thus (50) admits two branches in the Painlevé analysis. For the branch with $A_0 = -3/2$ the resonances are given by $r_1 = -1$ and $r_2 = 3/2$. Equation (49) admits an expansion of the form

$$A(\xi) = (\xi - \xi_1)^{-1}\sum_{j=0}^{\infty} A_j(\xi - \xi_1)^{j/2}$$

where at the resonance $r = 3/2$ the expansion coefficient is arbitrary. For the second branch we find the resonances $r_1 = -1$ and $r_2 = -3$. It follows that (50) passes the so-called weak Painlevé test. Owing to the two branches of equation (50) and since all terms are dominant in (50) we find two special solutions for equation (49) given by

$$g(\xi) = -2\xi^{-1/2}, \qquad g(\xi) = -\frac{1}{2}\xi^{-2}$$

so that

$$\phi(\xi, \eta) = \eta + 2\xi^{-1/2}, \qquad \phi(\xi, \eta) = \eta + \frac{1}{2}\xi^{-2}$$

satisfy conditions (47) and (53).

Equation (41) admits the Lie symmetry vector fields

$$\left\{ \frac{\partial}{\partial\xi},\ \frac{\partial}{\partial\eta},\ -\xi\frac{\partial}{\partial\xi} + \eta\frac{\partial}{\partial\eta},\ -\xi\frac{\partial}{\partial\xi} - \eta\frac{\partial}{\partial\eta} + u\frac{\partial}{\partial u} \right\}.$$

The first two Lie symmetry vector fields are related to the fact that (41) does not depend explicitly on η and ξ. The third Lie symmetry vector field is related to the

Lorentz transformation and the fourth is related to the scale invariance of (41), i.e., $\eta \to \epsilon^{-1}\eta$, $\xi \to \epsilon^{-1}\xi$, $u \to \epsilon u$. No Lie Bäcklund vector fields can be found for (41).

The Lie symmetry vector fields $\partial/\partial\xi$, $\partial/\partial\eta$ lead to the similarity ansatz

$$v(\xi, \eta) = f(\varsigma) \tag{51}$$

where the similarity variable ς is given by

$$\varsigma = c_1\xi + c_2\eta. \tag{52}$$

c_1 and c_2 are non-zero constants. Inserting (51) into (41) yields

$$\frac{d^2f}{d\varsigma^2} = \frac{1}{c_1 c_2}f^3. \tag{53}$$

Equation (53) passes the Painlevé test. This is in agreement with the fact that

$$\phi(\xi, \eta) = c_1\eta + c_2\xi$$

satisfies equation (47). Equation (53) can be solved in terms of Jacobi elliptic functions.

The Lie symmetry vector field $-\xi\partial/\partial\xi + \eta\partial/\partial\eta$ leads to the similarity ansatz

$$v(\xi, \eta) = f(\varsigma) \tag{54}$$

where the similarity variable ς is given by

$$\varsigma = \eta\xi. \tag{55}$$

Inserting (54) into (41) yields

$$\frac{d^2f}{d\varsigma^2} + \frac{1}{\varsigma}\frac{df}{d\varsigma} - \frac{1}{\varsigma}f^3 = 0. \tag{56}$$

Equation (56) does not pass the Painlevé test. This is in agreement with the fact that

$$\phi(\xi, \eta) = \eta\xi \tag{57}$$

does not satisfy condition (47).

The Lie symmetry vector field $-\xi\partial/\partial\xi - \eta\partial/\partial\eta + u\partial/\partial u$ leads to the similarity ansatz

$$v(\xi, \eta) = \frac{1}{\xi}f(\varsigma) \tag{58}$$

where the similarity variable ς is given by

$$\varsigma = \frac{\eta}{\xi}. \tag{59}$$

Inserting (57) into (41) yields

$$\frac{d^2 f}{d\varsigma^2} + \frac{2}{\varsigma}\frac{df}{d\varsigma} + \frac{1}{\varsigma}f^3 = 0. \tag{60}$$

Equation (60) passes the Painlevé test. This is in agreement with the fact that

$$\phi(\xi, \eta) = \frac{\eta}{\xi} \tag{61}$$

satisfies (47). From the Painlevé test we find a particular solution to (60)

$$f(\varsigma) = \frac{\sqrt{-2\varsigma_1}}{\varsigma - \varsigma_1}.$$

Let us discuss the truncated ansatz (57). Inserting the truncated ansatz into (54) yields

$$v_0^2 = 2\phi_\eta \phi_\xi \tag{62}$$

$$-\phi_{\xi\eta}v_0 - \phi_\xi v_{0\eta} - \phi_\eta v_{0\xi} = 3v_0^2 v_1 \tag{63}$$

$$v_{0\xi\eta} = 3v_0 v_1^2 \tag{64}$$

$$v_{1\xi\eta} = v_1^3. \tag{65}$$

It follows that

$$-4\phi_\xi \phi_\eta \phi_{\eta\xi} - \phi_\eta^2 \phi_{\xi\xi} - \phi_\xi^2 \phi_{\eta\eta} = 6v_0 v_1 \phi_\xi \phi_\eta \tag{66}$$

and

$$12\phi_\xi^2 \phi_\eta^2 v_1^2 = \phi_\eta \phi_\xi \phi_{\eta\eta}\phi_{\xi\xi} - \phi_\eta^2 \phi_{\eta\xi}\phi_{\xi\xi} - \phi_\xi^2 \phi_{\eta\xi}\phi_{\eta\eta} + 2\phi_\xi \phi_\eta^2 \phi_{\eta\xi\xi} + 2\phi_\xi^2 \phi_\eta \phi_{\eta\eta\xi} + \phi_\xi \phi_\eta \phi_{\eta\xi}^2 \tag{67}$$

where v_0 is given by (62). We see that the truncated expansion (44) leads to a different condition on ϕ compared with condition (47). If $\phi_{\eta\xi} = 0$, it follows that

$$-\phi_\eta^2 \phi_{\xi\xi} - \phi_\xi^2 \phi_{\eta\eta} = 6v_0 v_1 \phi_\xi \phi_\eta \tag{68}$$

and

$$12\phi_\xi^2 \phi_\eta^2 v_1^2 = \phi_\eta \phi_\xi \phi_{\eta\eta}\phi_{\xi\xi}. \tag{69}$$

Here, too, the condition on ϕ is different from condition (48). Solutions can be constructed when we insert a solution of (65). The simplest case is $v_1 = 0$. Then from (66), (67) and (44) it follows that

$$v(\eta, \xi) = \frac{\sqrt{-2c_1 c_2}}{c_1 \xi + c_2 \eta} \tag{70}$$

is a special solution to (41). Whether solution (70) can be used to construct another solution with the help of the Bäcklund transformation (44) and the conditions (66) and (67) is not obvious since we have to prove that (66) and (67) are compatible.

17.8 Integrability Concepts

In literature there are different conecpts of integrability. Here we list some of them.

1. Integrability through linearization, (sometimes called C-integrability) means that the nonlinear equation may be linearized through a local transformation. The standard prototype is Burgers equation which becomes linear by the Cole-Hopf transformation. Integrability through linearization is close to the original and somehow vague concept of an integrable dynamical system as a system for which one can find a suitable transformation to another system with known solutions.

2. Liouville integrability. Liouville's theorem relates integrability to the existence of constants or invariants or integrals of the motion for Hamiltonian systems. Integrals of motion can also be obtained for dissipative systems such as the Lorenz equation.

3. Poincaré integrability demands that the integrating transformation be analytic so that pertubation methods provide actual integration. Poincaré integrability extended to infinite systems is the starting point,

4. Normal form integrability is a generalization of Poincaré's idea referring to the possibility of reducing an equation to the simplest possible form.

5. Painlevé integrability refers to local analytic properties of differential equations (singularity structure). This notion has the advantage of allowing investigations to be made algorithmically (Painlevé test) and of providing criteria of "partial integrability". The Painlevé test was discussed in detail in this chapter.

6. IST-solvability of partial differential equation's, (sometimes called S-integrability) corresponds to the fact that the system is associated with a linear eigenvalue (scattering) problem and that it can be "linearized" through an inverse scattering transform.

7. Lax integrability of partial differential equation's, is closely related to the concepts 5, 6, 8 and refers to the possibility of expressing the equation as the consistency condition for a suitable pair of linear equations (Lax pair).

8. N-soliton integrability of partial differential equation's, $N = 1, 2, \ldots$ refers to the existence of multiple parameter families of special solutions (multisoliton solutions). This property is closely related to algebraic properties of the equation, such as the existence of an infinite sequence of conserved quantities and can also be tested (to some extent) in the framework of Hirota's bilinear forms. This criteria is referred to in contributions. The Hirota technique is discussed in detail in Steeb and Euler [126].

17.9   Computer Algebra Applications

We show that

$$\frac{d^2u}{dx^2} = u^3$$

passes the Painlevé test.

```
%Painleve test for u'' - u^3;
%Program name: Painlev.our;
linelength 72;
off nat$
out painlevl$

operator a;
for n:=0 step 1 until 8 sum (x**(n-1))*a(n);
u := ws;
b := df(u,x,2) - u*u*u;
c := num(b);
d0 := coeffn(c,x,0);
second(solve({d0},{a(0)}));
a(0) := part(ws,2);
d1 := coeffn(c,x,1);
solve({d1},{a(1)});
part(ws,1);
a(1) := part(ws,2);
d2 := coeffn(c,x,2);
solve({d2},{a(2)});
part(ws,1);
a(2) := part(ws,2);
d3 := coeffn(c,x,3);
solve({d3},{a(3)});
part(ws,1);
a(3) := part(ws,2);
d4 := coeffn(c,x,4);
solve({d4},{a(4)});
part(ws,1);
a(4) := part(ws,2);

write ";end"$
shut painlevl$
on nat$
```

Chapter 18

Lie Algebra Valued Differential Forms

18.1 Introduction

In chapter 7 we introduced differential forms. In this chapter we extend the differential forms over $\mathbf{R}$ or $\mathbf{C}$ to Lie algebra valued differential forms which play an important role in the derivation of the Yang-Mills and self-dual Yang-Mills equations. The Yang-Mills equations are an extension of Maxwell's equations. Whereas Maxwell's equations are linear the Yang-Mills equations are nonlinear. From the self-dual Yang-Mills equations we can derive soliton equations via exact reductions. We also show that a system of ordinary diffferential equation with chaotic behaviour can be derived from the Yang-Mills equations.

Let M be a C^∞ finite-dimensional oriented pseudo-Riemannian manifold of dimension m and pseudometric signature (k, q), $k + q = m$. The metric tensor field is denoted by g. Let us recall the notations that were previously introduced

$\alpha \wedge \beta$,   exterior product of the differential forms α and β
$d\alpha$,   exterior derivative of the differential form α
$Z \lrcorner \alpha$,   interior product (contraction)
$L_Z \alpha$,   Lie derivative of the differential form α with respect to the vector field Z
$f^* \alpha$,   pull back of a differential form α by a map f
$*\alpha$,   Hodge star operation
$[X, Y]$,   commutator of two vector fields .

Denote by $\bigwedge_p T^* M\big|_x$ the set of all C^∞ p-differential forms on M for each $p = 0, 1, \ldots, m$. Let L be a finite dimensional Lie algebra ($\dim L = n$) over the real field.

Definition 18.1 *A* **Lie algebra valued p-differential form** *on M is an element of the tensor product $\bigwedge_p T^*M\big|_{\mathbf{x}} \otimes L$. If $\{X_i, \ i = 1, \dots, n\}$ is a basis for L, then a Lie algebra valued p-differential form $\tilde{\alpha}$ can be written as*

$$\tilde{\alpha} := \sum_{i=1}^{n} \alpha_i \otimes X_i \tag{1}$$

where $\alpha_i \in \bigwedge_p T^*M\big|_{\mathbf{x}}$.

Throughout the Lie algebra valued differential forms are denoted by Greek letters with a tilde. Real valued differential forms are denoted by Greek letters without a tilde.

The actions of the Hodge operator $*$ and the exterior derivative d for a Lie algebra valued differential form may be consistently defined by

$$*\tilde{\alpha} := \sum_{i=1}^{n} (*\alpha_i) \otimes X_i \tag{2}$$

$$d\tilde{\alpha} := \sum_{i=1}^{n} (d\alpha_i) \otimes X_i \tag{3}$$

for any choice of a basis $X_1, \dots, X_n$ for L. Moreover, we define consistently

$$Z \lrcorner \tilde{\alpha} := \sum_{i=1}^{n} (Z \lrcorner \alpha_i) \otimes X_i \tag{4}$$

$$f^*\tilde{\alpha} := \sum_{i=1}^{n} (f^*\alpha_i) \otimes X_i \tag{5}$$

$$L_Z\tilde{\alpha} := \sum_{i=1}^{n} (L_Z\alpha_i) \otimes X_i \tag{6}$$

$$\beta \wedge \tilde{\alpha} := \sum_{i=1}^{n} (\beta \wedge \alpha_i) \otimes X_i \tag{7}$$

$$d(\beta \wedge \tilde{\alpha}) := (d\beta) \wedge \tilde{\alpha} + (-1)^p \beta \wedge d\tilde{\alpha} \quad (\beta : p - \text{form}). \tag{8}$$

As a consequence of the definition (3) we find that

$$dd\tilde{\alpha} = 0. \tag{9}$$

In the following let

$$\tilde{\alpha} := \sum_{i=1}^{n} \alpha_i \otimes X_i$$

and

$$\tilde{\beta} := \sum_{j=1}^{n} \beta_j \otimes X_j$$

be two Lie algebra valued differential forms, where α_i $(i = 1, \ldots, n)$ are p-forms and β_j $(j = 1, \ldots, n)$ are q-forms.

Definition 18.2 *The bracket* [,] *of Lie algebra valued differential forms* $\tilde{\alpha}$ *and* $\tilde{\beta}$ *is defined as*

$$[\tilde{\alpha}, \tilde{\beta}] := \sum_{i=1}^{n} \sum_{j=1}^{n} (\alpha_i \wedge \beta_j) \otimes [X_i, X_j]. \tag{10}$$

From the definition it follows that the bracket [,] has the following properties

$$[\tilde{\alpha}, \tilde{\beta} + \tilde{\delta}] = [\tilde{\alpha}, \tilde{\beta}] + [\tilde{\alpha}, \tilde{\delta}]$$

$$[\tilde{\alpha}, \tilde{\beta}] = (-1)^{pq+1} [\tilde{\beta}, \tilde{\alpha}] \tag{11}$$

$$(-1)^{pr} [\tilde{\alpha}, [\tilde{\beta}, \tilde{\gamma}]] + (-1)^{pq} [\tilde{\beta}, [\tilde{\gamma}, \tilde{\alpha}]] + (-1)^{rq} [\tilde{\gamma}, [\tilde{\alpha}, \tilde{\beta}]] = 0 \tag{12}$$

where $\tilde{\gamma}$ is a Lie algebra valued r-differential form and $\tilde{\delta}$ is a Lie algebra valued q-form. Property (11) is called the Z_2-graded anticommutativity law and property (12) is the Jacobi identity.

As a consequence of the definition of [,] and $*$ we find

$$[\tilde{\alpha}, *\tilde{\alpha}] = 0. \tag{13}$$

As a consequence of the Jacobi identity we find that

$$[\tilde{\alpha}, [\tilde{\alpha}, \tilde{\alpha}]] = 0. \tag{14}$$

Moreover, we obtain

$$f^*[\tilde{\alpha}, \tilde{\beta}] = [f^*\tilde{\alpha}, f^*\tilde{\beta}] \tag{15}$$

$$L_Z[\tilde{\alpha}, \tilde{\beta}] = [\tilde{\alpha}, L_Z\tilde{\beta}] + [L_Z\tilde{\alpha}, \tilde{\beta}] \tag{16}$$

$$d[\tilde{\alpha}, \tilde{\beta}] = [d\tilde{\alpha}, \tilde{\beta}] + (-1)^p [\tilde{\alpha}, d\tilde{\beta}]. \tag{17}$$

18.2   Covariant Exterior Derivative

Definition 18.3 *Let $\tilde{\alpha}$ be a Lie algebra valued one-form and let $\tilde{\beta}$ be a Lie algebra valued p-form. The* **covariant exterior derivative** *of a Lie algebra valued p-form $\tilde{\beta}$ with respect to a Lie algebra valued one-form $\tilde{\alpha}$ is defined as*

$$D_{\tilde{\alpha}}\tilde{\beta} := d\tilde{\beta} - g[\tilde{\alpha}, \tilde{\beta}] \tag{18}$$

where

$$g := \begin{cases} -1 & p \ \text{even} \\ -\frac{1}{2} & p \ \text{odd} \end{cases} \tag{19}$$

Thus $D_{\tilde{\alpha}}\tilde{\beta}$ is a Lie algebra valued $(p+1)$-form. In particular, we have

$$D_{\tilde{\alpha}}\tilde{\alpha} = d\tilde{\alpha} + \frac{1}{2}[\tilde{\alpha}, \tilde{\alpha}] \tag{20}$$

and

$$D_{\tilde{\alpha}}(D_{\tilde{\alpha}}\tilde{\alpha}) = 0. \tag{21}$$

This equation is called the **Bianchi identity**. The proof is as follows:

$$
\begin{aligned}
D_{\tilde{\alpha}}(D_{\tilde{\alpha}}\tilde{\alpha}) &= D_{\tilde{\alpha}}(d\tilde{\alpha} + \frac{1}{2}[\tilde{\alpha}, \tilde{\alpha}]) \\
&= d(d\tilde{\alpha} + \frac{1}{2}[\tilde{\alpha}, \tilde{\alpha}]) + [\tilde{\alpha}, d\tilde{\alpha} + \frac{1}{2}[\tilde{\alpha}, \tilde{\alpha}]] \\
&= \frac{1}{2}d[\tilde{\alpha}, \tilde{\alpha}] + [\tilde{\alpha}, d\tilde{\alpha}] \\
&= \frac{1}{2}[d\tilde{\alpha}, \tilde{\alpha}] - \frac{1}{2}[\tilde{\alpha}, d\tilde{\alpha}] + [\tilde{\alpha}, d\tilde{\alpha}].
\end{aligned}
$$

Since

$$[d\tilde{\alpha}, \tilde{\alpha}] = -[\tilde{\alpha}, d\tilde{\alpha}]$$

which is consequence of (12) it follows that $D_{\tilde{\alpha}}(D_{\tilde{\alpha}}\tilde{\alpha}) = 0$.   ♠

Theorem 18.1 *Let $\tilde{\alpha}$ be a Lie algebra valued one-form. Then*

$$d(D_{\tilde{\alpha}}\tilde{\alpha}) = \frac{1}{2}([D_{\tilde{\alpha}}, \tilde{\alpha}] - [\tilde{\alpha}, D_{\tilde{\alpha}}\tilde{\alpha}]). \tag{22}$$

Proof: Since

$$dd\tilde{\alpha} = 0$$

it follows from (20) that

$$d(D_{\widetilde{\alpha}}\widetilde{\alpha}) = \frac{1}{2}d[\widetilde{\alpha}, \widetilde{\alpha}] = \frac{1}{2}[d\widetilde{\alpha}, \widetilde{\alpha}] - \frac{1}{2}[\widetilde{\alpha}, d\widetilde{\alpha}].$$

Owing to

$$d\widetilde{\alpha} = D_{\widetilde{\alpha}}\widetilde{\alpha} - \frac{1}{2}[\widetilde{\alpha}, \widetilde{\alpha}] \tag{23}$$

and

$$[\widetilde{\alpha}, [\widetilde{\alpha}, \widetilde{\alpha}]] = 0$$

we obtain the identity. ♠

Example: Let

$$\widetilde{\alpha} = \frac{dx_1}{x_1} \otimes \begin{pmatrix} 1 & 0 \\ 0 & 0 \end{pmatrix} + \frac{dx_2}{x_1} \otimes \begin{pmatrix} 0 & 1 \\ 0 & 0 \end{pmatrix}$$

where $0 < x_1 < \infty$, $-\infty < x_2 < \infty$ and

$$g = -\frac{1}{2}.$$

Then

$$d\widetilde{\alpha} = -\frac{dx_1 \wedge dx_2}{x_1^2} \otimes \begin{pmatrix} 0 & 1 \\ 0 & 0 \end{pmatrix}$$

and

$$[\widetilde{\alpha}, \widetilde{\alpha}] = \frac{2dx_1 \wedge dx_2}{x_1^2} \otimes \begin{pmatrix} 0 & 1 \\ 0 & 0 \end{pmatrix}.$$

Consequently

$$D_{\widetilde{\alpha}}\widetilde{\alpha} = d\widetilde{\alpha} + \frac{1}{2}[\widetilde{\alpha}, \widetilde{\alpha}] = 0.$$

We used that

$$dx_1 \wedge dx_1 = 0, \qquad dx_2 \wedge dx_2 = 0.$$

♣

18.3   Yang-Mills and Self-Dual Yang-Mills Equations

Definition 18.4 *Let $\tilde{\alpha}$ be a Lie algebra valued one-form. With the definition*

$$\tilde{\beta} := D_{\tilde{\alpha}}\tilde{\alpha} \tag{24}$$

the **Yang-Mills equations** *are given by*

$$D_{\tilde{\alpha}}(*\tilde{\beta}) = 0. \tag{25}$$

Thus the Yang-Mills equations can be written as

$$D_{\tilde{\alpha}}(*D_{\tilde{\alpha}}\tilde{\alpha}) = 0. \tag{26}$$

Remark: $\tilde{\alpha}$ *is called the* **connection** *and $\tilde{\beta}$ the* **curvature form***. In physics $\tilde{\alpha}$ is referred to as the vector potential and $\tilde{\beta}$ the field strength tensor.*

Definition 18.5 *Let $\tilde{\alpha}$ be a Lie algebra valued one-form. The* **self-dual Yang-Mills equations** *are given by*

$$*D_{\tilde{\alpha}}\tilde{\alpha} = D_{\tilde{\alpha}}\tilde{\alpha}. \tag{27}$$

Besides the bracket $[\tilde{\beta}, \tilde{\gamma}]$ of two Lie algebra valued forms $\tilde{\beta}$ and $\tilde{\gamma}$, we introduce a further product of two Lie algebra valued forms. Let the Lie algebra be represented by $r \times r$ matrices. Let

$$\tilde{\beta} = \sum_{i=1}^{n} \beta_i \otimes X_i \qquad (\beta_i : \ p\text{-forms})$$

and

$$\tilde{\gamma} = \sum_{j=1}^{n} \gamma_j \otimes X_j \qquad (\gamma_j : \ q\text{- forms}).$$

We say that $\tilde{\beta}$, $\tilde{\gamma}$ are **matrix valued differential forms**.

Remark: *If X is an $r \times r$ matrix over* **R** *given by*

$$X = \begin{pmatrix} X_{11} & \cdots & X_{1r} \\ \vdots & \ddots & \vdots \\ X_{r1} & \cdots & X_{rr} \end{pmatrix}$$

and α a differential form. Then one also finds the notation

$$\begin{pmatrix} X_{11}\alpha & \cdots & X_{1r}\alpha \\ \vdots & \ddots & \vdots \\ X_{r1}\alpha & \cdots & X_{rr}\alpha \end{pmatrix}$$

instead of $\alpha \otimes X$.

We define

$$\tilde{\beta} \wedge \tilde{\gamma} := \sum_{i=1}^{n} \sum_{j=1}^{n} (\beta_i \wedge \gamma_j) \otimes (X_i X_j) \tag{28}$$

where $X_i X_j$ is the usual matrix product of X_i and X_j. Obviously $\tilde{\beta} \wedge \tilde{\gamma}$ is a matrix valued $(p+q)$-form. The product $\tilde{\beta} \wedge \tilde{\gamma}$ has the following properties

$$(\tilde{\beta}_1 + \tilde{\beta}_2) \wedge (\tilde{\gamma}_1 + \tilde{\gamma}_2) = \tilde{\beta}_1 \wedge \tilde{\gamma}_1 + \tilde{\beta}_1 \wedge \tilde{\gamma}_2 + \tilde{\beta}_2 \wedge \tilde{\gamma}_1 + \tilde{\beta}_2 \wedge \tilde{\gamma}_2$$

$$d(\tilde{\beta} \wedge \tilde{\gamma}) = (d\tilde{\beta}) \wedge \tilde{\gamma} + (-1)^p (\tilde{\beta} \wedge d\tilde{\gamma}).$$

Remark: *Let $\tilde{\alpha}$ be a Lie algebra valued differential one-form. Then*

$$\tilde{\alpha} \wedge \tilde{\alpha} = \sum_{i=1}^{n} \sum_{j=1}^{n} (\alpha_i \wedge \alpha_j) \otimes (X_i X_j).$$

Since $\alpha_i \wedge \alpha_j = -\alpha_j \wedge \alpha_i$ we find

$$\tilde{\alpha} \wedge \tilde{\alpha} = \sum_{i<j}^{n} (\alpha_i \wedge \alpha_j) \otimes [X_i, X_j].$$

Thus the covariant exterior derivative given by (20) can also be written as

$$D_{\tilde{\alpha}} \tilde{\alpha} := d\tilde{\alpha} + \tilde{\alpha} \wedge \tilde{\alpha}. \tag{29}$$

Let $X \in GL(n, \mathbf{R})$. Let $\Omega = X^{-1} dX$. Then

$$d\Omega + \Omega \wedge \Omega = 0$$

where we used the relation $X X^{-1} = I$ (I unit matrix) and $dI = 0$. Let

$$\Omega \to \Omega' = U^{-1} dU + U^{-1} \Omega U$$

where $\det U = 1$. Then $d\Omega' + \Omega' \wedge \Omega' = 0$. This is called the **gauge invariance**.

Example: Let

$$\Omega = \begin{pmatrix} \dfrac{dx_1}{x_1} & \dfrac{dx_2}{x_1} \\ 0 & 0 \end{pmatrix}$$

where $0 < x_1 < \infty$ and $-\infty < x_2 < \infty$. It follows that

$$d\Omega = \begin{pmatrix} 0 & -\dfrac{dx_1 \wedge dx_2}{x_1^2} \\ 0 & 0 \end{pmatrix}$$

and

$$\Omega \wedge \Omega = \begin{pmatrix} \dfrac{dx_1}{x_1} & \dfrac{dx_2}{x_1} \\ 0 & 0 \end{pmatrix} \wedge \begin{pmatrix} \dfrac{dx_1}{x_1} & \dfrac{dx_2}{x_1} \\ 0 & 0 \end{pmatrix} = \begin{pmatrix} 0 & \dfrac{dx_1 \wedge dx_2}{x_1^2} \\ 0 & 0 \end{pmatrix}.$$

Consequently $d\Omega + \Omega \wedge \Omega = 0$. The matrix X which satisfies the relation $\Omega = X^{-1}(dX)$ is given by

$$X = \begin{pmatrix} x_1 & x_2 \\ 1 & 0 \end{pmatrix}$$

since

$$X^{-1} = \frac{1}{x_1} \begin{pmatrix} 1 & -x_2 \\ 0 & x_1 \end{pmatrix}$$

$$dX = \begin{pmatrix} dx_1 & dx_2 \\ 0 & 0 \end{pmatrix}$$

it follows that

$$X^{-1}(dX) = \begin{pmatrix} \dfrac{1}{x_1} & -\dfrac{x_2}{x_1} \\ 0 & 1 \end{pmatrix} \begin{pmatrix} dx_1 & dx_2 \\ 0 & 0 \end{pmatrix} = \begin{pmatrix} \dfrac{dx_1}{x_1} & \dfrac{dx_2}{x_1} \\ 0 & 0 \end{pmatrix} = \Omega. \qquad \clubsuit$$

In physics the matrix valued n-form

$$\tilde{\alpha} \wedge (*\tilde{\alpha}) \tag{30}$$

is important. Let

$$\tilde{\alpha} = \sum_{i=1}^{n} \alpha_i \otimes X_i$$

be an arbitrary matrix valued p-form. Then

$$\tilde{\alpha} \wedge (*\tilde{\alpha}) = \sum_{i=1}^{n} \sum_{j=1}^{n} (\alpha_i \wedge *\alpha_j) \otimes (X_i X_j). \tag{31}$$

Let $\tilde{\alpha}$ be a Lie algebra valued p-form

$$\tilde{\alpha} = \sum_{i=1}^{n} \alpha_i \otimes X_i$$

where X_i $(i = 1, \ldots, n)$ are $r \times r$ matrices. Then we define the trace of a Lie algebra valued p-form as follows:

$$\operatorname{tr}\tilde{\alpha} := \sum_{i=1}^{n} \alpha_i(\operatorname{tr}X_i). \tag{32}$$

A special sort of Lie algebra valued differential forms is important. Let G be a Lie group whose Lie algebra is L. L is identified with the left invariant vector fields on G. Now suppose that $X_1, \ldots, X_n$ is a basis of L and that $\omega_1, \ldots, \omega_n$ is a dual basis of left invariant one-forms. There is a natural Lie algebra valued one-form $\tilde{\omega}$ on G which can be written as

$$\tilde{\omega} = \sum_{i=1}^{n} \omega_i \otimes X_i,$$

where

$$X_i \lrcorner \omega_j = \delta_{ij}.$$

It can be shown that

$$D_{\tilde{\omega}}\tilde{\omega} = 0 \tag{33}$$

or

$$d\tilde{\omega} + \frac{1}{2}[\tilde{\omega}, \tilde{\omega}] = 0. \tag{34}$$

This equation is called the **Maurer-Cartan equation**. Since

$$[X_i, X_j] = \sum_{k=1}^{n} C_{ij}^k X_k \tag{35}$$

where C_{ij}^k are the structure constants $(C_{ij}^k \in \mathbf{R})$ and

$$[\tilde{\omega}, \tilde{\omega}] = \sum_{i=1}^{n} \sum_{j=1}^{n} (\omega_i \wedge \omega_j) \otimes [X_i, X_j] \tag{36}$$

equation (34) can be written as

$$d\omega_k = -\frac{1}{2} \sum_{i=1}^{n} \sum_{j=1}^{n} C_{ij}^k \omega_i \wedge \omega_j \tag{37}$$

for all $k = 1, \ldots, n$.

Let $\omega_1, \ldots, \omega_r$ $(r < n)$ be one-forms defined on a neighbourhood U of the origin in $\mathbf{R}^n$ and assume that one-forms ω_i are linearly independent at each point of U. We put

$$\Omega = \omega_1 \wedge \ldots \wedge \omega_r.$$

The system

$$\omega_1 = 0, \quad \ldots \quad , \omega_r = 0$$

is called **completely integrable** if it fulfils any of the conditions of the following lemma.

Lemma *The following conditions are equivalent*

(i) *There exist one-forms θ_{ij} satisfying*

$$d\omega_i = \sum_{j=1}^{r} \theta_{ij} \wedge \omega_j.$$

(ii) $d\omega_i \wedge \Omega = 0$ \qquad $(i = 1, \ldots, r)$.

(iii) *There exist a one-form λ satisfying $d\Omega = \lambda \wedge \Omega$.*

For the proof we refer to Choquet-Bruhat *et al* [17] and Flanders [46].

We now apply the **Frobenius integration theorem**.

Theorem 18.2 *Let ω_i $(i = 1, \ldots, r)$ $(r \leq n)$ be one-forms defined on a neighbourhood U of the origin 0 in $\mathbf{R}^n$. Assume that the one-forms ω_i are linearly independent at each point of U and that there are r^2 one-forms θ_{ij} $(j = 1, \ldots, r)$ on U such that*

$$d\omega_i = \sum_{j=1}^{r} \theta_{ij} \wedge \omega_j. \tag{38}$$

Then there is a neighbourhood V of 0 contained in U, r-functions g_k defined on V, and an $r \times r$ matrix (f_{ik}) whose entries are r^2 functions defined on V such that $\det(f_{ik}) \neq 0$ at each point of V, and

$$\omega_i = \sum_{k=1}^{r} f_{ik} dg_k \tag{39}$$

holds on V.

For the proof we refer to Choquet-Bruhat *et al* [17] and Flanders [46].

In connection with gauge theory we need the following

Theorem 18.3 *Let $\tilde{\alpha}$ be a Lie algebra valued differential form. Let the Lie algebra be represented by $r \times r$ matrices. L is associated with the Lie group G, where G is a Lie subgroup of $GL(n, \mathbf{R})$. Let $U \in C^{\infty}(M, G)$. We define*

$$\tilde{\alpha} \cdot U := U^{-1}\tilde{\alpha}U + U^{-1}dU. \tag{40}$$

If

$$D_{\tilde{\alpha}}(*D_{\tilde{\alpha}}\tilde{\alpha}) = 0$$

then

$$D_{\tilde{\alpha} \cdot U}(*D_{\tilde{\alpha} \cdot U}\tilde{\alpha} \cdot U) = 0. \tag{41}$$

Note that dU is a matrix valued one-form. For the proof we need

$$d(I) = d(UU^{-1}) = (dU)U^{-1} + U(dU^{-1}) = 0$$

where I is the unit matrix. From this theorem it follows that the gauge group is an invariance group of the Yang-Mills equation.

Let us now give some applications of Lie algebra valued differential forms.

Example: Let

$$\tilde{\alpha} = \sum_{j=1}^{n} (a_j(x_1, x_2)\, dx_1 + A_j(x_1, x_2)\, dx_2) \otimes X_j$$

be a Lie algebra valued differential form where $\{X_1, \ldots, X_n\}$ form a basis of a Lie algebra. It follows from (3) that the exterior derivative is given by

$$d\tilde{\alpha} = \sum_{j=1}^{n} \left(-\frac{\partial a_j}{\partial x_2} + \frac{\partial A_j}{\partial x_1} \right) dx_1 \wedge dx_2 \otimes X_j. \tag{42}$$

We calculate the covariant derivative of $\tilde{\alpha}$ to obtain the equation which follows from the condition

$$D_{\tilde{\alpha}}\tilde{\alpha} = 0.$$

We have

$$D_{\tilde{\alpha}}\tilde{\alpha} = \sum_{i=1}^{n} \left[\left(-\frac{\partial a_i}{\partial x_2} + \frac{\partial A_i}{\partial x_1} \right) + \frac{1}{2} \sum_{k=1}^{n} \sum_{j=1}^{n} (a_k A_j - a_j A_k) C^i_{kj} \right] dx_1 \wedge dx_2 \otimes X_i.$$

Since $\{X_i \ : \ (i = 1, \ldots, n)\}$ form a basis of a Lie algebra, the condition

$$D_{\tilde{\alpha}}\tilde{\alpha} = 0$$

yields

$$\left(-\frac{\partial a_i}{\partial x_2} + \frac{\partial A_i}{\partial x_1} \right) + \frac{1}{2} \sum_{k=1}^{n} \sum_{j=1}^{n} (a_k A_j - a_j A_k) C^i_{kj}$$

for $i = 1, \ldots, n$. Since

$$C^i_{kj} = -C^i_{jk}$$

it follows that

$$\left(-\frac{\partial a_i}{\partial x_2} + \frac{\partial A_i}{\partial x_1} \right) + \sum_{k<j}^{n} (a_k A_j - a_j A_k) C^i_{kj} = 0$$

for $i = 1, \ldots, n$. We now consider the case where $n = 3$ and X_1, X_2 and X_3 satisfy the commutation relation

$$[X_1, X_2] = 2X_2, \qquad [X_1, X_3] = 2X_3, \qquad [X_2, X_3] = X_1.$$

We find

$$-\frac{\partial a_1}{\partial x_2} + \frac{\partial A_1}{\partial x_1} + a_2 A_3 - a_3 A_2 = 0 \qquad (43a)$$

$$-\frac{\partial a_2}{\partial x_2} + \frac{\partial A_2}{\partial x_1} + 2(a_1 A_2 - a_2 A_1) = 0 \qquad (43b)$$

$$-\frac{\partial a_3}{\partial x_2} + \frac{\partial A_3}{\partial x_1} - 2(a_1 A_3 - a_3 A_1) = 0. \qquad (43c)$$

A convenient choice of a basis $\{X_1, X_2, X_3\}$ is given by

$$X_1 = \begin{pmatrix} 1 & 0 \\ 0 & -1 \end{pmatrix}, \qquad X_2 = \begin{pmatrix} 0 & 1 \\ 0 & 0 \end{pmatrix}, \qquad X_3 = \begin{pmatrix} 0 & 0 \\ 1 & 0 \end{pmatrix}.$$

Consequently the Lie algebra under consideration is $sl(2, \mathbf{R})$.

We now let

$$a_1 = -\eta, \quad a_2 = \frac{1}{2}\frac{\partial u}{\partial x_1}, \quad a_3 = -\frac{1}{2}\frac{\partial u}{\partial x_1}$$

$$A_1 = -\frac{1}{4\eta}\cos u, \quad A_2 = A_3 = -\frac{1}{4\eta}\sin u$$

where $\eta \neq 0$ is an arbitrary constant. By inserting the above into (43) we obtain the equation of motion

$$\frac{\partial^2 u}{\partial x_1 \partial x_2} = \sin u$$

which is the sine-Gordon equation.   ♣

18.4 Yang-Mills Equation and Chaos

In this section we show how a Hamilton system with chaotic behaviour can be derived from the Yang-Mills equations.

We consider the Lie algebra valued differential one-form

$$\tilde{\alpha} = \sum_{i=1}^{3} \alpha_i \otimes X_i$$

with the Lie algebra $su(2)$ and

$$g = dx_1 \otimes dx_1 + dx_2 \otimes dx_2 + dx_3 \otimes dx_3 - dx_4 \otimes dx_4.$$

The differential one-form α_i is given by

$$\alpha_i = \sum_{j=1}^{4} A_{ij}(\mathbf{x})\, dx_j$$

with $\mathbf{x} = (x_1, \ldots, x_4)$. We choose the basis of $su(2)$ as

$$X_1 = \frac{1}{2}\begin{pmatrix} 0 & -i \\ -i & 0 \end{pmatrix}, \qquad X_2 = \frac{1}{2}\begin{pmatrix} 0 & -1 \\ 1 & 0 \end{pmatrix}, \qquad X_3 = \frac{1}{2}\begin{pmatrix} -i & 0 \\ 0 & i \end{pmatrix}.$$

We now give the explicit form of the Yang-Mills equation for the Lie algebra $su(2)$. The commutation relations of X_1, X_2 and X_3 are given by

$$[X_1, X_2] = X_3, \quad [X_2, X_3] = X_1, \quad [X_3, X_1] = X_2.$$

From

$$D_{\tilde{\alpha}}\tilde{\alpha} = d\tilde{\alpha} + \frac{1}{2}[\tilde{\alpha}, \tilde{\alpha}]$$

$$\frac{1}{2}[\tilde{\alpha}, \tilde{\alpha}] = (\alpha_1 \wedge \alpha_2) \otimes X_3 + (\alpha_2 \wedge \alpha_3) \otimes X_1 + (\alpha_3 \wedge \alpha_1) \otimes X_2$$

$$d\tilde{\alpha} = \sum_{i=1}^{3} d\alpha_i \otimes X_i$$

and

$$\begin{aligned} *(D_{\tilde{\alpha}}\tilde{\alpha}) &= (*d\alpha_1 + *(\alpha_2 \wedge \alpha_3)) \otimes X_1 + (*d\alpha_2 + *(\alpha_3 \wedge \alpha_1)) \otimes X_2 \\ &\quad + (*d\alpha_3 + *(\alpha_1 \wedge \alpha_2)) \otimes X_3 \end{aligned}$$

we obtain

$$D_{\widetilde{\alpha}}(D_{\widetilde{\alpha}}\widetilde{\alpha}) = d(*d\alpha_1 + *(\alpha_2 \wedge \alpha_3)) \otimes X_1$$

$$+ d(*d\alpha_2 + *(\alpha_3 \wedge \alpha_1)) \otimes X_2 + d(*d\alpha_3 + *(\alpha_1 \wedge \alpha_2)) \otimes X_3$$

$$+ (\alpha_2 \wedge (*d\alpha_3 + *(\alpha_1 \wedge \alpha_2)) - \alpha_3 \wedge (*d\alpha_2 + *(\alpha_3 \wedge \alpha_1))) \otimes X_1$$

$$+ (\alpha_3 \wedge (*d\alpha_1 + *(\alpha_2 \wedge \alpha_3)) - \alpha_1 \wedge (*d\alpha_3 + *(\alpha_1 \wedge \alpha_2))) \otimes X_2$$

$$+ (\alpha_1 \wedge (*d\alpha_2 + *(\alpha_3 \wedge \alpha_1)) - \alpha_2 \wedge (*d\alpha_1 + *(\alpha_2 \wedge \alpha_3))) \otimes X_3.$$

From the condition

$$D_{\widetilde{\alpha}}(*D_{\widetilde{\alpha}}\widetilde{\alpha}) = 0$$

it follows that

$$d(*d\alpha_1 + *(\alpha_2 \wedge \alpha_3)) + (\alpha_2 \wedge (*d\alpha_3 + *(\alpha_1 \wedge \alpha_2)) - \alpha_3 \wedge (*d\alpha_2 + *(\alpha_3 \wedge \alpha_1))) = 0$$

$$d(*d\alpha_2 + *(\alpha_3 \wedge \alpha_1)) + (\alpha_3 \wedge (*d\alpha_1 + *(\alpha_2 \wedge \alpha_3)) - \alpha_1 \wedge (*d\alpha_3 + *(\alpha_1 \wedge \alpha_2))) = 0$$

$$d(*d\alpha_3 + *(\alpha_1 \wedge \alpha_2)) + (\alpha_1 \wedge (*d\alpha_2 + *(\alpha_3 \wedge \alpha_1)) - \alpha_2 \wedge (*d\alpha_1 + *(\alpha_2 \wedge \alpha_3))) = 0.$$

Now, $*(\alpha_i \wedge \alpha_j)$ is a two-form and therefore $d * (\alpha_i \wedge \alpha_j)$ is a three form. We thus obtain 12 coupled partial differential equations. Let us now impose the gauge condition

$$A_{j4} = 0, \quad \sum_{i=1}^{3} \frac{\partial A_{ji}}{\partial x_i} = 0, \quad \frac{\partial A_{ji}}{\partial x_k} = 0 \quad (i, k = 1, 2, 3)$$

where $j = 1, 2, 3$. Thus we eliminate the space dependence of the fields. We thus arrive at the autonomous system of second order ordinary differential equations

$$\frac{d^2 A_{ji}}{dx_4^2} + \sum_{k=1}^{3} \sum_{l=1}^{3} (A_{kl} A_{kl} A_{ji} - A_{jl} A_{kl} A_{ki}) = 0$$

together with

$$\sum_{k=1}^{3} \sum_{l=1}^{3} \sum_{m=1}^{3} \epsilon_{jkm} A_{kl} \frac{dA_{ml}}{dx_4} = 0.$$

Further reduction leads to the Hamilton system

$$\frac{d^2 u_1}{dt^2} = -u_1(u_2^2 + u_3^2), \qquad \frac{d^2 u_2}{dt^2} = -u_2(u_1^2 + u_3^2), \qquad \frac{d^2 u_3}{dt^2} = -u_3(u_1^2 + u_2^2).$$

This system is nonintegrable (we can apply Ziglin theorem) and can be derived from the Hamilton function

$$H(\mathbf{u}, \dot{\mathbf{u}}) = \frac{1}{2}(\dot{u}_1^2 + \dot{u}_2^2 + \dot{u}_3^2) + \frac{1}{2}(u_1^2 u_2^2 + u_1^2 u_3^2 + u_2^2 u_3^2).$$

It also shows chaotic behaviour.

18.5 Soliton Equations

A large number of complete integrable (soliton) equations (such as the Korteweg de Vries equation and the nonlinear one-dimensional Schrödinger equation) can be derived from the self-dual Yang-Mills equations. Furthermore ordinary differential equations can be derived which are completely integrable (for example the Euler equation). Thus one can ask the following question: Is the self-dual Yang Mills equation the "master equation" of all complete integrable partial differential equations ? This question has been discussed by Ward [143], Mason and Sparling [82] and Chakravarty and Ablowitz [13]. Here we show that the Korteweg-de Vries equation and the nonlinear Scgrödinger equation can be derived from the self-dual Yang-Mills equations via exact reductions. We also describe an interesting connection with the Yang-Mills equations (Steeb *et al* [130]).

As an example we consider here the symmetry reduction of the self-dual Yang-Mills equation to the Korteweg-de Vries equation and nonlinear Schrödinger equation. Consider the space $\mathbf{R}^4$ with coordinates

$$x^a = (x, y, u, t)$$

and metric tensor field

$$g = dx \otimes dx - dy \otimes dy + du \otimes dt + dt \otimes du$$

and a totally skew orientation tensor

$$\epsilon_{abcd} = \epsilon_{[abcd]}.$$

Let

$$D_a := \partial_a - A_a$$

where the A_a are basis elements of the Lie algebra $sl(2, \mathbf{C})$ and $\partial_a \equiv \partial/\partial x^a$. The A_a are defined up to the gauge transformation

$$A_a \to h A_a h^{-1} - (\partial_a h) h^{-1}$$

where $h = h(x^a) \in SL(2, \mathbf{C})$. Then the self-dual Yang-Mills equation takes the form

$$\frac{1}{2} \epsilon_{ab}^{cd} [D_c, D_d] = [D_a, D_b] \tag{44}$$

where we used summation convention. This is equivalent to the following three commutator equations

$$[D_x + D_y, D_u] = 0 \tag{45a}$$

$$[D_x - D_y, D_x + D_y] + [D_u, D_t] = 0 \tag{45b}$$

$$[D_x - D_y, D_t] = 0 \tag{45c}$$

These equations also follow from the integrability condition of the linear system

$$(D_x - D_y + \lambda D_u)\phi = 0 \tag{46a}$$

$$(D_t + \lambda(D_x + D_y))\phi = 0 \tag{46b}$$

where λ is an affine complex coordinate on the Riemann sphere CP^1 (the so-called spectral parameter) and ϕ is a two component column vector. We set

$$D_x := \partial_x - A, \qquad D_u := \partial_u - B, \qquad D_t := \partial_t - C, \qquad D_y := \partial_y - D.$$

We require that the bundle and its connection possess two commuting symmetries which project to a pair of orthogonal spacetime translations one timelike and one null. These are along $\partial/\partial y$ and $\partial/\partial u$. Next we assume that A, B, C and D are independent of u and y. We also impose the gauge condition

$$A + D = 0.$$

The gauge transformations are now restricted to $SL(2, \mathbb{C})$ valued functions of t alone and which A and B transform by conjugation,

$$B \rightarrow hBh^{-1}.$$

System (45) reduces to

$$\partial_x B = 0, \qquad [\partial - 2A, \partial_t - C] = 0, \qquad 2\partial_x A - [B, C] = \partial_t B. \tag{47}$$

These equations follow from the integrability conditions on the reduction of the linear system

$$(\partial_x - 2A + \lambda B)\phi = 0. \tag{48a}$$

$$(\partial_t - C + \lambda\partial_x)\phi = 0. \tag{48b}$$

When (47a) holds, B depends only on the variable t, so the gauge freedom may be used to reduce B to a normal form. When B vanishes, the equations are trivially satisfied. Thus we assume that B is everywhere non-vanishing. The matrix B then has just two normal forms

$$B = \begin{pmatrix} 0 & 0 \\ 1 & 0 \end{pmatrix} \tag{49a}$$

$$B = \begin{pmatrix} 1 & 0 \\ 0 & 1 \end{pmatrix} \tag{49b}$$

The self-dual Yang-Mills equations are solved with B of (49a) by

$$2A = \begin{pmatrix} q & 1 \\ q_x - q^2 & -q \end{pmatrix}$$

$$2C = \begin{pmatrix} (q_x - q^2)_x & -2q_x \\ 2w & -(q_x - q^2)_x \end{pmatrix}$$

where

$$4w = \frac{\partial^3 q}{\partial x^3} - 4q\frac{\partial q}{\partial x} - 2\left(\frac{\partial q}{\partial x}\right)^2 + rq^2\frac{\partial q}{\partial x}$$

and q satisfies

$$4\frac{\partial q}{\partial t} = \frac{\partial^3 q}{\partial x^3} - 6\left(\frac{\partial q}{\partial x}\right)^2.$$

From

$$u := -\frac{\partial q}{\partial x} = \mathrm{tr}(BC)$$

we obtain the Korteweg-de Vries equation

$$4\frac{\partial u}{\partial t} = \frac{\partial^3 u}{\partial x^3} + 12u\frac{\partial u}{\partial x}.$$

To find the nonlinear Schrödinger equation, we set

$$2A = \begin{pmatrix} 0 & \psi \\ -\bar{\psi} & 0 \end{pmatrix}, \qquad 2\kappa C = \begin{pmatrix} \psi\bar{\psi} & \psi_x \\ \bar{\psi}_x & -\psi\bar{\psi} \end{pmatrix}$$

provided ψ and $\bar{\psi}$ satisfy

$$2\kappa\frac{\partial \psi}{\partial t} = \frac{\partial^2 \psi}{\partial x^2} + 2\psi^2\bar{\psi}, \qquad 2\kappa\frac{\partial \bar{\psi}}{\partial t} = -\frac{\partial^2 \bar{\psi}}{\partial x^2} - 2\bar{\psi}^2\psi$$

and $2\kappa = 1$ or $\kappa = -i$.

From the self-dual Yang Mills equation we also find, after exact reduction, the system of ordinary differential equation (Euler equation)

$$\frac{du_1}{dt} = u_2 u_3 \tag{50a}$$

$$\frac{du_2}{dt} = u_1 u_3 \tag{50b}$$

$$\frac{du_3}{dt} = u_1 u_2. \tag{50c}$$

This is an algebraically completely integrable differential equation. The first integrals are given by

$$I_1 = u_1^2 - u_2^2, \qquad I_2 = u_1^2 - u_3^2.$$

When we differentiate system (50) with respect to t and insert (50) into the new second order equation we arrive at

$$\frac{d^2u_1}{dt^2} = u_1(u_2^2 + u_3^2) \tag{51a}$$

$$\frac{d^2u_2}{dt^2} = u_2(u_1^2 + u_3^2) \tag{51b}$$

$$\frac{d^2u_3}{dt^2} = u_3(u_1^2 + u_2^2). \tag{51c}$$

Using Ziglin's theorem we can show that system (51) is not algebraic complete integrable. Furthermore it does not have the Painlevé property. From the Yang-Mills equation (26) we find, after reduction, the nonintegrable system

$$\frac{d^2u_1}{dt^2} = -u_1(u_2^2 + u_3^2) \tag{52a}$$

$$\frac{d^2u_2}{dt^2} = -u_2(u_1^2 + u_3^2) \tag{52b}$$

$$\frac{d^2u_3}{dt^2} = -u_3(u_1^2 + u_2^2). \tag{52c}$$

System (52) is nonintegrable (again we can apply Ziglin theorem) and can be derived from the Hamilton function

$$H(\mathbf{u}, \dot{\mathbf{u}}) = \frac{1}{2}(\dot{u}_1^2 + \dot{u}_2^2 + \dot{u}_3^2) + \frac{1}{2}(u_1^2u_2^2 + u_1^2u_3^2 + u_2^2u_3^2).$$

It shows chaotic behaviour and therefore is not integrable. Furthermore, we find that it does not pass the Painlevé test. The resonances are given by -1, 1 (twofold), 2 (twofold) and 4. Studying the behaviour at the resonances we find a logarithmic psi-series. For $u_3 = 0$ we find a chaotic system which is not purely ergodic since it exhibits a very small stable island. The total area of the elliptic region is very small. Now the systems (51) and (52) are equivalent up to the sign on the right hand side. When we consider the system in the complex domain for the Painlevé test we can find system (51) from system (52) via the transformation $t \to it$. System (51) has been derived from the self-dual Yang-Mills equation and system (52) from the Yang-Mills equation.   ♣

18.6 Computer Algebra Applications

Exterior differential forms and the corresponding Cartan calculus are increasingly applied in classical mechanics and classical field theory, as well as in gauge theories and in general relativity and its extensions. The Excalc package of REDUCE, which can be loaded according to the pattern

LOAD EXCALC;

was devised by E. Schrüfer exactly for this purpose. The package is able to handle scalar-valued differential forms, vectors and operations between them, as well as non-scalar valued, i.e. indexed forms. We have made extensive use of this package in our research, finding exact solutions in general relativity and gauge theories of gravitation.

A form is declared by specifying its rank and its valence.

PFORM CHRIST1(A,B)=1;

declares CHRIST1 as a 1-form with 2 indices A and B.

PFORM CURV2(A,B)=2;

declares CURV2 as a 2-form. It also has 2 indices. The names of the indices are arbitrary, that is, we could have declared PFORM CURV2(I,J)=2 instead. An ordinary function, a 0-form, must be declared as well:

PFORM PSI=0;

With

FDOMAIN PSI=PSI(R);

the function Ψ is declared to depend on the variable r. Hence @(PSI,THETA) ,@ being the partial differential sign in EXCALC in analogy to the df of REDUCE, will now evaluate to zero.

The exterior differential is denoted by D, and the exterior product sign by the wedge ^. Then, using the declared forms we can, for instance, formulate the following statement:

CURV2(-A,B) := D CHRIST1(-A,B)+CHRIST1(-C,B)^CHRIST1(-A,C);

The negative sign in front of an index signals that it is a subscript, i.e., a *co*-variant index, whereas a positive sign (or no sign at all) marks a superscript, or a *contra*-variant index. In EXCALC the Einstein summation convention applies automatically, with the result that EXCALC sums over repeated indices in different positions, like over -C and C in the example above. We do not need a FOR statement for displaying all components of CURV2(A,B). A, B, and C run over the set of the indices which were made known to the system by means of the declaration

INDEXRANGE T,R,THETA,PHI;

for example, or by the coframe-command.

The specific operators provided in EXCALC are the following

^	exterior multiplication	"nary" infix operator
D	exterior differentiation	unary prefix operator
	partial differentiation	"nary" prefix operator
_|	inner product	binary infix operator
|_	Lie derivative	binary infix operator
#	Hodge star operator	unary prefix operator

Unary means that there is one, binary that there are two, and nary that there is any number of arguments.

Suppose we declare two vectors (tangent vectors) and a 2-form:

TVECTOR V,W; PFORM F=2;

Then the inner product of V and W reads

V _| F;

(the blanks are optional). Then the Lie derivative

W |_ F;

evaluates to W _| D F + D(W _| F). In EXCALC we can also perfom a Lie derivative of a vector V with respect to another vector W, according to

W |_ V;

Similarly to operators in REDUCE, an indexed form like CHRIST1(A,B) can be declared to be antisymmetric under the exchange of its arguments:

ANTISYMMETRIC CHRIST1;

There is a corresponding command for SYMMETRIC.

In order to inform EXCALC of the dimension of the space we are working in, we declare

SPACEDIM 4;

(or any other positive number). Now EXCALC knows the range of the indices involved in the indexed p-forms declared above. If one works in 4-dimensional spacetime manifolds, the COFRAME-statement is enough to declare the dimension of the spacetime, the underlying 1-form basis and the signature of the metric.

Chapter 19

Bose Operators and Lie Algebras

19.1 Embedding and Bose Operators

In this chapter we describe a connection between nonlinear autonomous systems of ordinary differential equations, first integrals, Bose operators and Lie algebras. An extension to nonlinear partial differential equations is given (Steeb and Euler [128], Kowalski and Steeb [71]).

One of the basic tasks in the study of nonlinear dynamical systems is to find out whether or not the dynamical system is integrable. For systems of ordinary differential equations one has to find the first integrals (if any exist) and for partial differential equations one has to find the conservation laws (if any exist). Here we show that these questions can be investigated with the help of Bose operators and Bose field operators.

It is well known that nonlinear autonomous systems of first order ordinary differential equations

$$\frac{d\mathbf{u}}{dt} = V(\mathbf{u}) \tag{1}$$

can be embedded into a linear infinite system. It is assumed that $V : \mathbf{C}^n \mapsto \mathbf{C}^n$ is analytic. The analytic vector fields form a Lie algebra under the commutator.

Steeb [119] and Kowalski and Steeb [71] showed that the infinite system can be expressed with the help of Bose operators. We show that the associated Bose operators form an isomorphic Lie algebra.

Let

$$V := \sum_{j=1}^{n} V_j(\mathbf{u}) \frac{\partial}{\partial u_j} \tag{2}$$

be the corresponding analytic vector field of (1). Let W be the corresponding vector field of the system $d\mathbf{u}/dt = W(\mathbf{u})$. Then we define

$$[V, W] := \sum_{k=1}^{n} \sum_{j=1}^{n} \left(V_j \frac{\partial W_k}{\partial u_j} - W_j \frac{\partial V_k}{\partial u_j} \right) \frac{\partial}{\partial u_k} \cdot \tag{3}$$

It is known that the analytic vector fields form a Lie algebra with this composition (commutator).

We now describe the embedding. Consider a family of linear operators $b_j, b_j^\dagger$, $1 \le j \le n$ on an inner product space V, satisfying the commutation relations

$$[b_j, b_k] = [b_j^\dagger, b_k^\dagger] = 0, \qquad [b_j, b_k^\dagger] = \delta_{jk} I$$

where I is the identity operator. Such linear (unbounded) operators appear in the method of second quantization in quantum mechanics where they are defined on a Hilbert space $\mathcal{H}$. There $b_j^\dagger$ is called a **creation operator** for bosons and its adjoint b_j is called an **annihilation operator**.

The linear space V must be infinite-dimensional for the commutation relations to hold. For, if A and B are $n \times n$ matrices such that $[A, B] = \lambda I$ then $\mathrm{tr}\,([A, B]) = 0$ implies $\lambda = 0$. If V consists of analytic functions in m variables $u_1, \ldots, u_n$, a realization of the commutation relations is provided by the assignment

$$b_j = \frac{\partial}{\partial u_j}, \qquad b_j^\dagger = u_j,$$

It follows directly from the commutation relations that the operators

$$E_{jk} := b_j^\dagger b_k$$

satisfy relations

$$[E_{jk}, E_{hl}] = \delta_{kh} E_{jl} - \delta_{jl} E_{hk}, \quad 1 \le j, k, h, l \le n.$$

One can easily construct representations of each of the classical Lie algebras (see chapter 5) in terms of annihilation and creation operators for bosons. Furthermore one can use the models to decompose V into subspaces transforming irreducibly under these representations.

We define

$$M := \mathbf{b}^\dagger \cdot V(\mathbf{b}) \equiv \sum_{j=1}^{n} b_j^\dagger V_j(\mathbf{b}). \tag{4}$$

Let

$$|\mathbf{u}(t)\rangle = \exp\left(-\frac{1}{2} |\mathbf{u}(t)|^2 \right) \exp\left(\mathbf{u}(t) \cdot \mathbf{b}^\dagger \right) |0\rangle \tag{5}$$

be a **coherent state**, i.e. $\mathbf{b}$ satisfies the eigenvalue equation

$$\mathbf{b}|\mathbf{u}(t)\rangle = \mathbf{u}(t)|\mathbf{u}(t)\rangle$$

where

$$\mathbf{u}(t) \cdot \mathbf{b}^\dagger := \sum_{j=1}^{n} u_j(t) b_j^\dagger \tag{6}$$

and $\mathbf{u}$ satisfies (1). Furthermore,

$$\langle 0|0 \rangle = 1$$

$$b_j |0\rangle = 0$$

$$0 = \langle 0|b_j^\dagger$$

for $j = 1, \ldots, n$. If we define

$$|\bar{\mathbf{u}}(t)\rangle := \exp\left(\frac{1}{2}(|\mathbf{u}(t)|^2 - |\mathbf{u}_0|^2)\right) |\mathbf{u}(t)\rangle \tag{7}$$

then

$$\frac{d}{dt}|\bar{\mathbf{u}}(t)\rangle = M|\bar{\mathbf{u}}(t)\rangle \tag{8}$$

is the corresponding infinite system. To find an isomorphic Lie algebra we define

$$V \mapsto M^\dagger := V(\mathbf{b}^\dagger) \cdot \mathbf{b} \equiv \sum_{j=1}^{n} V_j(\mathbf{b}^\dagger) b_j \,. \tag{9}$$

Let

$$W \mapsto N^\dagger := W(\mathbf{b}^\dagger) \cdot \mathbf{b} \equiv \sum_{j=1}^{n} W_j(\mathbf{b}^\dagger) b_j \tag{10}$$

then

$$[M^\dagger, N^\dagger] = \sum_{k=1}^{n}\sum_{j=1}^{n} \left(V_j(\mathbf{b}^\dagger)\frac{\partial W_k(\mathbf{b}^\dagger)}{\partial b_j^\dagger} - W_j(\mathbf{b}^\dagger)\frac{\partial V_k(\mathbf{b}^\dagger)}{\partial b_j^\dagger} \right) b_k. \tag{11}$$

Comparing (11) and (3) we see that the Lie algebra for the analytic vector fields and the Lie algebra of the Bose operators defined by (9) are isomorphic.

19.2   Examples

We give three examples to show how the embedding is performed. Moreover we show how first integrals can be studied.

Example: First we consider the case $n = 1$ with the vector fields d/du, ud/du and $u^2 d/du$. These vector fields form a basis of a Lie algebra under the commutator. Owing to (9) we have

$$\frac{d}{du} \mapsto b, \qquad u\frac{d}{du} \mapsto b^\dagger b, \qquad u^2\frac{d}{du} \mapsto b^\dagger b^\dagger b. \qquad \clubsuit$$

Example: Consider the **Lotka-Volterra model**

$$\frac{du_1}{dt} = -u_1 + u_1 u_2 \tag{12}$$

$$\frac{du_2}{dt} = u_2 - u_1 u_2. \tag{13}$$

Here the associated vector field is given by

$$V = (-u_1 + u_1 u_2)\frac{\partial}{\partial u_1} + (u_2 - u_1 u_2)\frac{\partial}{\partial u_2}. \tag{14}$$

Owing to (9), the corresponding Bose operator is given by

$$M^\dagger = (-b_1^\dagger + b_1^\dagger b_2^\dagger)b_1 + (b_2^\dagger - b_1^\dagger b_2^\dagger)b_2. \tag{15}$$

The first integral of the Lotka-Volterra model takes the form

$$I(\mathbf{u}) = u_1 u_2 e^{-(u_1 + u_2)}. \tag{16}$$

In the formulation with Bose operators the first integral (16) is the state vector

$$b_1^\dagger b_2^\dagger e^{-(b_1^\dagger + b_2^\dagger)}|0\rangle. \tag{17}$$

Consequently,

$$[(-b_1^\dagger + b_1^\dagger b_2^\dagger)b_1 + (b_2^\dagger - b_1^\dagger b_2^\dagger)b_2]b_1^\dagger b_2^\dagger e^{-(b_1^\dagger + b_2^\dagger)}|0\rangle = 0 \tag{18}$$

since $L_V I = 0$, where L_V denotes the Lie derivative. $\clubsuit$

We now describe the extension to explicitly time-dependent first integrals for autonomous systems of first-order ordinary differential equations (1). We extend (1) to

$$\frac{d\mathbf{u}}{d\lambda} = V(\mathbf{u}) \tag{19}$$

$$\frac{dt}{d\lambda} = 1. \tag{20}$$

Then

$$W = \sum_{j=1}^{n} V_j(\mathbf{u})\frac{\partial}{\partial u_j} + \frac{\partial}{\partial t} \tag{21a}$$

is the corresponding vector field of system (19). An explicitly time-dependent smooth function $I(u(t), t)$ is a first integral of system (19) if $L_W I = 0$, where $L_W(.)$ denotes the Lie derivative. By (9) the corresponding Bose operator of the vector field W is

$$W \mapsto M^\dagger := \sum_{j=1}^{n} V_j(\mathbf{b}^\dagger)b_j + b_{n+1} \tag{21b}$$

where we have identified t with u_{n+1}.

Example: As an application, let us consider the Lorenz system

$$\frac{du_1}{dt} = \sigma u_2 - \sigma u_1 \tag{22a}$$

$$\frac{du_2}{dt} = -u_2 - u_1 u_3 + r u_1 \tag{22b}$$

$$\frac{du_3}{dt} = u_1 u_2 - b u_3. \tag{22c}$$

For $b = 0$, $\sigma = \frac{1}{3}$ and r arbitrary we find the explicitly time-dependent first integral

$$I(\mathbf{u}(t), t) = (-ru_1^2 + \frac{1}{3}u_2^2 + \frac{2}{3}u_1 u_2 + u_1^2 u_3 - \frac{3}{4}u_1^4)e^{4t/3} \tag{23}$$

and the Lorenz system takes the form

$$\frac{du_1}{dt} = \frac{1}{3}u_2 - \frac{1}{3}u_1 \tag{24a}$$

$$\frac{du_2}{dt} = -u_2 - u_1 u_3 + r u_1 \tag{24b}$$

$$\frac{du_3}{dt} = u_1 u_2. \tag{24c}$$

Consequently we find that the first integral expressed in Bose operators takes the form

$$\left(-r(b_1^\dagger)^2 + \frac{1}{3}(b_2^\dagger)^2 + \frac{2}{3}b_1^\dagger b_2^\dagger + (b_1^\dagger)^2 b_3^\dagger - \frac{3}{4}(b_1^\dagger)^4\right)e^{4b_4^\dagger/3}|0\rangle. \tag{25}$$

Therefore

$$M^\dagger \left(-r(b_1^\dagger)^2 + \frac{1}{3}(b_2^\dagger)^2 + \frac{2}{3}b_1^\dagger b_2^\dagger + (b_1^\dagger)^2 b_3^\dagger - \frac{3}{4}(b_1^\dagger)^4\right)e^{4b_4^\dagger/3}|0\rangle = 0 \tag{26}$$

where

$$M^\dagger = (\frac{1}{3}b_2^\dagger - \frac{1}{3}b_1^\dagger)b_1 + (-b_2^\dagger + b_1^\dagger b_3^\dagger + r b_1^\dagger)b_2 + (b_1^\dagger b_2^\dagger)b_3 + b_4. \tag{27}$$

♣

19.3   Embedding and Bose Field Operators

The embedding can be extended to partial differential equations. We now discuss the extension to (nonlinear) partial differential equations. Let

$$f, g : W \to W$$

be two maps, where W is a topological vector space ($u \in W$). Assume that the Gateaux derivative of f and g exists, i.e.,

$$f'(u)[v] := \left. \frac{\partial f(u + \varepsilon v)}{\partial \varepsilon} \right|_{\varepsilon = 0} \tag{28a}$$

$$g'(u)[v] := \left. \frac{\partial g(u + \varepsilon v)}{\partial \varepsilon} \right|_{\varepsilon = 0}. \tag{28b}$$

We introduced the Lie bracket (or commutator) for f and g

$$[f, g] := f'(u)[g] - g'(u)[f]. \tag{29}$$

The Bose field operators satisfy the commutation relations

$$[b_j(\mathbf{x}), b_k^\dagger(\mathbf{x}')] = \delta_{jk} \delta(\mathbf{x} - \mathbf{x}') \tag{30a}$$

$$[b_j(\mathbf{x}), b_k(\mathbf{x}')] = [b_j^\dagger(\mathbf{x}), b_k^\dagger(\mathbf{x}')] = 0 \tag{30b}$$

where $j = 1, \ldots, n$ and $\mathbf{x} = (x_1, \ldots, x_m, x_{m+1})$. Here $\delta(\mathbf{x} - \mathbf{x}')$ denotes the delta function and δ_{ij} the Kroneker symbol. In order to give the formal relations (30) a precise mathematical meaning, we need to smear out the field b_j by integrating it against suitable test functions and define the resulting smeared field

$$b_j(f) := \int d\mathbf{x}\, b_j(\mathbf{x}) f(\mathbf{x}) \tag{31}$$

as an operator in a Hilbert space.

We define both the Fock-Hilbert space $\mathcal{H}_F$ and the operators $\mathbf{b}(f)$ in $\mathcal{H}_F$ by the following standard specifications.

(1) For each square-integrable function $f(\mathbf{x})$ there is an operator $\mathbf{b}(f)$ in $\mathcal{H}_F$ such that

$$[\mathbf{b}(f), \mathbf{b}(g)^\dagger] = \int \mathbf{dx}\, \bar{g}(\mathbf{x})f(\mathbf{x})$$

$$[\mathbf{b}(f), \mathbf{b}(g)] = 0.$$

These relations may be derived by considering

$$\mathbf{b}(f) = \int d\mathbf{x}\, \mathbf{b}(\mathbf{x})f(\mathbf{x}).$$

(2) There is a vector Ψ_F in $\mathcal{H}_F$ such that

$$\mathbf{b}(f)\Psi_F = 0 \tag{32}$$

for all the square-integrable functions f.

(3) The vectors obtained by application to Ψ of all the polynomials in the $\mathbf{b}(f)^\dagger$'s form a dense subset of $\mathcal{H}_F$, i.e., this space is generated by applying these polynomials to Ψ_F. Thus, in a standard terminology, Ψ_F is a vacuum vector, the $\mathbf{b}(f)^\dagger$'s and $\mathbf{b}(f)$'s are creation and annihilation operators respectively and the various many-particle states are obtained by applying combinations of creation operators to the vacuum. We represent space translations by the group of unitary operators $\{V(\mathbf{x})\}$ which leave Ψ_F invariant and implement the formal relation

$$V(\mathbf{x})\mathbf{b}(\mathbf{x}')V(\mathbf{x})^\dagger = \mathbf{b}(\mathbf{x} + \mathbf{x}').$$

To be precise $V(\mathbf{x})$ is defined by the formula

$$V(\mathbf{x})\Psi_F = \Psi_F \tag{33a}$$

$$V(\mathbf{x})\mathbf{b}(f_1)^\dagger \ldots \mathbf{b}(f_k)^\dagger \Psi_F = \mathbf{b}(f_{1,\mathbf{x}})^\dagger \ldots \mathbf{b}(f_{k,\mathbf{x}})^\dagger \Psi_F \tag{33b}$$

where

$$f_{\mathbf{x}}(\mathbf{x}') = f(\mathbf{x}' - \mathbf{x}).$$

Likewise, we represent gauge transformations by the unitary group

$$\{W(\alpha) : \alpha \in \mathbf{R}\}$$

which leave Ψ_F invariant and implement the formal relation

$$W(\alpha)\mathbf{b}(\mathbf{x})W(\alpha)^\dagger = \mathbf{b}(\mathbf{x})e^{i\alpha}.$$

The precise definition of $W(\alpha)$ is given by the formula

$$W(\alpha)\Psi_F = \Psi_F \tag{34a}$$

$$W(\alpha)\mathbf{b}(f_1)^\dagger \ldots \mathbf{b}(f_k)^\dagger \Psi_F = e^{-ik\alpha}\mathbf{b}(f_1)^\dagger \ldots \mathbf{b}(f_k)^\dagger \Psi_F. \tag{34b}$$

Let

$$\frac{\partial \mathbf{u}}{\partial x_{m+1}} - \mathbf{F}(\mathbf{u}, D^\alpha \mathbf{u}; \mathbf{x}) = 0 \tag{35}$$

be a partial differential equation, where

$$\mathbf{u} : \mathbf{R}^m \times \mathbf{R} \to \mathbf{C}^n$$

and the function $\mathbf{F}$ is analytic. We set

$$D^\alpha \mathbf{u} := (D^{\alpha_1} u_1, \cdots, D^{\alpha_n} u_n) \tag{36}$$

where the α_j's are multi-indices and

$$D^\beta := \frac{\partial^{|\beta|}}{\partial x_1^{\beta_1} \ldots \partial x_m^{\beta_m}}. \tag{37}$$

The notation $\mathbf{F}(\ldots; \mathbf{x})$ indicates that the function $\mathbf{F}$ can explicitly depend on $\mathbf{x}$. Now we introduce the mappings from the right hand side of (35) to the Bose field operators, namely

$$M = \int d\mathbf{x}\, \mathbf{b}^\dagger(\mathbf{x}) \cdot \mathbf{F}(\mathbf{b}(\mathbf{x}), D^\alpha \mathbf{b}(\mathbf{x}); \mathbf{x}). \tag{38}$$

Then the mappings $\mathbf{F}$ and the corresponding Bose field operators M are isomorphic Lie algebras under the commutator.

Example: Consider the two mappings

$$f(u) := \frac{\partial u}{\partial x_2} - u\frac{\partial u}{\partial x_1} - \frac{\partial^3 u}{\partial x_1^3}$$

$$g(u) := \frac{\partial u}{\partial x_2} - \frac{\partial^2 u}{\partial x_1^2}$$

where

$$[f, g] = -2\frac{\partial u}{\partial x_1}\frac{\partial^2 u}{\partial x_1^2}.$$

Then we have

$$f \mapsto M := \int d\mathbf{x}\, b^{\dagger}(\mathbf{x}) \left(\frac{\partial b(\mathbf{x})}{\partial x_2} - b(\mathbf{x}) \frac{\partial b(\mathbf{x})}{\partial x_1} - \frac{\partial^3 b(\mathbf{x})}{\partial x_1^3} \right) \tag{39a}$$

$$g \mapsto N := \int d\mathbf{x}\, b^{\dagger}(\mathbf{x}) \left(\frac{\partial b(\mathbf{x})}{\partial x_2} - \frac{\partial^2 b(\mathbf{x})}{\partial x_1^2} \right) \tag{39b}$$

where $\mathbf{x} = (x_1, x_2)$ and

$$d\mathbf{x} = dx_1 dx_2.$$

The commutator is given by

$$[M, N] = -2 \int d\mathbf{x}\, b^{\dagger}(\mathbf{x}) \frac{\partial b(\mathbf{x})}{\partial x_1} \frac{\partial^2 b(\mathbf{x})}{\partial x_1^2}. \tag{40}$$

♣

To summarize: We have shown that dynamical systems given by ordinary differential or partial differential equations can be expressed with the help of Bose operators and Bose field operators, respectively. The technique can also be applied to difference equations. The first integrals and conservations laws can now be found within this approach. The soliton theory (Lax representation, conservation laws, recursion operators, master symmetries, etc.) can also be expressed with the help of Bose field operators. For a detailed discussion we refer to Kowalski and Steeb [71].

19.4   Computer Algebra Applications

Here we give an implementation of coherent states. The coherent state is denoted by cs
and the dual coherent state is denoted by ds. In REDUCE repart(z) denotes the real
part of a complex number z and impart(z) the imaginary part of a complex number z.
Furthermore, conj(z) finds the complex conjugate of z.

```
%Coherent states;
%Program name: coher.our;


operator b, bd, cs, ds;
noncom b, bd, cs, ds;

for all z let b*cs(z) = z*cs(z);
for all z let ds(z)*bd = ds(z)*z;
for all z, w let ds(z)*cs(w) =
exp(-(1/2)*(z*conj(z)+w*conj(w)-2*conj(z)*w));

%Example 1;
r1 := b*(b*cs(z));
z := 1;
r1;
clear z;

%Example 2;
ds(z)*cs(z);

%Example 3;
r2 := b*cs(z);
r3 := ds(w)*r2;
```

The output is

```
r1 := z**2*cs(z);
cs(1);
1;
r2 := z*cs(z);
r3 := (e**((impart(w)*i*w + impart(z)*i*z +
2*repart(w)*z)/2)*z)/e**((2*impart(w)*i*z +
repart(w)*w + repart(z)*z)/2);
```

Chapter 20

Computer Algebra

20.1 Computer Algebra Packages

Computer algebra systems now available on many computers make it feasible to perform a variety of analytical procedures automatically. In particular, the construction of Lie and Lie-Bäcklund symmetry vector fields, commutation relations of Lie algebras, calculation of the Gateaux derivative, performing the Painlevé test are a small number of applications.

The most important general purpose computer algebra systems currently available are MACSYMA (Math Lab Group MIT), REDUCE (A.C. Hearn, Rand Corporation), MAPLE (B. Char, University of Waterloo, Canada), DERIVE (D.R. Stoutemyer, The Software-house, Honolulu, Hawaii), MATHEMATICA (Wolfram Research, Inc.), AXIOM (R.D. Jenks and D. Yun, IBM Watson Laboratories) and SymbolicC++ (School for Scientific Computing, Rand Afrikaans University). MACSYMA, REDUCE, DERIVE and AXIOM are based on LISP. MAPLE is based on C. All packages also allow numerical manipulations. AXIOM is also a computer algebra system based on abstract data types, where the various types are constructed from elementary data types. SymbolicC++ is based on C++ and uses an object-oriented design based on abstract data types.

A survey about computer algebra is given by Champagne *et al* [14]. Davenport et al [22] describe algorithms for algebraic computations. Algebraic computation with REDUCE is discussed by MacCallum and Wright [80]. Discussions and applications on computer algebra can also be found in Zwillinger [149] as well as in Rogers and Ames [96]. A general introduction to applications of computer algebra methods in MACSYMA is given by Rand [92].

20.2   Programs for Lie Symmetries

A survey of programs calculating Lie symmetry vector fields is given by Champagne *et al* [14]. More recently, Hereman [57] gives a comprehensive review on programs for Lie symmetries. He also discusses other applications in this field. We give a short excerpt from these articles. The reader is referred to these articles for a more detailed discussion. For all computer algebra systems described above programs exist for finding Lie symmetry vector fields.

The REDUCE program SPDE developed by Schwarz [109] attempts to solve the determining equations with minimal intervention by the user. Based on the exterior calculus, Edelen [32] as well as Gragert, Kersten and Martini [54] did some pioneering work in using REDUCE to calculate the Lie symmetry vector fields of differential equations. Kersten [66] later developed a REDUCE software package for the calculation of the Lie algebra of Lie symmetry vector fields (and corresponding Lie-Bäcklund transformations) of an exterior differential system. Eliseev, Fedorova and Kornyak [33] wrote a REDUCE program to generate (but not solve) the system of determining equations for point and contact symmetries. Fedorova and Kornyak [45] generalized the algorithm to include the case of Lie-Bäcklund symmetries. Nucci [86] developed interactive REDUCE programs for calculating classical (Lie point), non-classical (generalized) and Lie-Bäcklund symmetry vector fields.

The program LIE by Head [56] is based on muMATH, and runs on IBM compatible PCs. Head's program calculates and solves the determining equations automatically. Interventions by the user are possible. The SYMCON package written by Vafeades [140] also uses muMATH to calculate the determining equations (without solving them). Furthermore, the program verifies whether the symmetry group is of variational or divergence type and computes the conservation laws associated with symmetries. Unfortunately these programs are confined to the 256 K memory accessible by muMATH and can therefore presently not handle very large systems of equations. This limitation motivated Vafeades to rewrite his SYMCON program in MACSYMA syntax (Vafeades [140]).

To the best of our knowledge, no package is available yet for the calculation of Lie symmetries with MAPLE and MATHEMATICA.

There are several MACSYMA programs for the calculation of Lie symmetry vector fields. MACSYMA is currently available for various types of computers, ranging from PCs to various work stations and main frame computers. Apart from a version by Champagne and Winternitz [15] and the work done by Rosencrans [99], there are three other MACSYMA-based symmetry programs. The MACSYMA version of SYMCON by Vafeades [140] was discussed above. Schwarzmeier and Rosenau [111] wrote a program that calculates the determining equations in their simplest form, but does not solve them automatically. The program SYM_DE by Steinberg [137] was recently added to the out-of-core library of MACSYMA. The program solves some (or all) of the determining equations automatically and, if needed, the user can (interactively) add extra information. Champagne *et al* [14] have written the program SYMMGRP.MAX in MACSYMA. This program is a modification of a package that has been extensively used over the last five years at the University of Montréal and elsewhere. It has been tested on hundreds of systems of equations and has thus been solidly debugged. The flexibility of this program and the possibility of using it in a partly interactive mode, allows us to find the symmetry group of, in principle, arbitrarily large and complicated systems of equations on relatively small computers. These are the main justifications for presenting yet another new symbolic program in a field where several programs already exist. The program SYMMGRP.MAX concentrates on deriving the determining equations. The program can also be used to verify calculated solutions of the determining equations.

Carminati *et al* [12] present their program LIESYMM in MAPLE for creating the determining equations via the Harrison-Estabrook technique. Within LIESYMM various interactive tools are available for integrating the determing equations, and for working with Cartan's differential forms.

Herod [58] developed the program MathSym for deriving the determining equations corresponding to Lie-point symmetries, including nonclassical (or conditional) symmetries. The program Lie.m of Baumann [5] written in MATHEMATICA follows the MACSYMA progam SYMMGRP.MAX closely.

Schwarz [109] rewrote SPDE for use with AXIOM.

Other applications of MACSYMA and REDUCE include the following: In Schwarz's paper [105] there is a description of a REDUCE program that will automatically determine first integrals of an autonomous system of ordinary differential equations. Schwarz [110] also write a factorization algorithm for linear ordinary differential equations. In Fateman [43] there is a description of a MACSYMA program that will automatically utilize the method of multiple scales to approximate the solution of differential equations. In Lo [79] there is a technique for calculating many terms in an asymptotic expansion. The computer language MACSYMA is used to perform the asymptotic matching at each stage. MACSYMA also provides us with a function called SERIES that computes the series expansion of a second order ordinary differential equation. In Rand and Winternitz [91] there is a MACSYMA program for determining whether a nonlinear ordinary differential equation has the Painlevé property. Obviously the differential equation must be a polynomial in both the dependent and independent variables and in all derivatives. A computer program in REDUCE for determining conservation laws is given in Ito and Kako [61]. In Gerdt, Shvachka and Zharkov [51] there is the description of a computer program in FORMAC that determines conservation laws, Lie-Bäcklund symmetries and also attempts to determine when an evolution equation is formally integrable. A REDUCE package called EXCALC can be used for calculations in differential geometry. The package is able to handle scalar-valued exterior forms, vectors and operations between them, as well as non-scalar valued forms.

To conclude, the availability of efficient computer algebra packages becomes crucial as the problem size increases. Thus, as the order, dimension and number of equations increases, so the total number N of the determining equations rises dramatically. This is readily illustrated for quasi-linear systems of first order where

$$N = \frac{1}{2}n(mn - n + 1)(mn - n + 2),$$

with m the number of independent variables and n the number of dependent variables. In the following table values of N are displayed to indicate how rapidly N increases with problem size.

The Number N of Determining Equations

n	2	3	3	4	4
m	2	3	4	4	5
N	12	84	180	364	680

Let us now describe methods for solving the determining equations. There are no algorithms to solve an arbitrary system of determining equations, which consists of linear homogeneous partial differential equations for the coefficients of the vector field. Most integration algorithms are based on a set of heuristic rules. Most commonly the following rules are used.

(1) Integrate single term equations of the form

$$\frac{\partial^{|I|} f(x_1, x_2, \ldots, x_n)}{\partial x_1^{i_1} \partial x_2^{i_2} \ldots \partial x_n^{i_n}} = 0,$$

where $|I| = i_1 + i_2 + \ldots + i_n$, to obtain the solution

$$f(x_1, x_2, \ldots, x_n) = \sum_{k=1}^{n} \sum_{j=0}^{i_k - 1} h_{kj}(x_1, x_2, \ldots, x_{k-1}, x_{k+1}, \ldots, x_n)(x_k)^j.$$

Thus introducing functions h_{kj} with fewer variables.

(2) Replace equations of type

$$\sum_{j=0}^{n} f_j(x_1, x_2, \ldots, x_{k-1}, x_{k+1}, \ldots, x_n)(x_k)^j = 0,$$

by $f_i = 0$ $(j = 0, 1, \ldots, n)$. More generally, this method of splitting equations (via polynomial decomposition) into a set of smaller equations is also allowed when f_j are differential equations themselves, provided the variable x_k is missing.

(3) Integrate linear differential equations of first and second order with constant coefficients. Integrate first order equations with variable coefficients via the integrating factor

technique, provided the resulting integrals can be computed in closed form.

(4) Integrate higher-order equations of type

$$\frac{\partial^n f(x_1, x_2, \ldots, x_n)}{\partial x_k^n} = g(x_1, x_2, \ldots, x_{k-1}, x_{k+1}, \ldots, x_n),$$

n successive times to obtain

$$\begin{aligned}
f(x_1, x_2, \ldots, x_n) &= \frac{(x_k)^n}{n!} g(x_1, x_2, \ldots, x_{k-1}, x_{k+1}, \ldots, x_n) \\
&+ \frac{x_k^{n-1}}{(n-1)!} h(x_1, x_2, \ldots, x_{k-1}, x_{k+1}, \ldots, x_n) \\
&+ \ldots + r(x_1, x_2, \ldots, x_{k-1}, x_{k+1}, \ldots, x_n),
\end{aligned}$$

where $h, \ldots, r$ are arbitrary functions.

(5) Solve any simple equation (without derivatives) for a function (or a derivative of a function) provided both (i) it occurs linearly and only once, (ii) it depends on all the variables which occur as arguments in the remaining terms.

(6) Explicitly integrate exact equations.

(7) Substitute the solutions obtained above in all the equations.

(8) Add differences, sums or other linear combinations of equations (with similar terms) to the system, provided these combinations are shorter than the original equations.

With these simple rules, and perhaps a few more, the determining system can often be drastically simplified. In many cases nothing beyond the above heuristic rules is needed to solve the determining equations completely. If that is not possible, after simplification, the program will return the remaining unsolved differential equations for further inspection. In most programs, the user can then interactively simplify and fully solve the determining equations on the computer, thereby minimizing human errors.

20.3   Basic REDUCE Commands

We now give a summary of the commands in REDUCE most commonly used in our programs. For a complete list of commands we refer to the REDUCE user manual.

First we remark that REDUCE does not distinguish between capital and small letters. Thus the commands sin(0), Sin(0) and SIN(0) are the same. The command most used in this book is differentiation. A differentiation can be performed as follows:

```
depend f, x;        % f depends on x

f := x*x+sin(x);    % the function f is declared

df(f,x);            % df is the differentiation operator

df(f,x,2);          % here we differentiate f twice with respect to x
```

An alternative option is to use operators wich we discuss later. To integrate we write

```
int(x**2+1,x);   % we integrate the function x**2+1 with respect to x
```

The command solve() solves a number of algebraic equations, and also systems of algebraic equations. For example, the command

```
solve(x**2+(a+1)*x+a= 0,x);
```

solves x**2+(a+1)*x+a=0 with respect to x. Another important command is the subsititution command, sub(). For example, the command

```
sub(x=2,x*y+x**2);
```

yields 2*y+4, i.e., x is replaced by 2. Amongst others, REDUCE includes the following mathematical functions: sqrt(x) (square root), exp(x) (exponential functions), log(x) (natural logarithm), and the trigonometric functions sin(x), cos(x), tan(x) with arguments in radians. REDUCE reserves i to represent $\sqrt{-1}$ and pi for the number π. Thus the input i*i gives -1 and sin(pi) gives 0. Other predefined constants are e, infinity, nil, t, where t stands for true.

Operators are the most general objects available in REDUCE. They are parametrized, and can be parametrized in a completely general way. Only the operator identifier is declared in and **operator** declaration, thus:

```
operator op1, op2, ... ;
```

The number of parameters is not declared, and operators with the same name but different numbers of parameters are not algebraically related. An operator needs to have at least one argument (or empty parentheses) to be recognized as an operator rather than a simple variable. This is because REDUCE allows an identifier to be used as a simple variable and as an algebraically-unrelated operator (or some other types of parametrized objects) at the same time. To do so is generally not good practice because of its potential for confusion, but it does allow identifiers of such parametrized objects to be passed as parameters themselves, i.e. passed as simple variables and then used as parametrized objects. A similar facility is provided in most other languages. REDUCE assumes that any unrecognized parametrized object is an operator; in batch mode it is automatically declared so, whilst in interactive mode the user is queried. Operators can be used to represent just about anything, although usually they will represent mathematical operators or functions. By default, operators remain purely symbolic, as do variables, and can be considered as generalized or parametrized variables. Operators can also be given values, either for all values of all their arguments, or for restricted ranges, so that for example they can be used to represent discontinuous functions, or functions like the factorial function which might be considered to be defined only over the non-negative integers. Operators are usually given values by using let-rules, e.g.

```
% derivatives of Jacobi elliptic functions


operator sn, cn, dn;


for all q let df(sn(q),q) = cn(q)*dn(q);
for all q let df(cn(q),q) = - sn(q)*dn(q);
for all q let df(dn(q),q) = -(k**2)*sn(q)*cn(q);


depend f, x;
f := sn(2*x)*dn(x**2);
df(f,x);
df(f,x,2);
```

For fixed argument values assignments can also be used. Operators are global in scope as are most objects in REDUCE, which means that once they have been defined they are available anywhere subsequently in the same REDUCE session. The declaration of an identifier to represent an operator is cleared by using **clear**, but its actual values mut all be cleared first. It is possible to use **SUB** to substitute one operator identifier for another within an expression.

20.4   Examples

Example: The following REDUCE program calculates the Lie bracket using the Gateaux derivative for the following two mappings

$$f(u) := \frac{\partial u}{\partial x_2} - u\frac{\partial u}{\partial x_1} - \frac{\partial^3 u}{\partial x_1^3}$$

$$g(u) := \frac{\partial u}{\partial x_2} - \frac{\partial^2 u}{\partial x_1^2}.$$

We find

$$[f, g] = -2\frac{\partial u}{\partial x_1}\frac{\partial^2 u}{\partial x_1^2}.$$

```
% Lie bracket and Gateaux Derivative
% GatCom.red


depend u, x1, x2;
depend A, ep;
depend B, ep;

V := df(u,x2) - u*df(u,x1) - df(u,x1,3);
W := df(u,x2) - df(u,x1,2);
A := sub(u=u+ep*W,V);
B := sub(u=u+ep*V,W);
R := df(A,ep);
S := df(B,ep);
ep := 0;
com := R - S;
com;
```

Example: The following REDUCE program finds the determining equations for the Lie symmetry vector fields of the inviscid Burgers' equation

$$\frac{\partial u}{\partial x_2} = u \frac{\partial u}{\partial x_1}.$$

Note that the program does not solve the determining equations.

```
% Determining equations
% for the Lie symmetries of
% u2 - u u1 = 0;
% invar.red


operator D1, D2, V;


depend xi1, x1, x2, u;
depend xi2, x1, x2, u;
depend eta, x1, x2, u;


% vertical vector field
uu := eta - xi1*u1 - xi2*u2;


% D1, D2 denote the total derivative operators
for all f let D1(f)=df(f,x1)+u1*df(f,u)+u11*df(f,u1)+u12*df(f,u2);
for all f let D2(f)=df(f,x2)+u2*df(f,u)+u22*df(f,u2)+u21*df(f,u1);


% V(f) denotes the prolongation of the vector field
for all f let V(f)=uu*df(f,u)+D1(uu)*df(f,u1)+D2(uu)*df(f,u2);


% mixed partial derivatives are the same
let u21 = u12;
```

```
% hs denotes the hypersurface
hs := u2 - u*u1;


% Lie derivative for symmetry
Lieder := V(hs);


% insertion of the differential consequences
u2 := u*u1;   u*u11 := u12 - u1*u1;   u22 := u*u1*u1+u*u12;
Lieder;


c1 := coeffn(Lieder,u1,1);
c2 := coeffn(Lieder,u1,0);
write "The first determining equation is ", c1, "=0";
write "The second determining equation is ",c2,"=0";
```

Appendix A

Differentiable Manifolds

In order to give a precise formulation of differentiable manifolds (Choquet-Bruhat *et al* [17], von Westenholz [141], Matsushima [83], we require some preliminary definitions.

Definition A.1   *1. A **topological space** is a pair (X, O) where X is a set and O a class of subsets, called open sets, such that*

   (i) The union of any collection of open sets is open;

   (ii) The intersection of a finite collection of open sets is open;

   (iii) The empty set ϕ and the set X are both open.

  *2. A collection B of open sets of X is a **basis** for the topology of X if every open set is the union of sets of B. (For instance, the open balls in $\mathbf{R}^n$ form a basis for the (metric) topology of $\mathbf{R}^n$; also the set of all open balls with rational radii and rational centre coordinates is also a basis for the topology of $\mathbf{R}^n$ which is actually countable.)*

  *3. If X has a countable basis, it is said to be **second countable** (alternatively, the space satisfies the second axiom of countability).*

  Remark: This implies that X is separable i.e., X contains a countable dense set.

  *4. The topological space X is a **Hausdorff space** if any two distinct points have disjoint open neighbourhoods.*

5. *Let* X, Y *be topological spaces. A map* $f : X \to Y$ *is* **continuous** *if the inverse image of any open set in* Y *is open in* X.

The map f *is a* **homeomorphism** *if it is bijective and both* f *and* f^{-1} *are continuous, in which case* $U \subset X$ *is open if and only if* $f(U) \subset Y$ *is open.*

Definition A.2 *An n-dimensional* **topological manifold** *is a Hausdorff space* M *with a countable basis that satisfies the following condition: Every point* $p \in M$ *has a neighbourhood* U *which is homeomorphic with an open subset of* $\mathbf{R}^n$

In this case we write $dim M = n$, *and* U *is called a coordinate neighbourhood. (Topological manifolds are often called Euclidean spaces.)*

A **chart** *for a topological manifold* M *is a pair* (U, h), *where* U *is an open subset of* $M, h : U \to h(U)$ *is a homeomorphism, with* $h(U)$ *open in* $\mathbf{R}^n$.

An **atlas** *on an n-dimensional topological manifold is a collection of charts*

$$\{(U_\alpha, h_\alpha) : \alpha \in A\}$$

where A *is some countable index set, such that the sets* U_α *constitute a cover of* M:

$$M = \bigcup_{\alpha \in A} U_\alpha.$$

The atlas is said to be finite if A is finite. Any **compact** topological manifold admits a finite atlas. In fact it can be shown that any topological manifold admits a finite atlas.

Let U_α, U_β be (coordinate) neighbourhoods of an atlas $\{U_\alpha, h_\alpha\}$ with $U_\alpha \cap U_\beta \neq \phi$. Then $h_{\beta\alpha} := h_\beta \circ h_\alpha^{-1}$ is a homeomorphism $h_{\beta\alpha} : h_\alpha(U_\alpha \cap U_\beta) \to h_\beta(U_\alpha \cap U)\beta)$ with inverse $h_{\alpha\beta} := h_\alpha \circ h_\beta^{-1}$.

The map $h_{\beta\alpha}$ is called the identification map for U_α and U_β. These are maps between open sets of $\mathbf{R}^n$ and can therefore be represented by n real-valued functions of n real variables. If $h_{\beta\alpha}$ is a C^k function, the charts (U_α, h_α), (U_β, h_β) are said to be C^k-**compatible**. If all overlapping charts of an atlas are C^k-compatible (respectively,

C^∞-compatible), the atlas is said to be C^k-compatible (respectively, C^∞-compatible or smooth). Thus an atlas is smooth if all of its identifiction maps are C^∞.

Let $\{(U_\alpha, h_\alpha)\}$, $\{(V_{\alpha'}, k_{\alpha'})\}$ be smooth atlases. These are said to be equivalent if their union is again a smooth atlas, that is, if all the maps

$$k_{\beta'} \circ h_\alpha^{-1} : h_\alpha(U_\alpha \cap V_{\beta'}) \to k_{\beta'}(U_\alpha \cap V_{\beta'})$$

and their inverses are smooth.

A differentiable (or smooth) structure on M is an equivalence class of smooth atlases on M.

A topological manifold endowed with a differentiable structure is called a **differentiable manifold**.

Remarks:

1. *Not every topological manifold can be endowed with a differentiable structure.*

2. *A topological manifold may carry differentiable structures that belong to distinct equivalence classes: for instance S^7.*

Example of differentiable manifolds:

1. $\mathbf{R}^n$ is an analytic manifold. For an open covering take $\mathbf{R}^n$ itself. Let r be the identity mapping from $\mathbf{R}^n$ to itself. Then it is clear that $\{(\mathbf{R}^n, r)\}$ is a C^ω coordinate neighbourhood system. The coordinates $x_1, \ldots, x_n$ of $\mathbf{R}^n$ form a (local) coordinate system for the whole $\mathbf{R}^n$. When $\mathbf{R}^n$ is considered as a differentiable manifold, it is called an affine space. Let U be an open set of $\mathbf{R}^n$, and let $f_1, \ldots, f_n$ be real valued C^r functions defined on U. Suppose the map $\mathbf{x} \mapsto \mathbf{f}(\mathbf{x}) = (f_1(\mathbf{x}), \ldots, f_n(\mathbf{x}))$ $(\mathbf{x} \in U)$ from U into $\mathbf{R}^n$ is a one-to-one map and that the Jacobian

$$\frac{\partial(f_1, \ldots, f_n)}{\partial(x_1, \ldots, x_n)}$$

is not 0 at each point of U. Then by the inverse function theorem, $\mathbf{f}(U)$ is an open set $\mathbf{R}^n$ and

$$\mathbf{f}^{-1} : \mathbf{f}(U) \to U$$

is of class C^r. This means that

$$\{(\mathbf{R}^n, r), \ (U, \mathbf{f})\}$$

is a C^r coordinate neighbourhood system for the affine space. Hence $(f_1, \ldots, f_n)$ is a local coordinate system of the affine space, and is called a **curvilinear coordinate system**. In particular, if each f_i is a linear function on $\mathbf{x}$, i.e., if

$$f_i(\mathbf{x}) = \sum_{j=1}^{n} a_{ij} x_j + b_i, \qquad i = 1, \ldots, n$$

then $\partial(f_1, \ldots, f_n)/\partial(x_1, \ldots, x_n)$ is the determinant $\det(a_{ij})$ of the matrix

$$\begin{pmatrix} a_{11} & a_{12} & \cdots & a_{1n} \\ a_{21} & a_{22} & \cdots & a_{2n} \\ \vdots & & & \\ a_{n1} & a_{n2} & \cdots & a_{nn} \end{pmatrix}$$

which is not 0 by assumption. In general, let

$$y_i = \sum_{j=1}^{n} a_{ij} x_j + b_i$$

where $i = 1, \ldots, n$. If $\det(a_{ij}) \neq 0$, then $(y_1, \ldots, y_n)$ is a (local) coordinate system defined on the whole affine space, and is called an **affine coordinate system**, or a **linear coordinate system**. We also call $(x_1, \ldots, x_n)$ the standard coordinate system of the affine space $\mathbf{R}^n$.

2. Let S^1 be the circle in the xy-plane $\mathbf{R}^2$ centered at the origin and of radius 1. Give S^1 the topology of a subspace of $\mathbf{R}^2$. Let

$$\begin{aligned}
U_1 &:= \left\{ p = (x,y) \in S^1 \ : \ y > 0 \right\} \\
V_1 &:= \left\{ p = (x,y) \in S^1 \ : \ y < 0 \right\} \\
U_2 &:= \left\{ p = (x,y) \in S^1 \ : \ x > 0 \right\} \\
V_2 &:= \left\{ p = (x,y) \in S^1 \ : \ x < 0 \right\}.
\end{aligned}$$

Then U_i and V_i are open sets of S^1, and $U_1 \cup U_2 \cup V_1 \cup V_2 = S^1$. Let ψ_i and φ_i be the maps from U_i and V_i, respectively, to the open interval

$$I := \{t \ : \ -1 < t < 1\},$$

given by $\psi_1(x,y) = x$, $\varphi_1(x,y) = x$, $\psi_2(x,y) = y$, $\varphi_2(x,y) = y$. The maps ψ_i and φ_i are homeomorphisms. Thus S^1 is a 1-dimensional topological manifold and

$$S := \left\{ (U_i, \psi_i), (V_i, \varphi_i) \right\}_{i=1,2}$$

is a coordinate neighbourhood system. Now

$$U_1 \cap U_2 = \left\{ p = (x,y) \in S^1 \ : \ x, y > 0 \right\}.$$

Thus $\psi_1(U_1 \cap U_2)$ and $\psi_2(U_1 \cap U_2)$ are both equal to

$$J := \{\, t \ : \ 0 < t < 1 \,\}.$$

For $t \in J$, we have

$$f_{21}(t) = \psi_2(\psi_1^{-1}(t)) = (1 - t^2)^{1/2}$$

and

$$f_{12}(t) = \psi_1(\psi_2^{-1}(t)) = (1 - t^2)^{1/2}.$$

$f_{21}(t)$ and $f_{12}(t)$ are analytic for $0 < t < 1$. Similarly the other coordinate transformations are also analytic. Thus S^1 is an analytic manifold.

3. The topological space obtained by giving the subspace topology to the set of points $(x_1, \ldots, x_{n+1})$ in $\mathbf{R}^{n+1}$ satisfying

$$(x_1)^2 + (x_2)^2 + \ldots + (x_{n+1})^2 = r^2$$

is denoted by S^n and called the n-**dimensional sphere**. The sphere S^n is an n-dimensional analytic manifold. To verify this we let

$$\begin{aligned} U_i &:= \{\, p = (x_1, \ldots, x_{n+1}) \in S^n \ : \ x_i > 0 \,\} \\ V_i &:= \{\, p = (x_1, \ldots, x_{n+1}) \in S^n \ : \ x_i < 0 \,\} \end{aligned} \quad (i = 1, \ldots, n+1)$$

and define maps φ_i and ψ_i from U_i and V_i to

$$B^n = \{\, y_1, \ldots, y_n) \ : \ (y_1)^2 + \ldots + (y_n)^2 < r^2 \,\},$$

the interior of the sphere S^{n-1} in $\mathbf{R}^n$, by

$$p = (x_1, x_2, \ldots, x_{n+1}) \to (x_1, \ldots, x_{i-1}, x_{i+1}, \ldots, x_{n+1}).$$

Then φ_i and ϕ_i are both homeomorphisms from U_i and V_i, respectively, onto the open set B^n in $\mathbf{R}^n$. Hence S^n is an n-dimensional topological manifold. As in the case of the circle, the coordinate transformation functions are analytic.

4. **Projective spaces.** Remove the origin $\mathbf{0} = (0, \ldots, 0)$ from $\mathbf{R}^{n+1}$, and define and equivalence relation on the set $\mathbf{R}^{n+1} \setminus \{\mathbf{0}\}$ as follows. Two points $\mathbf{x} = (x_i)$ and $\mathbf{y} = (y_i)$ are defined to be equivalent if there is a nonzero real number λ such that $x = \lambda y$, i.e.,

$$x_i = \lambda y_i$$

where $i = 1, \ldots, n+1$. The set of equivalence classes given by this equivalence relation is denoted by P^n. Let $\pi(\mathbf{x})$ denote the equivalence class containing the point $\mathbf{x}$ of $\mathbf{R}^{n+1} \setminus \{\mathbf{0}\}$. Then π is a map from $\mathbf{R}^{n+1} \setminus \{\mathbf{0}\}$ onto P^n. To an element $\pi(\mathbf{x})$ of P^n associate the line $t\mathbf{x}$ $(t \in \mathbf{R})$ through the origin $\mathbf{R}^{n+1}$. This correspondence is one-to-one. Thus we can regard P^n as the set of all lines passing through the origin in $\mathbf{R}^{n+1}$. A subset U of P^n is defined to be an open set if the inverse image $\pi^{-1}(U)$ in $\mathbf{R}^{n+1} \setminus \{0\}$ is an open set. In this way a topology is defined on P^n, and P^n becomes a Hausdorff space. Furthermore the projection π from $\mathbf{R}^{n+1} \setminus \{0\}$ onto P^n is continuous. Take a point p of P^n and let

$$p = \pi(\mathbf{x})$$

where $\mathbf{x} = (x_1, \ldots, x_{n+1})$. The property that the ith coordinate x_i of $\mathbf{x}$ is not 0 does not depend on the representative $\mathbf{x}$ for p. Let U_i $(i = 1, \ldots, n+1)$ be the set of all points p in P^n, which are represented by points $\mathbf{x}$ in $\mathbf{R}^{n+1} \setminus \{0\}$ whose ith coordinate x_i is not 0. Since

$$\pi^{-1}(U_a) = \{\mathbf{x} \ : \ x_a \neq 0\}$$

is an open set in $\mathbf{R}^{n+1} \setminus \{0\}$, U_a is an open set in P^n. It can be shown that P^n is an analytic manifold. P^n is called the n-dimensional (real) **projective space**.

5. **Open submanifolds.** Let M be a C^r manifold, D an open set of M, and $\{(U_\alpha, \psi_\alpha)\}_{\alpha \in A}$ a coordinate neighbourhood system of class C^r. Set

$$U'_\alpha = U_\alpha \cap D$$

and let ψ'_α be the restriction of ψ_α to U'_α. Then

$$\{ (U'_\alpha, \psi'_\alpha) \}_{\alpha \in A}$$

is a coordinate neighbourhood system of class C^r on D, and D is a C^r manifold. D is called an open submanifold of M.

6. **Product manifolds.** Let M and N be C^r manifolds of dimension m and n, respectively. Let

$$\{ (U_\alpha, \psi_\alpha) \}_{\alpha \in A}$$

and

$$\{ (V_i, \varphi_i) \}_{i \in I}$$

be coordinate neighbourhood systems of class C^r of M and N, respectively. First, if we give the direct product set $M \times N$ the topology of the product space of M and N, then $M \times N$ becomes a Hausdorff space. Note that

$$\{ U_\alpha \times V_i \}_{(\alpha, i) \in A \times I}$$

is an open cover of $M \times N$. Let $\psi_\alpha \times \varphi_i$ be the map that sends the point (p, q) of $U_\alpha \times V_i$ to the point $(\psi_\alpha(p), \varphi_i(p))$ of $\mathbf{R}^{m+n}$. Then $\psi_\alpha \times \varphi_i$ is a homeomorphism from $U_\alpha \times V_i$ onto an open set $\psi_\alpha(U_\alpha) \times \varphi_i(V_i)$ of $\mathbf{R}^{m+n}$. Hence $M \times N$ is an $(m+n)$-dimensional topological manifold. The coordinate transformation functions are of class C^r, so $M \times N$ is a C^r manifold. We call $M \times N$ the product manifold of M and N.

7. **The n-dimensional torus.** The product $S^1 \times \cdots \times S^1$ of n circles S^1 is an n-dimensional analytic manifold. We denote this by T^n, and call it the n-dimensional torus (or n-torus).

8. The group $GL(n, \mathbf{R})$ of all nonsingular $n \times n$ matrices with real entries is a differentiable manifold. In this case, if $g = (a_{hj})$, we can define a homeomorphism h by putting

$$h(g) = \{ \, a_{hj} \; : \; a_{hj} \in \mathbf{R}, \; \det a_{hj}, \neq 0 \, \} \in \mathbf{R}^{n^2}.$$

Let

$$\Delta : \mathbf{R}^{n^2} \to \mathbf{R}$$

be the determinant function: $\Delta(a_{hj}) = \det(a_{hj})$. Then $h(G) = \Delta^{-1}(\mathbf{R} \setminus \{0\})$ which is open in $\mathbf{R}^{n^2}$. ♣

Let $f : M \to \mathbf{R}$ be a function on M, and (U_α, h_α) be a chart on M such that $U_\alpha \subset \operatorname{dom} f$. The map $g_\alpha := f \circ h_\alpha^{-1} : h_\alpha(U_\alpha) \to \mathbf{R}$ is a real-valued function on $h_\alpha(U_\alpha) \subset \mathbf{R}^n$. If $h_\alpha(p) = (u_1, \ldots, u_n) \in \mathbf{R}^n$, then

$$g_\alpha(u_1, \ldots, u_n) = f \circ h_\alpha^{-1}(u_1, \ldots, u_n) = f(p).$$

In this manner the function $f : M \to \mathbf{R}$ is represented by the real-valued functions g_α on $h_\alpha(U_\alpha) \subset \mathbf{R}^n$.

The function f is said to be of class C^k (respectively C^∞ or smooth) on U_α if the functions g_α are of class C^k (respectively C^∞ or smooth) on $h_\alpha(U_\alpha)$. Clearly this definition is independent of the choice of the chart by virtue of the C^k-compatability of the charts of the atlases. The collection of all C^k functions defined on a neighbourhood of $p \in M$ is denoted by C_p^k.

Let (U, h) be a chart, with $h(p) = (u_1, \ldots, u_n)$. The assignment to p of the jth coordinate $u_j (1 \leq j \leq n)$ is a function $x_1 : U \to \mathbf{R}$, with

$$x_j(p) = u_j.$$

This function is called the jth coordinate function in the chart (U, h). For any $f : U \to \mathbf{R}$, we write $g = f \circ h^{-1} : h(U) \to \mathbf{R}$ as before, and we define the derivatives of f with respect to x_j as

$$\frac{\partial f}{\partial x_j} = \frac{\partial g}{\partial u_j} \circ h, \text{ which is a map: } U \to \mathbf{R}$$

so that

$$\frac{\partial f}{\partial x_j}(p) = \frac{\partial g}{\partial u_j}(h(p)) = \frac{\partial g(u_1, \ldots, u_n)}{\partial u_j}.$$

This definition entails the usual rules: if $f, g \in C_p^\infty$, and $a, b \in \mathbf{R}$:

$$\frac{\partial}{\partial x_j}(af + bg) = a\frac{\partial f}{\partial x_j} + b\frac{\partial g}{\partial x_j}, \quad \text{(linearity)}$$

$$\frac{\partial(fg)}{\partial x_j} = f\left(\frac{\partial g}{\partial x_j}\right) + \left(\frac{\partial f}{\partial x_j}\right)g. \quad \text{(derivation property)}$$

Let (U_α, h_α), (U_β, h_β) be a pair of overlapping charts whose coordinate functions are denoted by $\{x_h : h = 1, \ldots, n\}$ and $\{\bar{x}_j : j = 1, \ldots, n\}$, respectively. The identification map

$$h_{\beta\alpha} = h_\beta \circ h_\alpha^{-1}$$

being a map between open sets of $\mathbf{R}^n$, is represented by a system of n equations

$$\bar{x}_j = f_j(x_1, \ldots, x_n), \quad j = 1, \ldots, n$$

with inverse $h_{\alpha\beta} = h_\alpha \circ h_\beta^{-1}$ being given by

$$x_h = g_h(\bar{x}_1, \ldots, \bar{x}_n).$$

By hypothesis the functions $\{f_j\}$ and $\{g_h\}$ are smooth on $h_\alpha(U_\alpha \cap U_\beta)$, $h_\beta(U_\alpha \cap U_\beta)$ respectively. Thus the Jacobian

$$J = \frac{\partial(x_1, \ldots, x_n)}{\partial(\bar{x}_1, \ldots, \bar{x}_n)}$$

is defined on $h_\beta(U_\alpha \cap U_\beta)$. If $j > 0$, the charts (U_α, h_α), (U_β, h_β) are said to **be oriented** consistently. The differerentiable manifold M is said to **orientable** if it admits an atlas all of whose overlapping charts are consistently oriented.

Example: A 2-sphere S^2 is orientable and a Möbius band is not orientable. ♣

Bibliography

[1] Ablowitz M. J. and Segur H. (1981) *Solitons and the inverse scattering transform*, SIAM, Philadelphia.

[2] Ablowitz M. J., Fuchssteiner B. and Kruskal M. (1987) *Topics in Soliton Theory and Exact Solvable Nonlinear Equations*, World Scientific, Singapore.

[3] Ames W.F. (1965) *Nonlinear Partial Differential Equations in Engineering*, Academic Press, New York.

[4] Anderson L. and Ibragmov N.H. (1979) *Lie-Bäcklund Transformations in Applications*, SIAM, Philadelphia.

[5] Baumann G. (1992) *Lie Symmetries of Differential Equations:* A MATHEMATICA program to determine Lie symmetries. Wolfram Research Inc., Champaign, Illinois, MathSource 0202-622.

[6] Baumslag B. and Chandler B. (1968) *Group Theory*, Schaum's Outline Series, McGraw-Hill, New York.

[7] Bluman G.W. and Kumei S. (1989) *Symmetries and Differential Equations*, Applied Mathematical Science **81**, Springer, New York.

[8] Bluman G.W., Kumei S. and Reid G. J. (1988) *J. Math. Phys.* **29**, 806.

[9] Bocharov A.V. and Bronstein M.L. (1989) *Acta Appl. Math.* **16**, 143.

[10] Calogero F. and Degasperis A. (1982) *Spectral Transform and Solitons I*, Studies in Mathematics and its Applications, **13**.

[11] Cantwell B. J. (1985) *J. Fluid Mech.* **85**, 257.

[12] Carminati J., Devitt J. S. and Fee G. J. (1992) *J. Symb. Comp.* **14**, 103.

[13] Chakravarty S. and Ablowitz M. J. (1990) *Phys. Rev. Lett.* **65**, 1085.

[14] Champagne B., Hereman W. and Winternitz P. (1991) *Comput. Phys. Commun.* **66**, 319.

[15] Champagne B. and Winternitz P. (1985) *A MACSYMA Program for Calculating the Symmetry Group of a System of Differential Equations.* Report CRM-1278 (Centre de Recherches Mathématiques, Montréal, Canada).

[16] Chen H.H., Lee Y.C. and Lin J.E. (1987) *Physica* **26D**, 165.

[17] Choquet-Bruhat Y., DeWitt-Morette C. and Dillard-Bleick M. (1978) *Analysis, Manifolds and Physics* (revised edition) North-Holland, Amsterdam.

[18] Chowdhury A.R. and Nasker M. (1986) *J. Phys. A:Math. Gen.* **19**, 1775.

[19] Cicogna G. and Vitali D. (1989) *J. Phys. A: Math. Gen.* **22**, L453.

[20] Cicogna G. and Vitali D. (1990) *J. Phys. A: Math. Gen.* **23**, L85.

[21] Cook J.M. (1953) *Trans. Amer. Math. Soc.* **74**, 222.

[22] Davenport J. H., Siret Y. and Tournier E. (1988) *Computer Algebra*, Academic Press, London.

[23] David D., Kamran N., Levi D. and Winternitz P. (1986) *J. Math. Phys.* **27**, 1225.

[24] Davis Harold T. (1962) *Introduction to Nonlinear Differential and Integral Equations*, Dover Publication, New York.

[25] Dodd R.K. and Bullough R.K. (1976) *Proc. Roy. Soc. London. Ser.* **A351**, 499.

[26] Dodd R.K., Eilbeck J.C., Gibbon J.D. and Morris H.C. (1982) *Solitons and Nonlinear Wave Equations*, Academic Press, London.

[27] Duarte L.G.S., Duarte S.E.S. and Moreira I.C. (1987) *J. Phys. A: Math. Gen.* **20**, L701.

[28] Duarte L.G.S., Duarte S.E.S. and Moreira I.C. (1989) *J. Phys. A:Math. Gen.* **22**, L201.

[29] Duarte L.G.S., Euler N., Moreira I.C. and Steeb W.-H. (1990) *J. Phys. A: Math. Gen.* **23**, 1457.

[30] Duarte L.G.S., Moreira I.C., Euler N. and Steeb W.-H. (1991) *Physica Scripta* **43**, 449.

[31] Dukek G. and Nonnenmacher T. F. (1986) *Physica* **135A**, 167.

[32] Edelen D.G.B. (1981) *Isovector Methods for Equations of Balance*, Sijthoff & Nordhoff, Alphen an de Rijn.

[33] Eliseev V.P., Fedorova R.N. and Kornyak V.V. (1985) *Comput. Phys. Commun.* **36**, 383.

[34] Estévez P.G. (1991) *J. Phys. A: Math. Gen.* **24**, 1153.

[35] Euler N., Leach P.C.L., Mahomed F.M. and Steeb W.-H. (1988) *Int. J. Theor. Phys.* **27**, 717.

[36] Euler N. and Steeb W.-H. (1989) *Int. J. Theor. Phys.* **28**, 11.

[37] Euler N. and Steeb W.-H. (1989) *Aust. J. Phys.* **42**, 1.

[38] Euler N. and Steeb W.-H. (1990) *Phys. Scripta* **41**, 289.

[39] Euler N., Steeb W.-H. and Cyrus K. (1989) *J. Phys. A: Math. Gen.* **22**, L195.

[40] Euler N., Steeb W.-H., Duarte L.G.S. and Moreira I.C. (1991) *Int. J. Theor. Phys.* **30**, 8.

[41] Euler N., Steeb W.-H. and Mulser P. (1991) *J. Phys. Soc. Jpn.* **60**, 1132.

[42] Euler N., Steeb W.-H. and Mulser P. (1991) *J. Phys. A: Math. Gen.* **24**, L785.

[43] Fateman R.J., (1976) *An Approach to Automatic Asymptotic Expansions*, in R.D. Jenks (ed.) *SYMSAC '76*, ACM, New York, 365.

[44] Fedorova R.N. and Kornyak V.V. (1986) *Comput. Phys. Commun.* **39**, 93.

[45] Fedorova R.N. and Kornyak V.V. (1987) *A REDUCE Program for Computing Determining Equations of Lie-Bäcklund Symmetries of Differential Equations*, Report R11-87-19 (JINR, Dubna).

[46] Flanders H. (1963) *Differential Forms with Applications to the Physical Sciences*, Academic Press, New York.

[47] Fokas A.S. (1987) *Studies in Applied Mathematics* **77**, 253.

[48] Fokas A.S. and Fuchsteiner B. (1981) *Nonlinear Anal.* **5**, 423.

[49] Forsyth A.R. (1959) *Theory of Differential Equations*, **VI**, Dover, New York.

[50] Gagnon L. and Winternitz P. (1988) *J. Phys. A : Math. Gen.* **21**, 1493.

[51] Gerdt P., Shvachka A.B. and Zharkov A.Y. (1985) *J. Symbolic Comp.* **1**, 101.

[52] Gilmore R. (1974) *Lie Groups, Lie Algebras, and Some of Their Applications*, Wiley-Interscience, New York.

[53] Goldschmidt H. and Sternberg S. (1973) *Ann. Inst. Fourier, Grenoble*, **23**, 203.

[54] Gragert P., Kersten P.H.M. and Martini A. (1983) *Acta Appl. Math.* **1**, 43.

[55] Harrison B.K. and Estabrook F.B. (1971) *J. Math. Phys.* **12**, 653.

[56] Head A.K. (1990) *LIE : A muMATH Program for the Calculation of the Lie Algebra of Differential Equations*, (CSIRO Division of Material Sciences, Clayton, Australia).

[57] Hereman W. (1994) *Review of Symbolic Software for the Computation of Lie Symmetries of Differential Equations*, Euromath Bulletin **1**, 45.

[58] Herod S. (1992) *MathSym: A MATHEMATICA program for Computing Lie Symmetries*. Preprint, Program in Applied Mathematics (The University of Colorado).

[59] Ibragimov N.H. and Sabat A.B. (1980). *Func. Anal. Appl.* **14**, 19.

[60] Ince E. L. (1956) *Ordinary Differential Equations*, Dover, New York.

[61] Ito M. and Kako F. (1985) *Comput. Phys. Commun.* **38**, 415.

[62] Jacobson N. (1962) *Lie Algebras*, Interscience Publisher, New York.

[63] Kapitanskii L.V. (1979) *Group-theoretical Analysis of the Navier-Stokes Equations in the Rotational Symmetric Case and Some New Exact Solutions*, Translated from Zapiski nauchrykh, Seminarov Leningradskogo Otdeleniya matematichesk.

[64] Kersten P.H.M. (1983) *J. Math. Phys.* **24**, 2374.

[65] Kersten P.H.M. (1987) *Infinitesimal Symmetries: A Computational Approach*, CWI Tract 34 (Center for Mathematics and Computer Science, Amsterdam).

[66] Kersten P.H.M. (1989) *Acta Appl.. Math.* **16**, 207.

[67] Kiiranen K. and Rosenhaus V. (1988) *J. Phys. A:Math. Gen.* **21**, L681.

[68] Kovalevskaya S. (1978) *A Russian childhood*, Springer, New York.

[69] Kowalevski S. (1989) *Sur le Problème de la Rotation d'ùn Corps Solide Autour d'un Point Fixe*, Acta Mathematica **12**, 177.

[70] Kowalski K. (1987) *Physica* **145A**, 98.

[71] Kowalski K. and Steeb W.-H. (1991) *Nonlinear Dynamical Systems and Carleman Linearization*, World Scientific, Singapore.

[72] Kumei S. (1977) *J. Math. Phys.* **18**, 256.

[73] Lamb G. L. Jr. (1974) *J. Math. Phys.* **15**, 2157.

[74] Lamb G. L. Jr. (1980) *Elements of Soliton Theory*, John Wiley, New York.

[75] Lax P.D. (1968) *Comm. Pure Appl. Math.* **21**, 467.

[76] Leach P.G.L. and Mahomed F.M. (1985) *Quast. Math.* **8**, 241.

[77] Leo R.A., Martina L. and Doliani G. (1986) *J. Math. Phys.* **27**, 2623.

[78] Lloyd S.P. (1981) *Acta Mechanica* **38**, 85.

[79] Lo L.L. (1985) *J. Comput. Physics* **61**, 38.

[80] MacCallum M. and Wright F. (1991) *Algebraic Computing with REDUCE*, Clarendon Press, Oxford.

[81] Mack G. and Salam A. (1969) *Ann. Phys. N.Y.* **53**, 174.

[82] Mason L. J. and Sparling G. A. J. (1989) *Phys. Lett. A* **137** 29.

[83] Matsushima Y. (1972) *Differential Manifolds*, Translated by Kobayashi E.T., Marcel Dekker Inc., New York.

[84] Miller W. Jr. (1972) *Symmetry Groups and their Applications*, Academic Press, New York.

[85] Milnor J. (1956) *J. Ann. Math.* **64**, 399.

[86] Nucci M. C. (1990) *Interactive REDUCE programs for Calculating, Classical, Non-Classical and Lie-Bäcklund Symmetries of Differential Equations.* Preprint GT Math. 062090-051, School of Mathematics (Georgia Institute of Technology, Atlanta, Georgia.

[87] Olver P.J. (1986) *Applications of Lie Groups to Differential Equations*, Springer, New York.

[88] Olver P.J. (1977) *J. Math. Phys.* **18**, 1212.

[89] Ovsiannikov L.V. (1982) *Group Analysis of Differential Equations,* Translation edited by W.F. Ames, Academic Press, New York.

[90] Popov M.D. (1985) Izvestiya AN Belor. SSR, Ser. Phys.-Math. **2**, 33.

[91] Rand D.W. and Winternitz P. (1986) *Comput. Phys. Commun.* **42**, 359.

[92] Rand R. H. (1984) *Computer Algebra in Applied Mathematics: An Introduction to MACSYMA*, Pitman, New York.

[93] Reid G. J. (1990) in: V. Hussin, *Lie Theory, Differential Equations and Representation Theory*, Proc. Annual Seminar of the Canadian Math. Soc., Les Publications de Centre de Recherches Mathématiques, Montréal, Canada.

[94] Reid G. J. (1990) *Finding Symmetries of Differential Equations Without Integrating Determining Equations*, Technical Report 90-4, Inst. of Appl. Math., The University of British Columbia, Vancouver, Canada.

[95] Reid G. J. (1990) *J. Phys. A:Math. Gen.* **23**, L853.

[96] Rogers C. and Ames W. F. (1989) *Nonlinear Boundary Value Poblems in Science and Engineering,* Academic Press, New York.

[97] Rogers C. and Shadwick W.F. (1982) *Bäcklund Transformations and their Applications,* Academic Press, New York.

[98] Rosenau P. and Schwarzmeier J.L. (1979) *Similarity solutions of Systems of Partial Differential Equations using MACSYMA,* Report/COO-3077-160 MF-94, Courant Institute of Mathematical Sciences, New York University, New York.

[99] Rosencrans S. I. (1985) *Comput. Phys. Commun.* **38**, 347.

[100] Rosenhaus V. (1986) *The Unique Determination of the Equation by its Invariant and Field-space Symmetry,* Hadronic Press.

[101] Rosenhaus V. (1988) *J. Phys. A:Math. Gen.* **21**, 1125.

[102] Rudra R. (1986) *J. Phys. A;Math. Gen* **19**, 2947.

[103] Sattinger D.H. and Weaver O.L. (1986) *Lie Groups and Algebras with Applications to Physics, Geometry, and Mechanics,* Applied Mathematical Science, **61**, Springer, New York.

[104] Schwarz F. (1982) *Comput. Phys. Commun.* **27**, 179.

[105] Schwarz F. (1985) *J. Symbolic Comp.* **1**, 229.

[106] Schwarz F. (1987a) *The Package SPDE for Determining Symmetries of Partial Differential Equations: User's Manual,* Distributed with REDUCE.

[107] Schwarz F. (1987b) *Programming with Abstract Data Types: The Symmetry Packages AIDE and SPDE in SCRATCHPAD,* Int. Symp. "Trends in Computer Algebra" Bad Neuenahr, Germany

[108] Schwarz F. (1988) in: R. Janssen *Trends in Computer Algebra,* Lecture Notes in Comput. Sci. 296, Springer, New York.

[109] Schwarz F. (1988) *SIAM Review* **30**, 450.

[110] Schwarz F. (1989) *A Factorization Algorithm for Linear Ordinary Differential Equations*, From: Proceedings of the ACM-SIGSAM 1989 International Symposium of Symbolic and Algebraic Computation, ISSAC'89.

[111] Schwarzmeier J.L. and Rosenau P. (1988) *Using MACSYMA to Calculate Similarity Transformations of Partial Differential Equations*, Report LA-UR 88-4157, Los Alamos National Laboratory, Los Alamos.

[112] Sen T. and Tabor M. (1990) *Physica D* **44**, 313.

[113] Sewell G.L. (1986) *Quantum Theory of Collective Phenomena*, Clarendon Press, Oxford.

[114] Shadwick W.F. (1980) *Lett. Math. Phys.* **4**, 241.

[115] Skierski M., Grundland A.M. and Tuszyński J.A. (1988) *Phys. Lett. A*, **133**, 213.

[116] Steeb W.-H. (1978) *Z. Naturforsch.* **33a**, 724.

[117] Steeb W.-H. (1980) *J. Math. Phys.* **21**, 1656.

[118] Steeb W.-H. (1982) *Int. J. Theor. Phys.* **24**, 237.

[119] Steeb W.-H. (1983) *Hadronic J.* **6**, 68.

[120] Steeb W.-H. (1985) *Int. J. Theor. Phys.* **24**, 237.

[121] Steeb, W.-H. (1993) *Invertible Point Transformation*, World Scientific, Singapore.

[122] Steeb W.-H., Brits S.J.M. and Euler N. (1990) *Int. J. Theor. Phys.* **29**, 6.

[123] Steeb W.-H., Erig W. and Strampp W. (1982) *J. Math. Phys.* **23**, 145.

[124] Steeb W.-H. and Euler N. (1987) *Prog. Theor. Phys.* **78**, 214.

[125] Steeb w.-H. and Euler N. (1987) *Lett. Math. Phys.* **13**, 234.

[126] Steeb W.-H. and Euler N. (1988) *Nonlinear Field Equations and Painlevé Test*, World Scientific, Singapore.

[127] Steeb W.-H. and Euler N. (1990) *Found. Phys. Lett.* **3**, 367.

[128] Steeb W.-H. and Euler N. (1990) *Z. Naturforsch.* **45a**, 929.

[129] Steeb W.-H. and and Lewien D. (1992) *Algorithms and Computation with REDUCE*, B.-I. Wissenschaftsverlag, Mannheim.

[130] Steeb W.-H., Euler N. and Mulser P. (1991) *IL NUOVO CIMENTO* **106B**, 1059.

[131] Steeb W.-H., Kloke M., Spieker B.M. and Kunick A. (1985) *Found. Phys.* **15**, 6.

[132] Steeb W.-H. (1994) *Chaos und Quanten Chaos in Dynamischen Systemen*, Spektrum Verlag, Heidelberg.

[133] Steeb W.-H., Schröter J. and Erig W. (1982) *Found. Phys.* **12**, 7.

[134] Steeb W.-H. and Strampp W. (1982) *Physica* **114 A**, 95.

[135] Steeb W.-H. and Wilhelm F. (1980) *J. Math. Appl.* **77**, 601.

[136] Steinberg S. (1979) in V.E. Lewis, Proc. of the 1979 MACSYMA User's Conference, MIT Press, Boston, 408.

[137] Steinberg S. (1990) MACSYMA Newsletter **7**, 3.

[138] Strampp W. (1984) *J. Phys. Soc. Jpn.* **53**, 11.

[139] Strampp W. (1986) *Prog. Theor. Phys.* **76**, 802.

[140] Vafeades P. (1990) *SYMCOM: A MACSYMA Package for the Determiniation of Symmetries and Conservation Laws of PDEs*, in: Proc. ISMM Int. Symposium on Computer Algebra Application in Design, Simulation and Analysis, New Orleans, Louisiana.

[141] Von Westenholz C. (1981) *Differential Forms in Mathematical Physics*, (Revised edition) North-Holland, Amsterdam.

[142] Wadati M. *Stud. Appl. Math.* **59**, 153.

[143] Ward R.S. (1981) *Commun. Math. Phys.* **80**, 563.

[144] Ward R.S. (1984) *Phys. Lett.* **102 A**, 279.

[145] Ward R.S. (1985) *Phil. Trans. R. Soc. Lond. A* **315**, 451.

[146] Weiss J. (1984) *J. Math. Phys.* **25**, 13.

[147] Weiss J., Tabor M. and Carnevale G. (1983) *J. Math. Phys.* **24**, 522.

[148] Winternitz P., Grundland A.M. and Tusyński J.A. (1987) *J. Math. Phys.* **28**, 2194.

[149] Zwillinger D. (1990) *Handbook of Differential Equations*, Academic Press, Inc. Boston.

Index